FLORA OF MIDDLE-EARTH

FLORA OF MIDDLE-EARTH

Plants of J. R. R. Tolkien's Legendarium

Walter S. Judd
and
Graham A. Judd

OXFORD
UNIVERSITY PRESS

Oxford University Press is a department of the University of Oxford. It furthers the University's objective of excellence in research, scholarship, and education by publishing worldwide. Oxford is a registered trade mark of Oxford University Press in the UK and certain other countries.

Published in the United States of America by Oxford University Press
198 Madison Avenue, New York, NY 10016, United States of America.

Library of Congress Cataloging-in-Publication Data
Names: Judd, Walter S., author. | Judd, Graham A.
Title: Flora of Middle-Earth : plants of J.R.R. Tolkien's legendarium /
Walter S. Judd and Graham A. Judd.
Description: New York: Oxford University Press, 2017. |
Includes bibliographical references and index.
Identifi ers: LCCN 2016049648 | ISBN 9780190276317 (hardback)
Subjects: LCSH: Tolkien, J. R. R. (John Ronald Reuel), 1892–1973—Criticism and interpretation. | Tolkien, J. R. R. (John Ronald Reuel), 1892–1973. Hobbit. | Plants in literature. | Trees in literature. | Botany in literature. | Ecology in literature. | BISAC: SCIENCE / Life Sciences / Botany. | SCIENCE / Life Sciences / Ecology.
Classifica tion: LCC PR6039.O32 Z6664 2017 | DDC 823/.912—dc23
LC record available at https://lccn.loc.gov/2016049648

The manufacturer's authorised representative in the EU for product safety is Oxford University Press España S.A. of El Parque Empresarial San Fernando de Henares, Avenida de Castilla, 2 – 28830 Madrid (www.oup.es/en or product.safety@oup.com). OUP España S.A. also acts as importer into Spain of products made by the manufacturer.

Printed by Integrated Books International, United States of America

Contents

List of Figures

Acknowledgments

This book is a team effort. Although each of us was responsible for certain sections, we both benefited from helpful comments from the other. Walter Judd had primary responsibility for Chapters 1–7 and Graham Judd for Chapter 8 and all of the illustrations. We thank our wives, Beverly and Danielle, for their emotional support and for their forbearance with their husbands' preoccupation with the world of J. R. R. Tolkien (and also the worlds of botany and fine art). We thank Paul and Cherith Davenport, who read and commented on Chapter 1. We express our sincere appreciation to the staff of the University of Florida Herbarium (of the Florida Museum of Natural History), especially Norris Williams and Kent Perkins, and also to Anita Cholewa of the University of Minnesota Herbarium (of the Bell Museum of Natural History) for their assistance in the use of herbarium material representative of the plants of Middle-earth. Finally, we thank the production staff of Oxford University Press for their assistance in bringing our work to its finished form.

We, the authors, assume all editorial responsibility for this book. We would greatly appreciate receiving any comments or corrections that readers may wish to send.

Walter S. Judd, Gainesville, Florida
Graham A. Judd, St. Paul, Minnesota

ONE

Introduction

The Importance of Plants in J. R. R. Tolkien's Legendarium

THE GENESIS OF THIS BOOK BEGAN YEARS AGO, WHEN WE, AS FATHER AND son, enjoyed reading *The Hobbit* and then *The Lord of the Rings* together—since then, we both have re-read these books and also *The Silmarillion* (SILM) many times—and each time the experience is not so much merely reading words on a page, but actually finding ourselves immersed in Middle-earth, as if we had awoken from sleep and found ourselves transported to a wondrous land. We share a love of the sub-created world of J. R. R. Tolkien's Middle-earth, although otherwise our interests are quite different. One of us (Graham) is an artist, specializing in printmaking, while the other (Walter) is a biological scientist, specializing in botany and plant systematics. This book is thus interdisciplinary, and in it we have joined forces to introduce the reader to the plants of Tolkien's legendarium—and to provide illustrations showing these plants, along with vignettes (in a woodcut style) of events in the history of Middle-earth in which these plants have played a key role.

Many readers of *The Hobbit* or *The Lord of the Rings* believe that the events of these books occur in an imaginary world and thus have no connection with the world around us. However, Tolkien sought to correct this misconception, stating that Middle-earth "is just the use of Middle English *middle-erde* (or *erthe*), altered from Old English *Middangeard*: the name for the inhabited lands of Men 'between the seas.'" He went on to say that "imaginatively this 'history' is supposed to take place in a period of the actual Old World of this planet" (Letters: No. 165). His writings, therefore, in no way encourage an escape from reality but are instead meant to reconnect us to important elements of our internal and cultural landscape and also to impact how we interact with other individuals and with the world in which we live—including the landscapes of our natural environment. This book focuses on one of the major components of our environment—the green plants—organisms to which many in our modern, highly technological world, have become blind. Plants are ecologically diverse and range dramatically in size—from microscopic, aquatic, green algae to the tallest flowering trees or conifers. They are critically important in maintaining a healthy biosphere—and in fact, without plants, animal (and, of course, human) life would be impossible. They provide our food, construction materials for our homes, add beauty to our surroundings, and even provide the air we breathe. In Tolkien's legendarium, plants are the primary concern of Yavanna Kementári, the Giver of Fruits and wife of Aulë, who has lordship over all the substances of which the Earth is made. As related in *The Silmarillion*, she is the "lover of all things that grow in the earth, and all their countless forms she holds in her mind, from the trees like towers in forests . . . to the moss upon stones or the small and secret things in the mould" (Valaquenta). Mythologically, the connection between Yavanna and Aulë is clear because her plants (and animals) depend upon the lands of Middle-earth, which are fashioned by her husband. Understandably, she is held in great reverence by the elves, as are the natural environments she oversees. We think Tolkien's reverence was comparable.

Tolkien's descriptions of Middle-earth are richly detailed, including succinct verbal sketches of many of its plants, and thus create a realistic stage for his dramas. His detailed treatment of plants plays a major role in the creation of this stage—providing the distinctive landscapes and natural locales of Middle-earth—from the tundra and ice-fields of the north, to the extensive prairies of Rohan, and the coniferous forests of Dorthonion, as well as the broad-leaved forests of Doriath or Fangorn and wetlands such as the Gladden Fields. The dominant species within each plant community are always mentioned, especially the trees, which Tolkien, like Yavanna, held most dear (see SILM 2). Thus, it is critical for our appreciation and understanding of Middle-earth to envision these scenes accurately. These plants, however, do more than merely provide descriptive detail, enhancing the veracity of the tales of Middle-earth. The plants within Tolkien's legendarium are actually part of the story and in ways that are more deeply significant than merely evident in the actions of Ents—anthropomorphized trees—that "speak on behalf of all things that have roots, and punish those that wrong them" (SILM 2). Their significance can be seen in the numerous connections between plants and important individuals in the myths and history of Middle-earth. For example, in the First Age (and earlier), how are we to understand the Two Trees of Valinor, fashioned by Yavanna, and why is it important that Thingol, the elven ruler of Doriath, was called the king of beech, oak, and elm? Why was his daughter, Lúthien, when first observed by Beren, dancing among the hemlock-umbels under the beeches of Neldoreth? And what is the link between her feet and the leaves of lindens? Why did hawthorns obscure the entrance to the Hidden Kingdom of Gondolin? During the Second Age, why did the elves give Aldarion, soon to become the sixth king of Númenor, a White Tree—Nimloth—and what is the connection between this tree and the White Trees of Gondor? Why did the elves bring to Númenor several different fragrant trees from Eressëa—and what did these trees look like? In the Third Age, how was pipe-weed integral to the culture of the Shire, and why was athelias (kingsfoil) useful in the hands of the king of Gondor? How did these two herbs get to Middle-earth? What is the connection of willows and the Withywindle valley (in the Old Forest), and should willows, therefore, be viewed negatively? Why does Quickbeam love rowan-trees, and why were mallorn-trees important to Galadriel and the elves of Lothlórien?

And finally, how should we envision the herbs elanor and niphredil, and what made these two plants so sacred to the elves? Of course, many more questions come quickly to mind. We will, therefore, deal with these questions, among many others, in the following pages (and especially in Chapters 6 and 7, in which the plants of Middle-earth are considered in detail).

It is obvious from even a cursory reading of *The Lord of the Rings* that the book was written by a person who was botanically knowledgeable—but more than that—a writer who really loved plants! But we don't need to merely accept this from our interpretations of his writings. Tolkien tells us of his appreciation of plants. He said in his letter to the Houghton Mifflin Co.: "I am (obviously) much in love with plants and above all trees, and always have been; and I find human maltreatment of them as hard to bear as some find ill-treatment of animals" (Letters: No. 164). We agree: his love of plants is obvious, and it is apparent on nearly every page of *The Hobbit* or *The Lord of the Rings*. Only a writer whose eyes were open to the diversity of the natural world could have accomplished such a task—closely integrating plants into his imagined world, and, as a result, including nearly all of the trees of England (and also most European trees) within the Middle-earth of the First through the Third Ages. Since the species of trees (as well as shrubs and herbs) growing in England and other European regions are for the most part members of widely distributed genera that also occur in temperate North America and Asia, especially eastern and southeastern Asia, we can find the plants of Tolkien's Middle-earth in the forests and fields around our homes. Thus, a major goal of this book, in addition to increasing our appreciation of the imagined landscapes of Middle-earth, is to increase our respect for and understanding of the plants that grow in the natural environments that exist around us. Tolkien appreciated the beauty and diversity of the natural world, and its destruction through urbanization and industrialization angered him (unfortunately, modern followers of Saruman are not hard to find!). Thus, one of our goals is to increase the visibility of and love for plants in our modern culture. And, taking the Ents as our role-models, we hope to foster the desire to protect the forests and meadows near our homes (and across the world). Finally, the wild plants of forest and field are not our only concern. In this book, we have also described the cultivated plants of vegetable and flower gardens as well as agricultural fields, addressing the interesting and long history of plants and people (or hobbits and elves!). We should appreciate not only wild plants (as do the Ents) but also the plants of orchards and cultivated fields (like the Entwives). In the end, the fact that an investigation of the plants of Tolkien's Middle-earth reconnects us with the plants of our own world should not be surprising. Tolkien, in his essay *On Fairy-Stories*, said that "Recovery" is one of the goals of fantasy, and by this he meant "a re-gaining—regaining of a clear view" and "seeing things as we are (or were) meant to see them." Thus, in "experiencing the fantastic, we recover a fresh view of the unfantastic, a view too long dulled by familiarity" (see Verlyn Flieger, 2002, *Splintered Light*, chapter 3). Tolkien's fantasy allows us to see oaks, beeches, and pines in a fresh light.

If the plants of Tolkien's legendarium are the trees, shrubs, and herbs of our own world, one might ask: What about plants such as elanor, niphredil, alfirin, simbelmynë, pipe-weed, or the White Tree of Gondor? Are these simply the creation of Tolkien's imagination, or do they also have links to our own world. The answer, we think, is both—certainly these plants, as Tolkien explained, "are lit by a light that would not be seen ever in a growing plant" (Letters: No. 312) in our world—so they arise, some more and others less, out of his imagination and are used in specific ways in the story in order to clarify aspects of elven, human, or hobbit culture. They are artistic creations, enhancing the wonder and mystery of Tolkien's imaginative world. But it is also important to keep in mind that perhaps all of the imaginative plants of Middle-earth are based, at least in part, on species of our own world. For example, Tolkien suggested that niphredil—if seen in the light of our world—would be "simply a delicate kin of a snowdrop," while elanor would be "a pimpernel (perhaps a little enlarged) growing sun-golden flowers and star-silver ones on the same plant" (Letters: No. 312). As early as 1956, Tolkien commented that "Botanists want a more accurate description of the mallorn, of elanor, niphredil, alfirin, mallos, and simbelmynë" (Letters: No. 187), and we trust that many readers have a similar desire. We have, therefore, done the necessary detective work to connect these imaginative plants with their sources and provide such accurate descriptions. We believe that this botanical knowledge will enrich the experience of those who have read (or are reading) Tolkien's works. This book explores the interactions between plants and the speaking-peoples of Middle-earth—such as humans, hobbits, elves, or ents—whether such plants are the common oaks, pines, or grasses found in the sunlight of our world or are those plants lit by a more imaginative light, such as niphredil or elanor. Thus, we attempt in this book to synthesize information from diverse realms: Tolkien's writings, etymology (the evolution of words), botany and plant systematics (study of plants and their evolutionary relationships), and artistic endeavors.

In Chapter 2, we consider the plant communities of Middle-earth and how these were distributed in the landscapes of the First through the Third Ages. We also compare these vegetation types to those of Europe, temperate North America, and northern China today. Chapter 3 is focused on the major groups of photosynthetic organisms, especially the green plants. The characteristics of these organisms and their evolutionary relationships are briefly considered. The language of descriptive botany is the topic of Chapter 4, and here we define (and illustrate) the technical terms used in our descriptions of the plants of Middle-earth, which in Chapter 5 are included in an identification key (highlighting the distinctive features of each plant). The Two Trees of Valinor (i.e., Telperion and Laurelin, the most important trees of Tolkien's legendarium) are the subject of Chapter 6. In Chapter 7, we provide detailed treatments for all of the 141 plants of Middle-earth. For each of the 100 most important plants of Tolkien's imaginative world, we include (1) the common and scientific names, along with an indication of the family to which the plant belongs; (2) a brief quote from one of Tolkien's works in which the plant is referenced; (3) a discussion of the significance of the plant in the context of Tolkien's legendarium; (4) the etymology, relating to both the English common name and the Latin (or Latinized) scientific name, and, where relevant, the name in one or more of the languages of Middle-earth; (5) a brief description of the plant's geographical distribution and ecology; (6) its economic importance; and (7) a brief description of the plant. Most of these are also provided with a woodcut-style illustration (as an aid to identification), along with an inset illustrating one of the events in the history of Middle-earth in which the plant played a role. The remaining less important species are gathered into one of four categories: Plants of Ithilien, Food Plants of Middle-earth, Bree Names, and Hobbit Names. Finally, in Chapter 8, we consider the topic of the art of Tolkien's Middle-earth, with an emphasis on how the artwork in this book should be interpreted.

The 141 plants considered in this book are those that are mentioned in any of the following works by J. R. R. Tolkien: *The Hobbit* (Hobbit), *The Lord of the Rings* (LotR), *The Adventures of Tom Bombadil* (TATB), *The Silmarillion* (SILM), *Unfinished Tales of Númenor and Middle-earth* (UT), *The Lays of Beleriand* (Lays; i.e., the Third Volume of *The History of Middle-earth*), *The Epilogue* (i.e., chapter 11 in the Ninth Volume of *The History of Middle-earth*), *The Wanderings of Húrin* (in the Eleventh Volume of *The History of Middle-earth*), and *The Children of Húrin* (CoH). Quotations from any of these works are indicated by the abbreviations (within parentheses) and indicated by chapter (for *The Hobbit, The Silmarillion,* and *Children of Húrin*), by poem number (for the *Adventures of Tom Bombadil*), by poem number and line (for the *Lays of Beleriand*), by book number and chapter (for *The Lord of the Rings*), and by part number and chapter (for the *Unfinished Tales*). Of course, a voluminous literature on the writings of J. R. R. Tolkien exists, and a few of these are also cited where appropriate. We hope that our detailed treatment of these plants will create a visual reference—and legitimacy—for both the plants growing in our forests, meadows, and marshes, as well as those that we have received as gifts from Tolkien's imagination.

TWO

Plant Communities of Middle-earth

THE LANDSCAPE OF MIDDLE-EARTH FROM THE FIRST TO THE THIRD AGES, and especially during the period covered by *The Hobbit* and *The Lord of the Rings*, supported an array of vegetation types similar to those of Europe, temperate North America, and, to a lesser extent, northern China today. We will introduce each of these major plant zones or biomes of Middle-earth, in turn, giving their general distribution and comparing them to the similar biomes of north-temperate regions today. Each of the plant zones or biomes is mapped in this chapter (see Figures 2.1 through 2.11), and, in addition, a useful map of the vegetation of Middle-earth can be found in *The Atlas of Middle-earth*, by K. W. Fonstad (1991). Following this overview, we will discuss the environmental modifications brought about by the activities of humans, hobbits, elves, and the like.

ARCTIC TUNDRA AND POLAR DESERT

The first of these biomes is arctic tundra (including polar desert communities), which in the First Age occurred in the far north where Aman approached Middle-earth at the Helcaraxë, a region of grinding ice, and in the vast regions north of the Ered Engrin (Iron Mountains). In the Third Age, such vegetation occurred on the Cape of Forochel, other territories surrounding the Ice Bay, and across the Northern Waste, far north of the Grey Mountains (Figure 2.1). These regions were dominated by mosses, lichens, grasses, sedges, and low shrubs, but, in more favorable localities, tall-shrub tundra developed, which was dominated by dwarf willows (*Salix*), birches (*Betula*), and low alders (*Alnus*). However, many regions were barren, nearly devoid of plant life, with the ground consisting only of bare soil or frost-shattered rocks and pebbles. In such localities, the few low herbs were restricted to cracks in the cold, dried soil. Northward, the land became completely desolate and covered by snow and ice. In the First Age, these regions were largely controlled by Morgoth, and very few of the events related in the *Silmarillion* took place in them. However, Fëanor abandoned Fingolfin, his sons, as well as Galadriel and Finrod, and their followers, in the far north of Aman, forcing

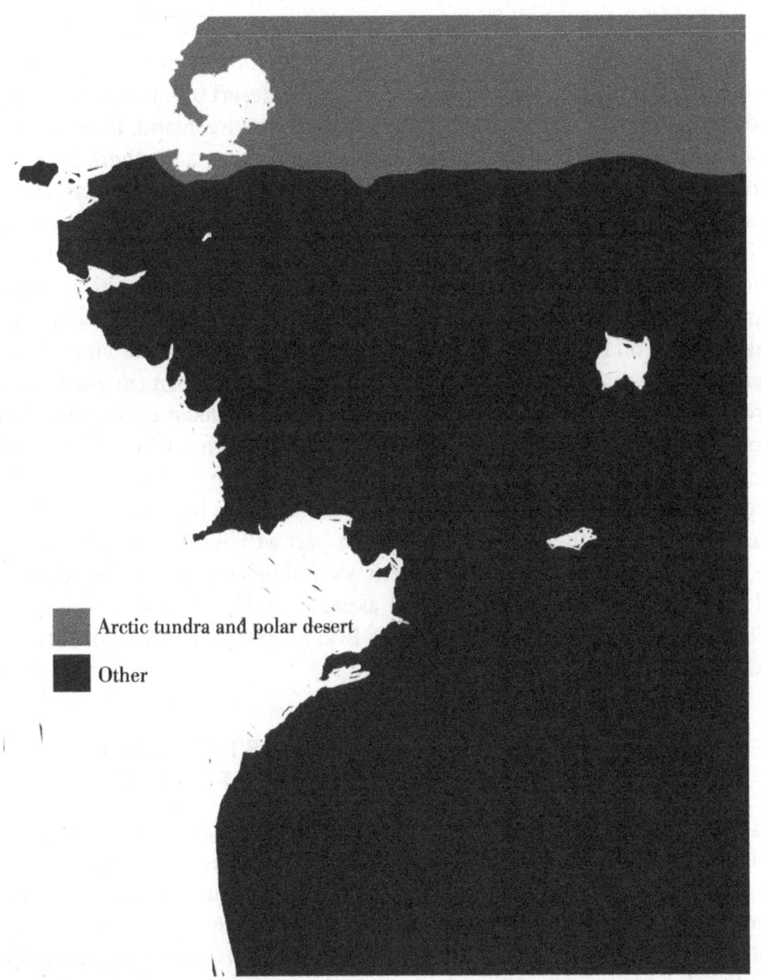

FIGURE 2.1 *Arctic tundra and polar deserts in western Middle-earth of the Third Age.*

them to cross the Helcaraxë, with its "cruel hills of ice" (SILM 9) in order to reach Middle-earth. During the Third Age, Arvedui, the last king of Arnor (the Northern Kingdom), was forced into the lands just south of the Ice Bay by the Witch-king of Angmar. He was there aided by the Snowmen of Forochel, with whom he left the Ring of Barahir before departing on an elven ship (sent by Círdan) that tragically soon sank in a blinding snowstorm, being blown back on and crushed against the ice-sheet that extended far out from the lands around the Ice Bay (see LotR, Appendix A (iii)). The bitter cold of the region of the Ice Bay can only partly be explained by its high latitude. Tolkien noted that the cold had lingered there since the First Age, when the region had been close to Morgoth's stronghold of Angband. Even during the Third Age it was close to Carn Dûm, controlled by the Witch-king whose power increased in winter and waned in summer. Mythologically, therefore, the bitter cold of this region can be seen as part of an attempt by Morgoth, and later Sauron, to distort and damage Middle-earth through extreme cold (although these evil beings also employed extreme heat, as we will see later). It is appropriate that "the light of the eyes of Melkor"—as related in *The Music of the Ainur*—"was like a flame that withers with heat and pierces with a deadly cold." In North America, the tundra and polar desert biome extends from western and far northern Alaska across northern Canada to Greenland. In Europe, tundra occurs in northern regions of Scandinavia and Russia (across northern Siberia). This vegetation type is absent from southeastern Asia. Tundra regions are characterized by more or less continuous darkness in the middle of winter, and they are snow-covered for much of the year. The water in the soil and rock below it is frozen to a great depth (i.e., the permafrost), with only the surface layer thawing during the brief summer. In this region, we certainly see extreme cold in action! However, it would be a mistake to think that in Tolkien's legendarium the tundra biome should be viewed as somehow defective or evil. This is clearly evident in Ilúvatar's statement (in the Ainulindalë) that even the themes of Melkor (= Morgoth) have been taken up into the Music of Creation, resulting in the "devising of things more wonderful, which he himself hath not imagined." Thus Ulmo's conception of water is in the end enhanced, with the cold of Morgoth resulting in the wonder of the snowflake and his heat leading to the beauty of clouds and rain, neither of which had been part of Ulmo's original theme in the Music. Therefore, this biome, however harsh, should be viewed as a valued part of our natural world: it is both good and beautiful, as are all of the vegetational biomes (being part of a good creation; see also Dickerson & Evans, 2006, *Ents, Elves, and Eriador*, and Bernthal, 2014, *Tolkien's Sacramental Vision*).

BOREAL CONIFEROUS FOREST (TAIGA) AND MONTANE CONIFEROUS FORESTS

Forests dominated by conifers (i.e., various cone-bearing plants) occurred in the First Age primarily in Dorthonion (Pine-land, in Sindarin), a highland region dominated by an extensive pine forest (*Pinus*), and the Ered Luin (Blue Mountains) and Ered Wethrin (Shadowy Mountains), the higher slopes of which supported both pine and fir (*Abies*) forests (Figure 2.2). In the Second Age, forests of fir and larches (*Larix*) occurred in the mountains of Forostar and Andustar (in Númenor; see Figure 2.3). In the Third Age, coniferous forests existed in southern Mirkwood, which was dominated by a dense fir forest, but most such forests occurred in the major mountain ranges (i.e., the Ered Luin [Blue Mountains], Hithaeglir [Misty Mountains], and Ered Nimrais [White Mountains]) (see Figure 2.4). In these montane forests, pines, firs, and larches predominated (mixed with a few broadleaved species, such as birches [*Betula*] and rowans [*Sorbus*]). Of course, some conifers also grew intermixed with broadleaved trees in various predominantly broadleaved, temperate forests (see the later section "Cool- to Warm-Temperate, Deciduous Forests"). In fact, the regions of highest conifer diversity seemed to be the mixed deciduous forests of northern Ithilien, on the western slopes of the Ephel Dúath (the western fence of Mordor), which in the time of *The Lord of the Rings* had only recently fallen under the control of Sauron. These diverse forests included many broadleaved species intermixed with pines, firs, true cedars (*Cedrus*), junipers (*Juniperus*), and cypresses (*Cupressus*). In North America, a band of boreal coniferous forests occurs immediately south of the tundra biome, and this constitutes the boreal forest biome (or taiga). A similar taiga zone occurs across northern Europe from Scandinavia and northern Russia,

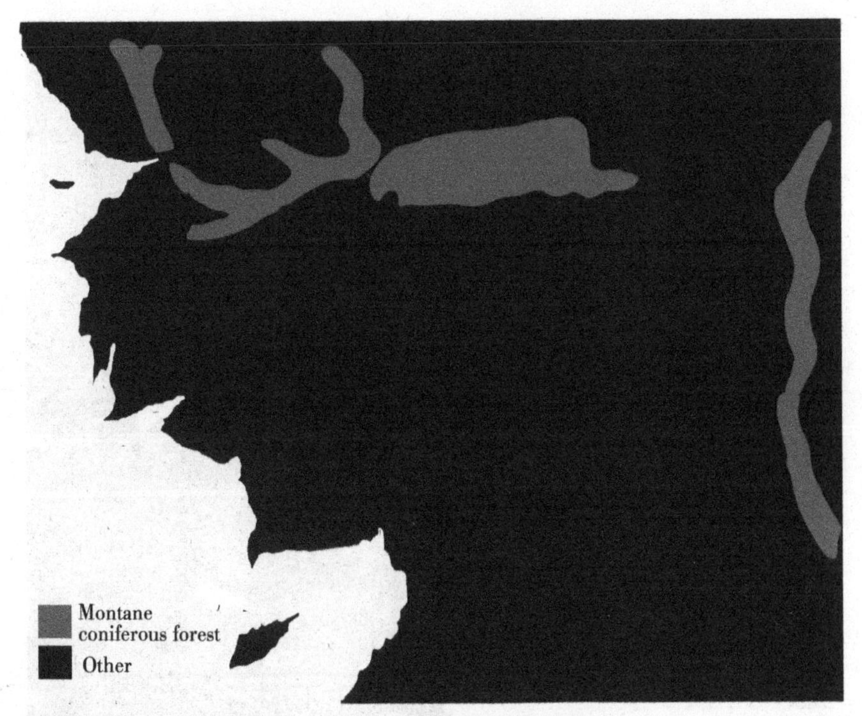

FIGURE 2.2 *Montane coniferous forests in Beleriand (First Age).*

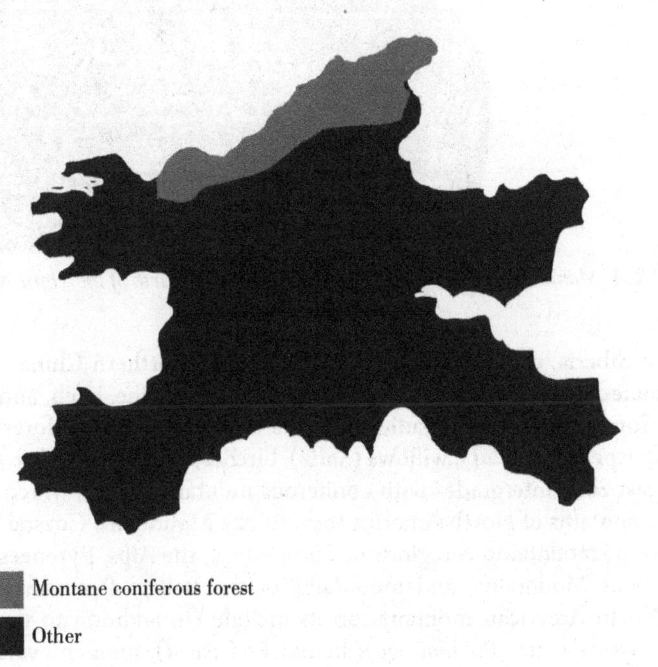

FIGURE 2.3 *Montane coniferous forests of Númenor (Second Age).*

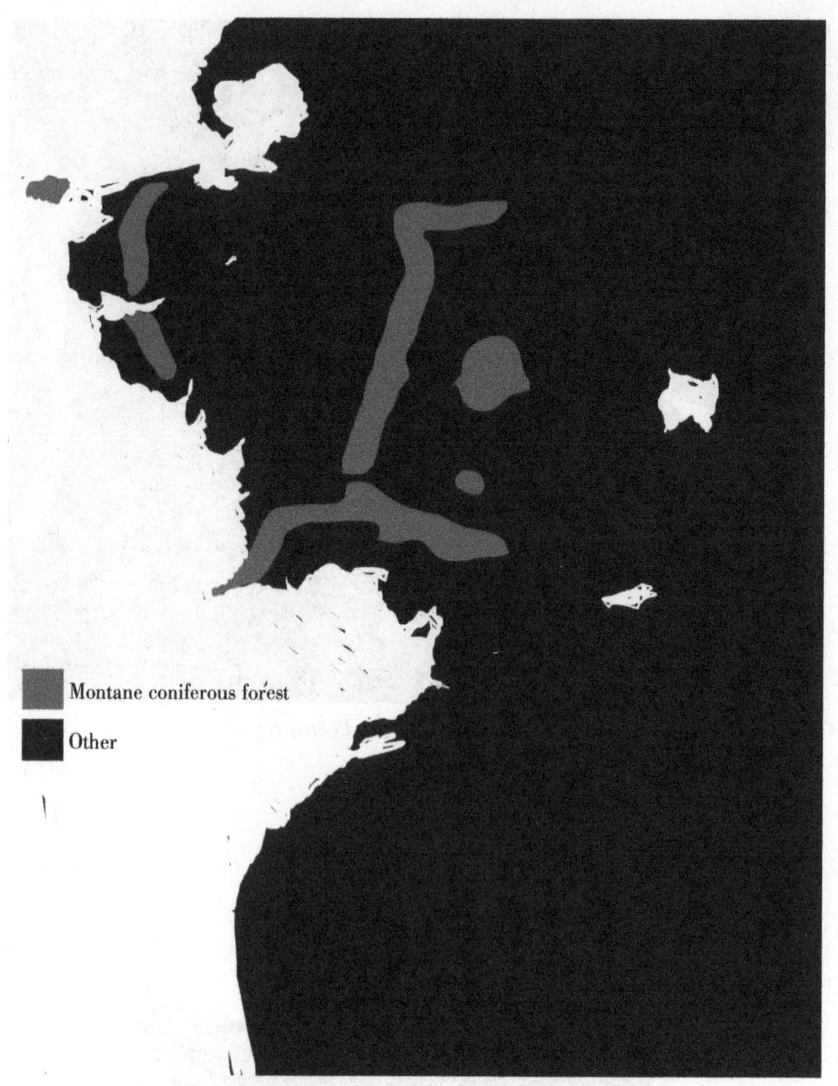

FIGURE 2.4 *Montane coniferous forests of western Middle-earth of the Third Age.*

and across Siberia, extending eastward into extreme northern China. These forests are dominated by various species of spruce (*Picea*), fir, pine, larch, and white-cedar (*Thuja*). Some deciduous, broadleaved trees also occur in such forests, especially poplars or aspens (*Populus*), willows (*Salix*), birches, and alders (*Alnus*). This broad, boreal forest zone intergrades with coniferous montane forests, which occur in the western mountains of North America (e.g., Rocky Mountains, Coastal Ranges) and in the various mountainous regions of Europe (e.g., the Alps, Pyrenees, Carpathian and Caucasus Mountains, and mountains of the Balkan Peninsula). Coniferous trees of North American montane forests include (in addition to those listed for the taiga) Douglas fir (*Pseudotsuga*), hemlocks (*Tsuga*), junipers, various "cedars" and cypresses (*Callitropsis, Calocedrus, Chamaecyparis, Thuja*), redwoods (*Sequoia*), and giant sequoias (*Sequoiadendron*). European montane conifers—both boreal and montane—are less diverse, including only firs, spruces, larches, pines, cypresses (*Cupressus*), and junipers. Some of the intermixed broadleaved species include alders, birches, ironwoods (*Ostrya*), maples (*Acer*), oaks (*Quercus*), poplars and aspens, willows, and mountain-ashes or rowans. Finally, above the tree line in mountainous regions, especially on high mountain summits, slopes, and ridges, there exists a region dominated by herbs and low shrubs—an alpine vegetational zone. It is interesting that the boreal forest zone is absent in Middle-earth, with barren mountainous regions, grasslands, and/or scrublands (i.e., a zone occupied by stunted trees and shrubs) occurring to the south of the zone of ice and tundra. In contrast, montane coniferous forests seem to be as well-developed in Middle-earth as they are in the mountains of Europe and North America today. Many of the key events in the history of Middle-earth occurred in such conifer-dominated forests. For example, in the First Age, Treebeard walked in the winter "among the pine-trees upon the highland of Dorthonion" (LotR 3: IV). Barahir, Beren, and their followers fought the orcs who had invaded Dorthonion during and after the Battle of Sudden Flame, and their struggles occurred both in pine forests and in high alpine meadows. And, later, Beleg wandered in the haunted pine forests of Dorthonion, searching for Túrin, who had been captured by orcs, and there, with the help of Gwindor, rescued him. During the elves' escape from Gondolin, Glorfindel fought with a balrog in a high pass of the Encircling Mountains. He died there, allowing the refugees (led by Tuor and Idril) to escape. And, in that high place, a "green turf came" and "yellow flowers bloomed upon it amid the barrenness of stone" (SILM 23)—certainly a reference to an alpine meadow. In the Third Age, it was in pine forests, intermixed with larches and firs, high on the eastern slopes of the Misty Mountains, that Gandalf, Bilbo, and the dwarves were trapped by orcs and wargs. While recovering in Rivendell, Frodo, on the very day of the Council of Elrond, told Gandalf that he would love to go walking and get into the pine woods covering the higher elevations north of Elrond's home. Firs dominated the forests near the eastern gates of Moria where the company, now led by Aragorn, briefly rested, and firs dominated the landscape of Dunharrow, where the muster of Rohan occurred and where Aragorn led those who followed him under the Dwimorberg, the Haunted Mountain. And Sam and Frodo struggled in the ravines of the Emyn Muil among twisted birch and gaunt fir. There are numerous illustrations of Middle-earth landscapes by Alan Lee and Ted Nasmith, and one of our favorites shows Túrin Turambar trudging through the snow in Dor-lómin, seeking Morwen, with the fir-clad slopes of the Ered Wethrin in the background (see *The Silmarillion*, opposite page 221, illustrated by Ted Nasmith). Tolkien's illustration "The Misty Mountains looking west from the eyrie toward Goblin Gate" shows well the characteristic form of fir trees (see illustration no. 38, in W. G. Hammond and C. Scull's *The Art of the Hobbit*). Finally, the haunted pine forest of Taur-na-Fúin was beautifully illustrated by Tolkien (see no. 48, in *The Art of the Hobbit*).

GRASSLANDS, PRAIRIES, AND STEPPES

Grass-dominated landscapes were common in Middle-earth. In the First Age, tall grasses covered the beautiful plain of Ard-galen (Figure 2.5), which was used for grazing horses by the men of Hithlum but perished in the Battle of Sudden Flame.

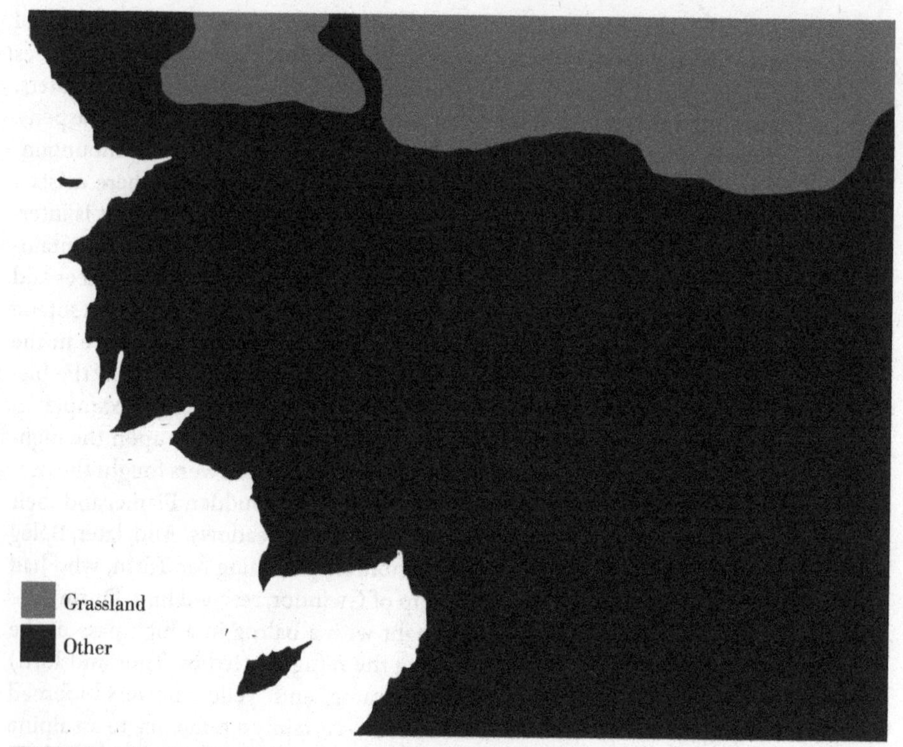

FIGURE 2.5 *Grasslands of western Middle-earth of the First Age.*

It became a burned and desolate waste and was thereafter called Anfauglith, the Gasping Dust. Shorter grasses dominated the plains east of Ard-galen and north of Maglor's Gap, and these were destroyed in the same battle. Finally, the northern portion of Hithlum, west of Ard-galen and on the eastern side of the Ered Wethrin, was also grass-dominated (Figure 2.5). In the Second Age, grasslands dominated the central portion of Númenor (which was called Mittalmar), and these prairies surrounded the tall mountain Meneltarma, sacred to the worship of Eru Ilúvatar (Figure 2.6). The grazing of sheep was centered in the region of Emerië, which was beloved by Erendis. The Orrostar (Eastlands) peninsula was originally dominated by taller grasses but later became the major grain-producing region of Númenor. In the Third Age, prairies characterized Rohan, with shorter grasses dominating in the East Emnet and taller grasses in the West Emnet—a large region east and south of Fangorn and extending southward to the White Mountains (Figure 2.7). Through the midst of this broad grassland flowed the Entwash, with its floodplain covered by sedge- and reed-dominated wetlands. Grasslands also dominated the adjacent northern portion of Gondor (Anórien), and, just to the south in Lebennin, grasslands occurred among scattered woodlands. Smaller and more discontinuous grassland regions occurred farther north, in the valley of the Anduin River, east of the Misty Mountains, and in the Barrow Downs and South Downs, located east of the Old Forest. Finally, a dry short-grass prairie occurred just south of the Sea of Rhûn, and this expansive prairie also occupied the southeastern portion of Mordor (i.e., Nurn and region around the Sea of Núrnen, which was

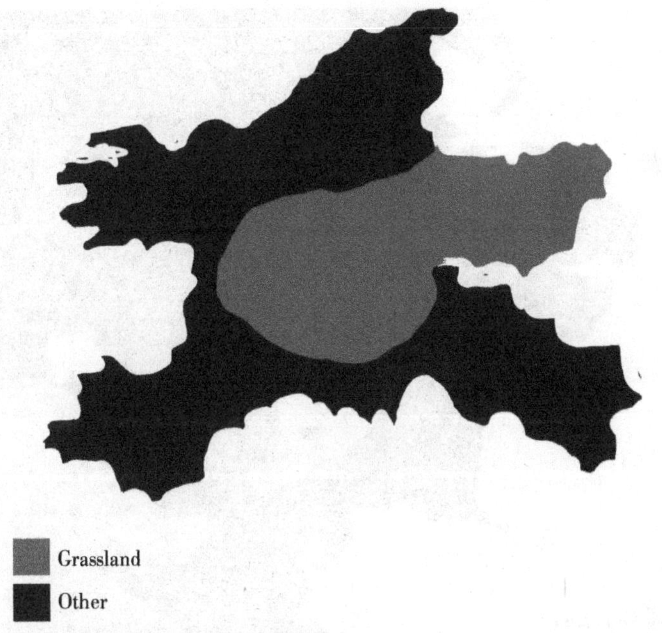

FIGURE 2.6 *Grasslands of Númenor (Second Age).*

surrounded by saline flats), extending as far south as Khand (Figure 2.7). The grasslands of Nurn were, however, much degraded, and mainly occupied by vast fields tended by Sauron's slaves. All these grasslands were undoubtedly similar to those occurring in our prairies and steppes today. Grasslands dominate the central portion of North America, extending from southern Canada, through the Great Plains of the United States, and ending in central Mexico. In these regions, limited rainfall and frequent fires limit the growth of trees. The dominant grasses in North American prairies are *Agropyron, Andropogon, Bouteloua, Buchloe, Calamovilfa, Koeleria, Muhlenbergia, Panicum, Poa, Shisachyrium, Sorghastrum, Sporobolus*, and *Stipa*. Eurasian grasslands occur from eastern Romania and Ukraine, eastward across Asia to northern China. The dominant grasses in these regions are *Agropyron, Cleistogenes, Festuca, Koeleria, Leymus, Poa*, and *Stipa*. The grasslands of both North America and Eurasia also support various sedges (*Carex*) as well as a diverse array of broadleaved herbs and low shrubs. It is interesting that grasslands are characteristic of the central regions of both North America and Eurasia. Finally, this biome has been much altered on all continents by agricultural activities. Although individual prairie grasses are not mentioned in Tolkien's writings, we suspect that the grasslands of Middle-earth were likewise dominated by a diverse array of species. Grasslands are a conspicuous part of our mental image of Middle-earth due to the importance of the Rohirrim (Riders of Rohan) in the events at the end of the Third Age—who can forget the moment when Aragorn, Gimli, and Legolas first encountered the grass of Rohan, which "swelled like a green sea up to the very foot of the Emyn Muil" or their meeting with Éomer, who exclaimed, "Have you sprung out of the grass?" (LotR 3: II). The expansive prairies of Rohan are shown in an illustration by Alan Lee (see *The Lord of the Rings* illustrated by Alan Lee,

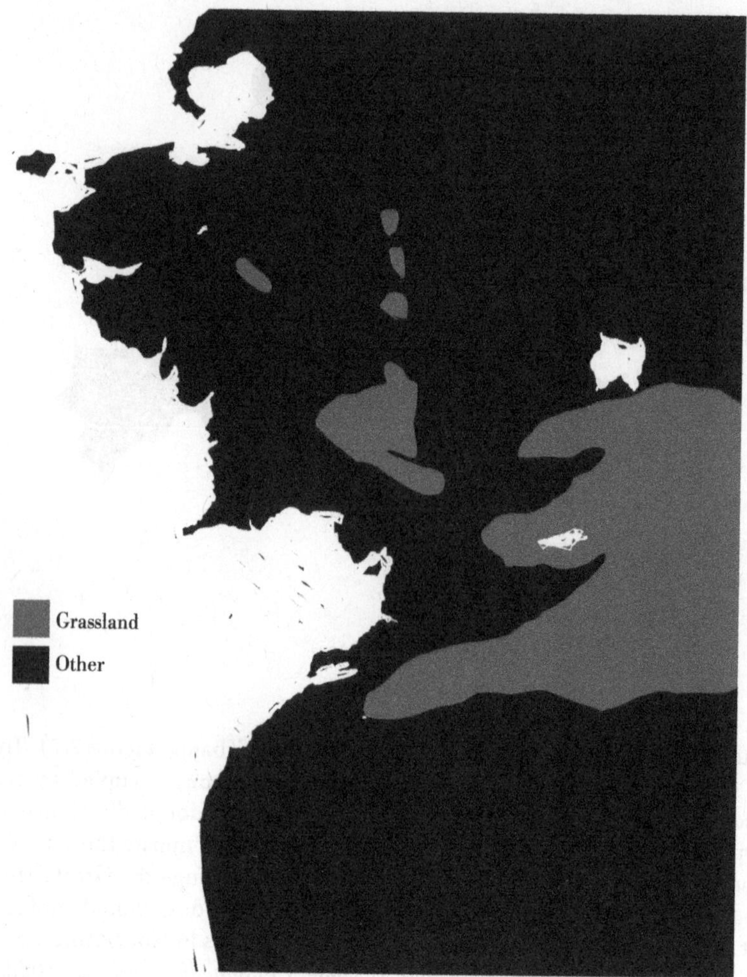

FIGURE 2.7 *Grasslands of western Middle-earth in the Third Age.*

opposite page 448). Grasslands were much less conspicuous in the First Age, and their destruction by "rivers of flame" (SILM 18) running down from Thangorodrim in the Battle of Sudden Flame is probably the most memorable event connected with these early grassy plains; Ard-galen then became Anfauglith. The crossing of this plain, now a place of "ashes and dust and thirsty dune" (Lays III: line 3478), by Lúthien and Beren in the fell forms of monstrous bat and werewolf is perhaps just as dramatic, especially as related in *The Lay of Leithian* (see lines 3456–3485 and painting by Ted Nasmith in the illustrated *Silmarillion*, opposite page 176).

DESERTS

The largest expanse of desert habitat in Middle-earth occurs in the far south, in the lands of the Haradrim (i.e., the Haradwaith) (Figure 2.8). However, these warm to hot deserts (and their unique plants) are nowhere described in Tolkien's writings.

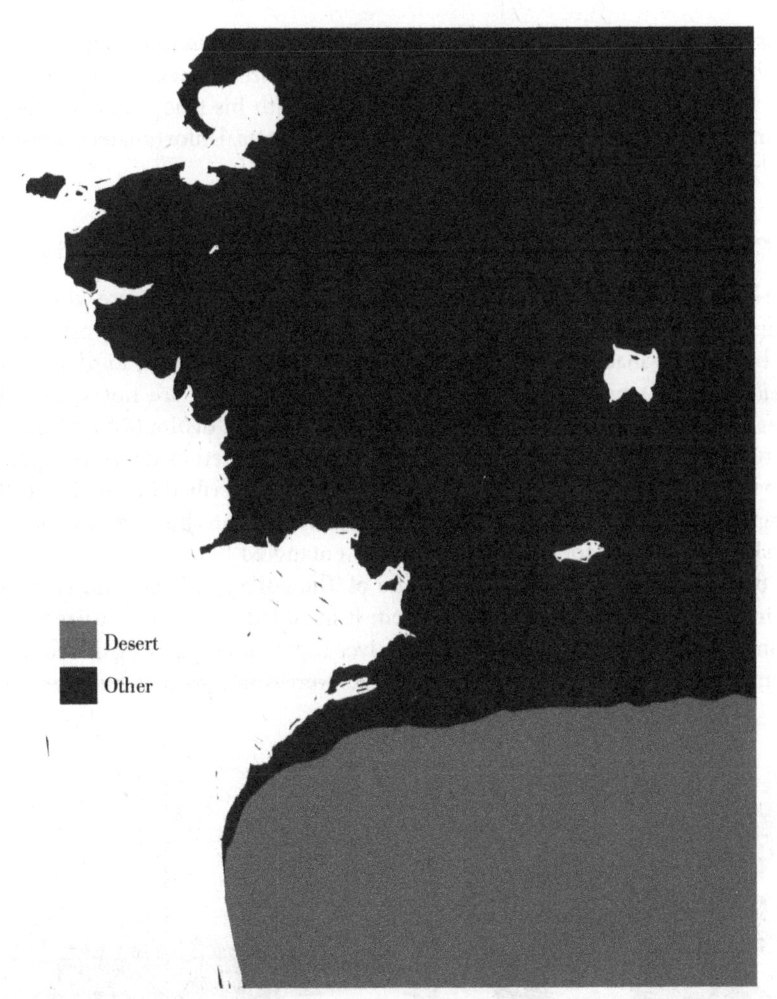

FIGURE 2.8 *Deserts of western Middle-earth in the Third Age.*

In North America, deserts occur in the extreme southwest—in California, Nevada, Arizona, New Mexico, and western Texas—and represent warm areas with low rainfall. Such regions are dominated by a diverse array of arid-adapted plants, including many with very small, leathery leaves, succulent stems, and thorns or spines. Deserts, with similarly adapted plants, occur in Eurasia from north and east of the Caspian Sea and extend eastward to northern China. Deserts and semi-deserts also occur just south of Europe, in northern Africa. During the Third Age of Middle-earth, for more than 2,000 years, the Haradrim (or Southrons) frequently attacked Gondor but were repeatedly defeated. Their troops used giant war-elephants called mûmakil, one of which Sam was delighted to see in Ithilien. They were one of the major forces allied with Mordor in the battle of the Pelennor Fields, during which Théoden attacked and threw down their chieftain and standard bearer. But these events occurred in Gondor, so, in *The Lord of the Rings*, as mentioned earlier, we do not find any description of these southern deserts, and thus none of the plants of Middle-earth treated in this book is representative of this common biome. Deserts are thus quite peripheral to the history of Tolkien's Middle-earth; yet they are not entirely absent, as in Appendix A, I (iv), we read of Ciryaher, king of Gondor, who with his troops crossed the River Harnen and defeated the people of these southern deserts. Unfortunately, these events are only very briefly related.

COOL- TO WARM-TEMPERATE, DECIDUOUS FORESTS

Dense broadleaved, deciduous forests as well as open woodlands were widespread in Middle-earth (Figures 2.9, 2.10, and 2.11). The former (i.e., dense forests) are indicated by treelike shading in the maps associated with *The Hobbit, Lord of the Rings,* and *Silmarillion,* whereas the latter (i.e., open woodlands) are not shown, being represented merely by a white background (and thus not distinguished from grassland, tundra, or desert biomes). Broadleaved forests, whether dense or open, predominate in the history of Middle-earth, and they are described in detail in Tolkien's writings. We here briefly describe the major forested regions but mention only a few of the key events that took place under their entangled branches. In the First Age, the protected region of Doriath, the home of Thingol and Melian and enclosed by the Girdle of Melian, was densely forested: it included the Forest of Brethil in the west and Neldoreth in the north, with the River Esgalduin separating this forest from Region in the south (Figure 2.9). These forests were dominated by beeches (*Fagus*),

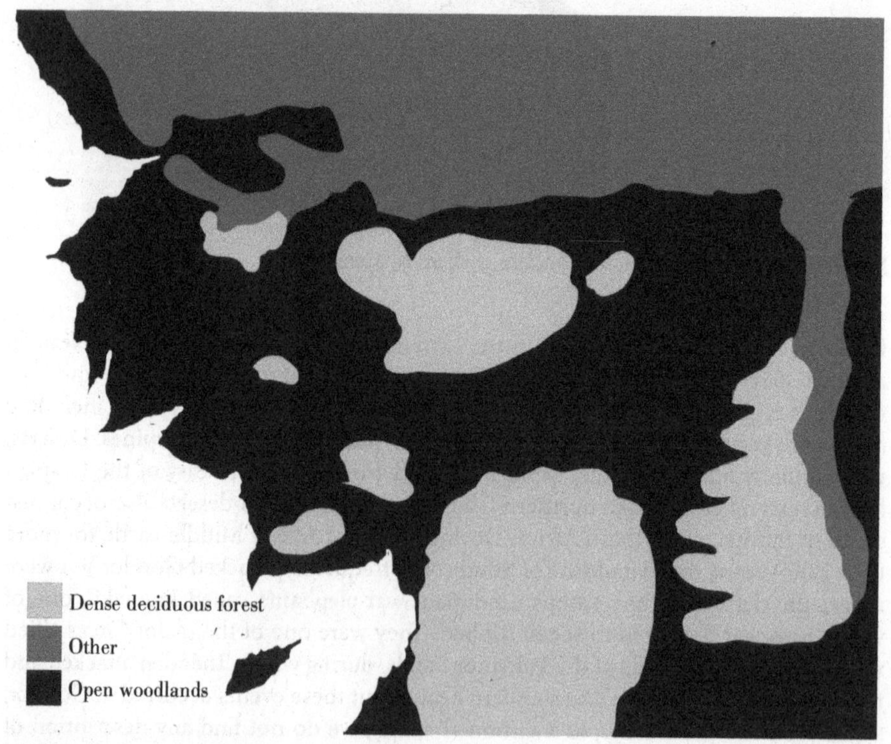

FIGURE 2.9 *Temperate, deciduous forests of Beleriand (First Age).*

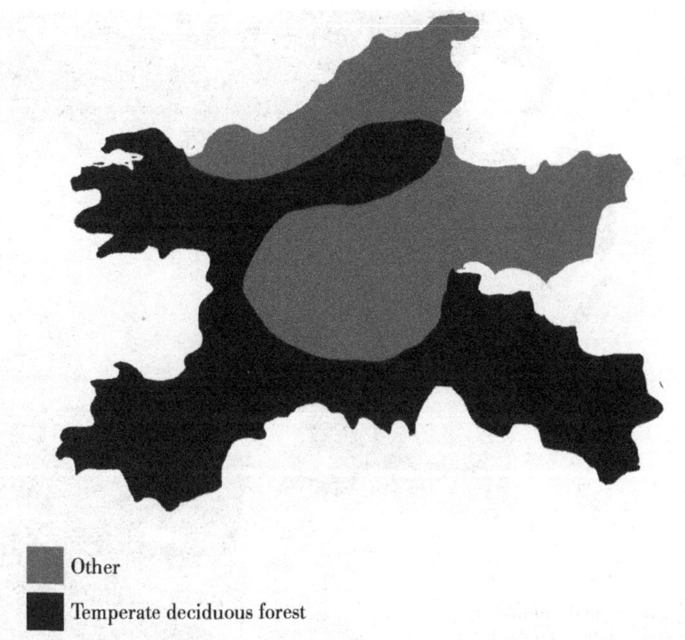

FIGURE 2.10 *Temperate, deciduous forests of Númenor (Second Age).*

oaks (*Quercus*), elms (*Ulmus*), and horse-chestnuts (*Aesculus*). In fact, Thingol is called "the king of beech and oak and elm" (Lays III: line 72). Beeches, especially, dominated the forest of Neldoreth—and Treebeard sang of walking in "the beeches of Neldoreth" (LotR 3: IV) in the autumn. After Lúthien rescued Beren from the dungeons of Sauron, they returned to the woods of Doriath:

> resting in deep and mossy glade;
> there lay they sheltered from the wind
> under mighty beeches silken-skinned
> and sang of love that still shall be,
> though earth be foundered under sea. (Lays III: lines 3221–3225)

Beeches grew near the gates of Menegroth (as shown in the illustration by Alan Lee, opposite page 96, in *The Children of Húrin*), and even the pillars of this great underground dwelling were "hewn in the likeness of the beeches of Oromë" (SILM 10). The forest of Brethil was dominated by beeches (as indicated by its name), along with numerous oaks and birches (see illustration opposite page 192 by Alan Lee, in *The Children of Húrin*, and illustration opposite page 224 by Ted Nasmith, in *The Silmarillion*). However, Region was a mixed forest, with beeches, oaks, elms, hollies (*Ilex*), lindens (*Tilia*), and yews (*Taxus*), among other trees. The nearby forest of Nan Elmoth, just east of Doriath (and south of the River Celon) was floristically similar to that of Region, but its trees were "the tallest and darkest in all Beleriand" (SILM 16). Melian had walked there in the twilight of Middle-earth, and this forest was also the home of the Dark Elf, Eöl. Just to the west of Doriath and along the southern slopes of

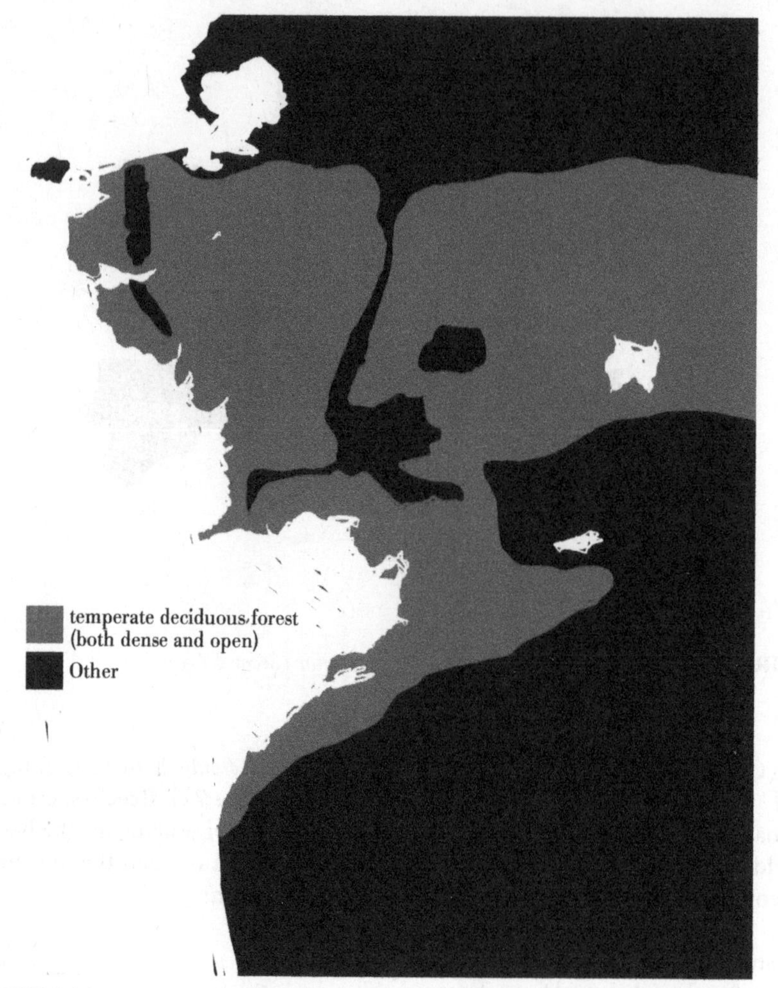

FIGURE 2.11 *Temperate, deciduous forests of western Middle-earth in the Third Age.*

the Ered Wethrin was another beech forest that surrounded the beautiful Lake Ivrin—the woods of Núath. The beautiful and ancient beeches of this forest, many of which grew near the shores of this lake, are described in the *Lay of the Children of Húrin*; their gray trunks "marched in majesty" and their "translucent" and "golden russet" leaves were blown in the morning breezes (Lays I: lines, 1512–1514). It was in this forest, on the shores of Lake Ivrin, that Túrin received healing from grief after the death of his friend Beleg.

Much farther south, where the River Narog joined the Sirion, was the beautiful forest of Nan-Tathren, which was swampy, dominated by willows (*Salix*), and under the protection of Ulmo. Nan-Tathren (Sindarin for Willow-vale) was visited by Voronwë, who told Tuor that the willows of this forest were the fairest of all the region's plants, describing them as "pale green, or silver in the wind" and stating that "the rustle of their innumerable leaves" was like a "spell of music." Time seemed to nearly stand still, and in this forest he stood for days "knee-deep in the grass and listened" and "wandered, naming new flowers" (UT 1: I). This forest was also described by Treebeard, who sang of walking "in the willow-meads of Tasarinan" in the spring, exclaiming: "Ah! The sight and the smell of the Spring in Nan-tasarion!" (LotR 3: IV). South of Nan-Tathren and just north of the Bay of Balar was Nimbrethil, a woods dominated almost exclusively by white birches (i.e., various species of *Betula* having lovely, smooth, and white trunks, as indicated by its name, which is Sindarin for white birch). It was timber from these trees that Eärendil, with the help of Cirdan, used in making Vingilot, his ship, the *Foam-Flower*, in which he and Elwing sailed to Valinor. Eastward, across the Sirion from Nimbrethil, was the great forest Taur-im-Duinath, which filled all the land between the River Sirion in the west and the River Gelion in the east. The dominant trees of this forest, which was peripheral to the major events in the history of the First Age, are not described. The last of the major densely forested regions of Beleriand in the First Age occurred on the western slopes of the Ered Luin and was dissected by the seven rivers of Ossiriand. This region was given by Thingol to Denethor, the lord of the Nandor (Green-elves), and these elves long lived among these trees. Treebeard sang of wandering in the summer "in the elm-woods of Ossiriand" (LotR 3: IV), although undoubtedly other trees also occurred there. The dominance of elms in this region is not surprising, given that most elms prefer the low, wet soils of river floodplains. This large forest is not indicated on the map of Beleriand that accompanies *The Silmarillion* (but it is shown in the vegetation map on page 185 in K. W. Fonstad's, 1991, *The Atlas of Middle-earth*).

In Númenor (during the Second Age; Figure 2.10) the peninsulas of Andustar, Hyarnustar, and Hyarrostar were largely covered by a diverse, mixed broadleaved forest (in which grew birch, beech, oaks, and elms, among many others). Beginning in the days of Tar-Aldarion, there were many tree plantations in the Hyarrostar, which provided timber for ship-building.

In the First Age and early Second Age, the dense forests of Eriador, west of the Ered Luin, were much more extensive than those shown in the map accompanying *The Lord of the Rings*. As Treebeard much later told Merry and Pippin, "there was all one wood once upon a time from here [Fangorn] to the Mountains of Lune, and this was just the East End" (LotR 3: IV). In fact, this plant biome was so common and widespread that it was used to represent Middle-earth itself, as seen in the elven song to Elbereth that Frodo heard in the Hall of Fire, which pictures the singer as gazing at Elbereth's stars "*o galadhremmin ennorath*" (LotR 2: I; "from tree-tangled Middle-earth"; see also discussion in T. Shippey's, 2002, *J. R. R. Tolkien: Author of the Century*). By late in the Third Age, these woods were much dissected, and large areas formerly occupied by dense forest now instead supported a landscape of open woodland or scrub. This destruction of the original virgin forest of the region, described by Treebeard as similar to the forest of Lothlórien, "only thicker, stronger, younger" (LotR 3: IV) was the result of years of forest clearing, especially by humans and dwarves, and also the burning and cutting of trees associated with the military campaigns and destructive environmental policies of Morgoth and later Sauron (and, to a lesser extent, Saruman). There is, therefore, some justification for the hatred of the trees of the Old Forest for those "that go free upon the earth, gnawing, biting, breaking, hacking, burning": as related by Tom Bombadil, the trees considered orcs, humans, dwarves, and even hobbits to be "destroyers and usurpers" (LotR 1L VII; see also discussion in *Taking the Part of Trees: Eco-Conflict in Middle-earth*, by V. Flieger, 2000, and *Representations of Nature in Middle-earth*, edited by M. Simonson, 2015).

By the time of the war against Sauron, the dense, broadleaved forests of Middle-earth (see areas indicated by tree-shading in the map of the West of Middle-earth in *The Lord of the Rings*) were few and far between and were represented west of the Misty Mountains only by (1) the remnants of the forests of Ossiriand (just west of the Ered Luin), occupying the small portion of Beleriand still extant; (2) the forests of the North Farthing, the Green-hill Country, and the Woody End of the Shire; (3) the Old Forest, just east of the Shire; (4) the Chetwood near Bree; and (5) the forests of the Trollshaws and nearby Rivendell, near the western slopes of the White Mountains. These forests were quite diverse, with different species dominating depending upon exposure, soil conditions, and hydrology. The dominant trees, however, were alders (*Alnus*), ashes (*Fraxinus*), beeches, birches (*Betula*), elms, firs (*Abies*), hawthorns (*Crataegus*), hazels (*Corylus*), horse-chestnuts, lindens, oaks, pines (*Pinus*), sloes (*Prunus*), wild apples (*Malus*), and willows, and, as the hobbits noted in the Old Forest, "strange and nameless trees of the denser wood" (LotR 1: VI). Holly trees (*Ilex*) seem to have occurred mainly in the region of the former elven land of Eregion (Hollin). Floristically similar forests also occurred east of White Mountains, and the most important of these were (1) the forest of northern and central Mirkwood, (2) the forest near the northeastern shore of the Sea of Rhûn, and (3) Fangorn forest, just east of the southern portion of the Misty Mountains. Of these forests, Fangorn was the most diverse, probably having all of the trees listed earlier as well as rowans (*Sorbus*) and strange holly-like trees with stiff upright flower-spikes (possibly *Banksia*). Nearly all of these forests are intimately connected to the events related in *The Hobbit* and *The Lord of the Rings*—from Bilbo and the dwarves' hike along the elven path through Mirkwood (see watercolor by Alan Lee opposite page 128 in illustrated edition of *The Hobbit*) and the first meeting of Frodo, Sam, and Pippin with the elves in the Woody End (see watercolor by Alan Lee opposite page 96 in illustrated edition of *The Lord of the Rings*), to the hobbits meeting Tom Bombadil along the Withywindle, discovering three stone trolls in the Trollshaws while being led by Aragorn, or Merry and Pippin meeting Treebeard and Quickbeam in Fangorn (see watercolor of Merry and Pippin in Fangorn forest by Alan Lee, opposite page 480 in the illustrated *Lord of the Rings*). These and many other events take place under a canopy of mixed, broadleaved hardwoods.

In North America, such deciduous, broadleaved forests are characteristic of the eastern half of the United States (although in many regions the forest has been much fragmented by agricultural activities or urban growth). This region is dominated by broadleaved, mainly deciduous species of trees, although evergreen species become more common in the southern portions of this widespread forest. In addition, on the sands of the southeastern US coastal plain, pines often become a dominant element in the landscape. The forest composition is diverse and varies with climatic regime, hydrological conditions, and soil type. Dominant trees include ashes, bays (*Persea*), beeches, birches, chestnuts (*Castanea*), cherries (*Prunus*), dogwoods (*Cornus*), maples (*Acer*), hickories (*Carya*), hollies, horse-chestnuts, elms, lindens or basswoods, magnolias (*Magnolia*), mulberries (*Morus*), oaks, pines, sugarberries or hackberries (*Celtis*), sweetgums (*Liquidambar*), tupelos (*Nyssa*), tulip-poplars (*Liriodendron*), and walnuts (*Juglans*). Such cool to warm temperate forests are extremely widespread in Europe, occurring in a broad east–west zone from Spain eastward to Russia and Kazakhstan, often bounded by boreal, coniferous forests to the north and (at least in eastern Europe) grasslands to the south. The deciduous forests of Europe are quite similar to those of eastern North America but are somewhat less diverse and are dominated by broadleaved species such as ashes, beeches, birches, chestnuts, horse-chestnuts, elms, lindens, maples, and oaks, but also include needle-leaved species such as pines and hemlocks (*Tsuga*). Finally, temperate broadleaved forests also occur in eastern Asia, especially central and northern China, Korea, and Japan. The species diversity of these forests is much greater than comparable forests of Europe and North America (as is evident from the species numbers provided in the plant treatments; see Chapter 7).

Two of the forested regions of Middle-earth were rather distinctive and so are discussed separately later. These are the forests of Lothlórien and Ithilien. The forest of Lothlórien, the dwelling place of Galadriel and Celeborn, was unique in Middle-earth, being dominated by majestic mallorn trees (see treatment of this tree in Chapter 7 and illustration no. 157 in *J. R. R. Tolkien: Artist and Illustrator*, by W. G. Hammond and C. Scull, 1995). Mallyrn were brought to Middle-earth by the elves with the assistance of Tar-Aldarion, the sixth king of Nûmenor, and this was the only region where they were common. This forest here is considered as kind of temperate broadleaved forest because the mallorn leaves turn golden yellow in the autumn. However, the trees were only tardily deciduous, with the leaves being held during the winter and only dropping in the spring when the trees came into bloom. Of course, mallorn forests have no counterpart in the forests of North America or Europe today (although they are somewhat reminiscent of mature beech forests—but more open and park-like). They represented an attempt by the elves (essentially horticultural in nature) to enhance the natural beauty found in temperate forests. We, like the elves of Lothlórien, appreciate the beauty of nature. Thus, we also attempt to enhance the beauty of our native trees and shrubs, for example, through the selection and breeding of genotypes with red leaves even during the growing season or those with weeping or twisted branches. And, of course, a similar impulse underlies the breeding of plants with enlarged, doubled, or otherwise more striking flowers.

The forests and open woodlands of Ithilien were almost equally distinctive (see watercolor showing Sam and the Oliphaunt in illustrated *The Lord of the Rings* opposite page 688). They were exceptionally diverse and contained many broadleaved evergreens, such as bays (*Laurus*), boxwoods (*Buxus*), olives (*Olea*), holm oaks (*Quercus ilex*), and myrtles (*Myrtus*), as well as conifers (e.g., true cedars [*Cedrus*], cypresses [*Cupressus*], junipers [*Juniperus*], and pines). Many resinous and aromatic species, such as terebinths (*Pistacia*), myrtles, cedars, and various medicinal or culinary herbs, also grew in these forests. They were floristically very similar to Mediterranean forests and woodlands, which characterize southern Europe today, especially in regions very close to the Mediterranean Sea. In fact, the forest of Ithilien appears especially similar to the forests of Greece and coastal Turkey, which, like all Mediterranean forests, develop in regions with dry summers and rainy winters. Of course, many key events in Tolkien's legendarium occurred in Lothlórien and Ithilien. We may first think of the meeting of Frodo and Galadriel or the visions revealed to Frodo and Sam in her Mirror. Our mental image of Cerin Amroth, the heart of Lothlórien, a hill crowned by two rings of trees—the outer of birches "leafless, but beautiful in their shapely nakedness," and the inner of mellyrn, "still arrayed in pale gold" (LotR 2: VI)—also is enduring. Of course, in Ithilien, Frodo and Sam are revived after their experience at the Gate of Mordor, and it is also where they first met Faramir, receiving unexpected help in their quest.

Forests and woodlands frequently are interspersed with more open habitats that develop as a result of edaphic or hydrological conditions. Two of these communities are conspicuous in Tolkien's descriptions of Middle-earth and are discussed here: (1) heathlands and moorlands and (2) various other wetland habitats. Heaths and moorlands are moist to wet open habitats (i.e., without trees or with only a few scattered trees or shrubs). They develop on quite acidic and low-nutrient soils and are dominated by grasses and grasslike plants such as sedges (*Carex, Scirpus,* and related genera), cotton-grass (*Eriophorum*), rushes (*Juncus*), and mat-grass (*Nardus*); occasional herbs, such as blackberries (*Rubus*); and ericaceous shrubs, such as heather (*Calluna*), heath (*Erica*), bilberry, whortleberry, and relatives (*Vaccinium*), and crowberry (*Empetrum*). Mosses are common, especially peatmoss (*Sphagnum*), and bracken (*Pteridium*), a widely distributed fern, occasionally forms dense brakes. In the First Age, the largest regions dominated by moors were the highlands of Dorthonion, especially the vicinity of Tarn Aeluin (see illustration opposite page 161 by Ted Nasmith, in the illustrated *Silmarillion*), where Barahir, Beren, and their few followers made their lair when they were hunted by the forces of Morgoth, and much of the vast plain between the Rivers Sirion and Narog, in which was situated Amon Rûdh, a prominent hill that for a time was the home of Túrin Turambar. In the Second Age, high moors occurred in the northern peninsula of Númenor. In the Third Age, moorlands are mentioned as occurring in the Shire (the North Moors) and on the summit of Dol Baran, the southernmost of the foothills of the Misty Mountains, but they probably also occurred in other hills of this range and are recorded from the highlands near Rivendell. Several of the previously listed plants are included in Tolkien's writings, but it is interesting that he did not mention either cotton-grass, which is striking when in fruit, or crowberry, with its showy, purple-black, subglobose fruits—both are common moorland species in the British Isles.

It is clear that wetlands were close to Tolkien's heart, and many are depicted in great detail in his writings. Major wetlands of the First Age include (1) the marshes of Nevrast; (2) the fens of Serech, north of Tol Sirion and between the Ered Wethrin and the mountains of western Dorthonion; (3) the Aelin-uial (Meres of Twilight), near the southwestern corner of Doriath; and (4) the marshes of the delta of the River Sirion. The species of these wetlands are not described in detail; instead, the presence of reeds (*Phragmites australis*) is emphasized. The brothers Húrin and Huor fought the orcs of Morgoth beside the fens of Serech at the end of the Nirnaeth Arnoediad (Battle of Unnumbered Tears), and it was there that their heroism allowed Turgon and his forces to escape, although in the end Huor was slain and Húrin taken captive. Later Tuor, Huor's son, wandered in the beautiful marshes of Nevrast before finding in Vinymar the armor left for him by Turgon, and he there received his quest from Ulmo to find the Hidden Kingdom of Turgon and deliver the message of Ulmo. Nienor and her mother Morwen (Húrin's wife) were ferried over the Meres of Twilight on their search for Túrin, Nienor's brother and Morwen's son, and, soon thereafter, Nienor was bewitched by the dragon Glaurung. Finally, near the end of the First Age, when all of the elf kingdoms had fallen, Tuor and his wife Idril, the daughter of Turgon, moved from the refuge of Nan-Tathren southward to the marshes of the Mouths of the Sirion. They dwelt there for a while in peace until they set sail "into the sunset and the West" (SILM 23). Why are so many events in the lives of Húrin and Huor and their children (of the House of Hador), connected with wetlands? Perhaps because Ulmo, the

Lord of Waters, never abandoned humans and elves, even when they lay under the wrath of the Valar; his power was focused in rivers and wetlands, and he had chosen the House of Hador to accomplish his purpose. In Númenor, during the Second Age, the only major wetland occurred near the mouths of the River Siril, where there were "wide marshes and reedy flats" (UT 2: I). In the Third Age, extensive wetland habitats occurred (1) in the Shire (Overbourn Marshes); (2) the Midgewater Marshes east of Bree; (3) the Nîn-in-Eilph (Swanfleet), where the Mitheithel joined the Glanduin River; (4) the Loeg Ningloron (Gladden Fields), along the Anduin River, north of Lórien; (5) along the Entwash, especially where it joined the Anduin River to form the Nindalf (Wetwang); and (6) in the adjacent Dead Marshes (see watercolor by Alan Lee opposite page 656 in illustrated *The Lord of the Rings*). Characteristic species of these wetlands—in order of their conspicuousness—include reeds (*Phragmites australis*), flag-lilies or yellow iris (*Iris pseudacorus*), rushes (*Juncus*), and sedges (*Carex*, and relatives). Undoubtedly many other plants occurred in these regions as well, especially waterlilies (*Nymphaea*), which are mentioned as occurring in several rivers and ponds. Again, these localities are connected with key events in Tolkien's narrative. The One Ring was lost in the Gladden Fields, and Isildur died there as he struggled among "great rushes and clinging weeds" (UT3: I). There the Ring lay there for well over 2,000 years until found by Déagol, and it was then almost immediately taken by Sméagol (Gollum). Aragorn led the four hobbits through the Midgewater Marshes on their way to Rivendell, and, in these marshes, they were tormented by flies and "came upon pools, and wide stretches of reeds and rushes" (LotR 1: XI). Much later, when the fellowship had become divided, Gandalf, Aragorn, Legolas, and Gimli passed through the wetlands adjacent to the Entwash, with grass "so high that it reached above the knees of the riders" and "broad acres of sedge waving above wet and treacherous bogs" (LotR 3: V). Frodo and Sam, led by Gollum, struggled through the "dead grasses and rotting reeds" of the Dead Marshes (LotR 4: II). As Gandalf and Galadriel returned north after the war of the ring, they crossed a river and saw to their west the Swanfleet wetlands, where "countless swans housed in a land of reeds" (LotR6: VI). Except for the Dead Marshes, such wetlands, although they can be difficult to traverse, are viewed as places of beauty—and in both *The Silmarillion* and *The Lord of the Rings* they are often associated with birds.

COMPARISON OF THE TREES OF MIDDLE-EARTH WITH THOSE OF ENGLAND

Contrary to what some readers of the Hobbit or the Lord of the Rings may think, Middle-earth is our own world. As Tolkien stated in a letter sent to Houghton Mifflin Co., "'Middle-earth' . . . is not a name of a never-never land without relation to the world we live in . . . It is just a use of Middle English *middle-erde* (or *erthe*), altered from Old English *Middangeard*: the name for the inhabited lands of Men 'between the seas'" (Letters: No. 165). Thus, we should not be surprised, as outlined earlier, that the plant biomes and communities of Middle-earth bear a striking resemblance to those of Europe and especially England. And since the flora of Europe is closely related to that of temperate regions of Asia and North America, we can find the plants of Middle-earth in the meadows and forests near our homes—at least if we live in the temperate regions of the Northern Hemisphere.

It is useful, therefore, to compare the trees native to England with those occurring in the landscapes of Middle-earth. Detailed presentations of each of the trees of Middle-earth can be found in Chapter 7, along with a discussion of their importance in Tolkien's legendarium. Here, we merely summarize some interesting distributional patterns. Of the 34 genera of trees occurring in Tolkien's legendarium, almost all are European natives (although most of these are quite broadly distributed in the Northern Hemisphere, so they also grow in temperate Asia and North America). The only exceptions are oranges (*Citrus* x *aurantium*, Sweet Orange Group) and camellias (species of *Camellia*, including tea, *C. sinensis*, and ornamentals, such as *C. japonica*), and both of these are native to eastern Asia. Both plants also are of minor importance in Tolkien's writings—with oranges only mentioned in the poem *Errantry*, and the genus *Camellia* noted only in the name Camellia Sackville, included among the Baggins of Hobbiton (see LotR, Appendix C) and in the reference to tea in Chapter 1 of *The Hobbit* and *The Lord of the Rings*. Both are edibles, and we can assume that they were imported into the Middle-earth of the Third Age. Of those Middle-earth trees occurring in Europe, 20 are native to England, and the remaining 13 do not occur natively there (although some of these have been introduced and have become naturalized, and many are grown as cultivated ornamentals). Those Middle-earth trees occurring natively in the British Isles include alders (*Alnus*), ashes (*Fraxinus*), beeches (*Fagus*), birches (*Betula*), box or boxwood (*Buxus*), elms (*Ulmus*), gorse (or whin; *Ulex*), hawthorns (*Crataegus*), hazels (*Corylus*), hollies (*Ilex*), junipers (*Juniperus*), lindens (*Tilia*), oaks (*Quercus*), pines (*Pinus*), poplars (*Populus*), rowans (*Sorbus*), sloe (and plums, cherries; *Prunus*), wild apples (*Malus silvestris*, with *M. domestica* presumably introduced at an early date), willows (*Salix*), and yews (*Taxus*). Amazingly, this list includes all of the trees native to England except for maples (*Acer*, represented by *A. campestre*) and hornbeams (*Carpinus*, represented by *C. betulus*). Both maples and hornbeams are widely distributed in temperate regions, occurring in Europe, Asia, and North America. It is a mystery why these two trees were omitted from Tolkien's legendarium—especially maples, which are widely used as shade or ornamental trees. Maples have, however, made their way into Middle-earth lore in a peripheral way: Alan Lee's illustration of Cerin Amroth, which clearly shows mellyrn and birches, is decorated by a border showing the smooth gray trunks of mallorn-trees and the autumn-yellow leaves of sugar maples (*Acer saccharum*)! The imaginative *Beyond Middle Earth* (Williams, 2014) suggests, on the basis of no actual data, that *A. campestre* and other species of *Acer* (e.g., *A. platanoides* and *A. pseudoplatanus*) grew in the Middle-earth of the Third Age. In addition, two related species, buckthorn (*Rhamnus catharticus*) and alder-buckthorn (*Frangula alnus*), occasionally reach tree size and are native to England but are not mentioned in Tolkien's writings. Tolkien states (again in Letter No. 165) that he is "much in love with plants and above all trees," and this is easily seen in the fact that he manages to include all but two (or four, if we include the largely shrubby buckthorns) of the trees native to England in his legendarium. Only someone whose eyes were open to the wonderful diversity of temperate plant communities—and the plants that comprise them—could have done this and done it as well as Tolkien has, including in his writings succinct but salient characterizations of nearly all of these trees. But additionally, as noted earlier, Tolkien has included within his legendarium many other European trees that are not native in England. These include bays (*Laurus*), cedars (*Cedrus*, which occurs in Turkey, adjacent to the European continent), ebony and relatives (*Diospyros lotus*, with other species of tropical Africa and Asia), firs (*Abies*), horse-chestnuts (*Aesculus*), laburnums (*Laburnum*), larches (*Larex*), olives (*Olea*), cornel (*Cornus mas*, but other species of *Cornus* are native to England), cypresses (*Cupressus*), myrtles (*Myrtus*), tamarisks (*Tamarix*), and terebinths (*Pistacia terebinthus*). He included such European trees because the history of Middle-earth is geographically wide-ranging, occurring across hundreds of miles and in an array of vegetational zones. The climates and floras of Rohan, Lothlórien, or Ithilien are much different from that of the Shire, and this vegetational diversity, therefore, is reflected in the plants encountered by those involved in these historical events. Finally, the plants of Middle-earth are not merely descriptive detail—part of the background of various scenes—but, as stressed in Chapter 1, are actually involved in the story.

PLANT COMMUNITIES, AS ALTERED BY HOBBITS, HUMANS, ELVES, OR ANGELIC BEINGS (MAIAR OR VALAR)

It is often important (and ecologically useful) to consider the changes in plant communities resulting from the activities of humans: for example, many of the forests of the Northern Hemisphere have been cut for timber or destroyed in connection with conversion of the land to agricultural use, and an even larger percentage of grassland habitats have been lost to agricultural uses (converted to croplands or pastures). Human activities also have modified fire regimes in many regions, with resulting changes in floristic composition. Finally, certain economically valuable species have been selectively eliminated from the landscape through commercial extraction or introduced diseases. Thus, the landscapes of Europe or North America are much different today than they were before these regions were occupied by large human populations. The same is true of the landscapes of Middle-earth—except that those landscapes also had been impacted by hobbit, elven, and dwarven cultures, as well as by angelic beings (such as the Valar or their assistants, the Maiar). Thus, we here briefly discuss such vegetational changes to the plant biomes and communities introduced earlier. We will also describe the agricultural landscapes of Middle-earth.

The vegetation of the Shire was undoubtedly originally a temperate, mixed hardwood forest dominated by deciduous broadleaved trees (such as oaks, ashes, elms, lindens, alders, birches, and willows) along with a few conifers (e.g., mainly pines and firs). However, by the time of the War of the Ring, many of the forests had been cut, and a significant proportion of the land had been converted to pastures and fields. Hobbits were good agriculturalists; they were close to the land—in fact living in burrows—and worked hard to achieve a sustainable agricultural practice. Their fields and pastures were intermixed with hedgerows and small woods (providing a source of firewood but also protecting habitats for wildlife). This mixture of villages, agricultural land, and wild habitats resulted in the beauty of the Shire. Their major crops were the grains wheat (*Triticum*) and barley (*Hordeum*), and their garden vegetables included cabbages and turnips (*Brassica*), potatoes (*Solanum*), carrots (*Daucus*), onions (*Allium*), beans (*Phaseolus*), peas (*Psium*), cucumbers (*Cucumis*), and lettuce (*Lactuca*). Pipe-weed (*Nicotiana*) was also grown in the Shire, especially in the Southfarthing, and was used for smoking (an invention of hobbits and introduced into the Shire by Tobold Hornblower). They grew cotton (*Gossypium*), flax (*Linum*), and hemp (*Cannabis*) for their fibers, which were valuable in making cloth and ropes. Clovers (*Trifolium*) and sweet clovers (*Melilotus*), along with various grasses, were grown for fodder for grazing animals. In their gardens, they certainly grew various culinary herbs and spices, such as peppermint and spearmint (*Mentha*), thyme (*Thymus*), sage (*Salvia*), and parsley (*Petroselinum*), although some spices (e.g., pepper [*Piper*] and cardamom [*Elettaria*]) were received in trade. They also maintained vineyards and orchards—growing grapes (*Vitis*), hops (*Humulus*), apples (*Malus*), and plums and cherries (*Prunus*). And their gardens probably also contained strawberries (*Fragaria*). Their crops were largely from the Old World, with a few significant exceptions such as pipe-weed (see treatment of this issue in Chapter 7). Hobbits maintained lawns and gardens in order to beautify their homes, and common plants of their flower gardens included roses (*Rosa*), sunflowers (*Helianthus*), marigolds (*Calendula*), nasturtians[1] (*Tropaeolum*), snapdragons (*Antirrhinum*), gillyflowers (*Dianthus*), pansies (*Viola*), peonies (*Paeonia*), poppies (*Papaver*), lilies (*Lilium*), and daffodils (*Narcissus*). Of course their yards, gardens, and fields also contained weeds—and a few of these are mentioned in Tolkien's writings (e.g., amaranths [*Amaranthus*], dandelions [*Taraxacum*], daisies [*Bellis*], fireweeds [*Chamerion*], mugworts [*Artemisia*], sweet clovers [*Melilotus*], and thistles [*Cirsium*]—undoubtedly there were many others!).

Agricultural activities in human communities are described in less detail in Tolkien's writings, but their crops were probably similar to those of the hobbits. As described in the *Children of Húrin*, many of the folk of Haleth, during the First Age, lived in the forest of Brethil and in the surrounding woodlands; they lived by hunting and farming, keeping pigs in the forests (where they often ate wild roots, nuts, and acorns) and growing crops in clearings. A striking contrast is seen when these scattered homesteads are compared to the much more developed agriculture of Gondor, in the Third Age. When Merry looked out from the top level of Minas Tirith, he saw "the Pelennor laid out before him, dotted into the distance with farmsteads and little walls, barns and byres" (LotR 5: I). Clearly, this was an area of intense agriculture supporting those living in the city. The agriculture of the people of Rohan depended more on pasture lands than intensely cultivated fields, although these also were present. Just before the battle of Helm's Deep, Théoden looked back down the valley toward the Westfold, telling Aragorn, "This was a rich vale and had many homesteads" (LotR 3: VII) that were then being burned by the forces of Saruman. We should imagine many of the Rohirrim living in scattered homesteads not that different from those of the folk of Haleth, except that they occupied a grassland landscape—not a forested one. In the Third Age, however, discounting densely settled regions such as the vicinity of Minas Tirith (or, to a much lesser extent, Edoras), most of Rohan and Gondor were occupied by prairies, woodlands, and forests not much altered by human hands. Again, part of the appeal of both Rohan and Gondor is that the environment is mixed—including towns and villages, agricultural lands, and wild habitats.

What about the regions occupied by elves? The major elven kingdoms of the First Age, such as Doriath, Nargothrond, and Gondolin, must have been supported by vast fields and orchards, and although these are nowhere described in the *Silmarillion*, they are glimpsed in *The Lay of the Children of Húrin* (see Lays I: lines 1794–1801).

[1] We follow J. R. R. Tolkien's preferred spelling here, instead of the more common "nasturtiums"; see Chapter 7.

Likewise, the agricultural activities of the elves of Mirkwood or Lothlórien in the Third Age are not emphasized. Certainly, the elves provided Frodo, Sam, and Pippin a fine feast in the Woody End—Pippin remembered the "bread, surpassing the savor of a fair white loaf to one who is starving; and fruits sweet as wildberries and richer than the tended fruits of gardens" (LotR 1: III). The elves must have grown the grain used in this bread and cultivated these fruits. We are just not told how. Tolkien does not dwell on the agriculture of the elves because he considered other aspects of their culture to be more important. A major emphasis in elven culture was (as discussed in N. Dickerson and J. Evan's 2006 book, *Ents, Elves, and Eriador*) a focus on the aesthetic qualities of the natural world. The elves sought, through an artistic application of horticulture, to increase beauty in the world. Thus, at this meeting of hobbits and elves, Tolkien instead tells us that their singing went to Sam's heart and that Frodo, even though he knew little of their language, listened eagerly. Elven aesthetic sensibility is also seen in the choice of their place of encampment—a locality "roofed by boughs of trees" and where "great trunks ran like pillars down each side"—lit by torches of gold and silver light (LotR 1: III; as seen in the watercolor by Alan Lee; see illustrated *Lord of the Rings*, opposite page 96). For the elves, ornamental plants (i.e., plants that would not merely add beauty, but would also increase the quality of the life for those who lived among them) were of great importance. Such ornamentals echoed the quality of the Two Trees of Valinor—and we see this in plants such as Nimloth (and its descendants, the various White Trees of Gondor) and the fragrant trees of Eressëa, both of which were originally gifts of the elves to the people of Númenor, or lissuin, with its fragrant white blossoms that bring heart's ease, and the mallorn-trees that graced Lothlórien, with their silver trunks and golden leaves. This aesthetic quality is seen not only in trees, but also in flowering herbs such as the lovely elanor and niphredil. Plant life, as enhanced by the elves, had a quality that cannot be seen in our everyday world; as Tolkien said in one of his letters, "those imagined flowers are lit by a light that would not be seen ever in a growing plant" (Letters: No. 312)—a light that can, perhaps, only be fully appreciated by those who have experienced the light of the Two Trees. This elven artistic impulse, however, which had the power to enhance the beauty of the natural world—spreading the light of the Two Trees—even after their death is also present (in a lesser degree) in hobbits and humans. Otherwise there would not have been elf-friends such as Bilbo and Sam, and we could not appreciate the beauty of our flower gardens and arboreta.

The good stewardship of the land practiced by hobbits, elves, and the human populations of Gondor (and formerly Arnor, as well) is sharply contrasted by the environmental destruction caused first by Morgoth and later by his lieutenant Sauron and the wizard Saruman. As Frodo and Sam approached the Gates of Mordor, they found nothing living—"not even the leprous growths that feed on rottenness." They saw only oxygen-starved pools that "were choked with ash and crawling muds, sickly white and grey, as if the mountains had vomited the filth of their entrails upon the lands" interspersed with "great cones of earth fire-blasted and poison-stained" that "stood like an obscene graveyard in endless rows" (LotR 4: II). The land was defiled and "diseased beyond all healing—unless the Great Sea should enter in and wash it with oblivion" (LotR 4: II), as had been the only cure to the earlier defilement of Beleriand caused by Morgoth (or the societal defilement of Numenor brought about by Sauron). We, like Sam, feel sick at seeing such destruction, and it reminds us of toxic waste dumps, the refuse left after strip-mining, the pollution resulting from uncontrolled industrialization, and the environmental horrors of war (as Tolkien stated "the Dead Marshes and the approaches to the Morannon owe something to Northern France after the Battle of the Somme"; Letters: No. 226; see also *Tolkien and the Great War*, by J. Garth, 2003, and *Ents, Elves, and Eriador*, by M. Dickerson and J. Evans, 2006). Later, as Frodo and Sam entered Mordor via Cirith Ungol, they actually encountered a few plants on the eastern slopes of the Ephel Dúath and realized that, although Mordor was a dying land, along its outer margins it was not yet dead. These plants, however, were harsh and twisted, and they struggled for life. The vegetation consisted of low scrubby thorn trees (*Crataegus*, and perhaps also other thorny shrubs, such as *Rhamnus*) that "lurked and clung" among the rocks of the Morgai. And living trees were far outnumbered by dead wood. They saw that "coarse grey grass-tussocks fought with the stones, and withered mosses crawled on them; and everywhere great writhing, tangled brambles [*Rubus* spp., perhaps also *Rosa* spp.] sprawled." Finally, the brambles were well-defended: "Some had long stabbing thorns, some hooked barbs that rent like knives" (LotR 6: II). Like Morgoth before him, it is clear that Sauron is intent upon the destruction of the natural world—the good creation of Eru. Finally, let's consider the wizard Saruman because his actions at Isengard clarify those more mythological activities of both Morgoth and Sauron. It is significant that Treebeard described the landscape of Isengard as a waste "of stump and bramble where once there were singing groves" (LotR 3: IV). The dominance of brambles, an early successional plant group, is a message that the landscape had been devastated. But why? Because, according to Treebeard, Saruman was greedy for power, had a "mind of metal and wheels," and did "not care for growing things, except as far as they can serve him for the moment" (LotR 3: IV). As has been noted by others (see T. Shippey, 1992, *The Road to Middle-earth*; M. Dickerson and J. Evans, 2006, *Ents, Elves and Eriador*; D. Hazell, 2006, *The Plants of Middle-earth*; S. Jeffers, 2014, *Arda Inhabited;* and *Representations of Nature in Middle-earth*, M. Simonson, ed., 2015), Saruman is linked to our all too frequent use of destructive technology, which Tolkien witnessed as the landscapes of his childhood became increasingly industrialized. In Saruman's opinion, growing things have no inherent value—they are merely things that can and should be used to gain wealth and power. Thus, as related by Quickbeam, the beautiful groves of rowan trees, whose red fruits were a beauty and wonder, had been cut by Saruman's orcs in order to feed the fires of Isengard. He then "called them by their long names, but they did not quiver, they did not hear or answer" (LotR 3: IV)—they were dead. Saruman viewed trees as things (mere commodities) to be used, whereas Quickbeam considered them beautiful friends with cool and soft voices.

THREE

The Diversity of Life, with a Focus on the Green Plants

SINCE THIS IS A BOOK ON THE PLANTS OF J. R. R. TOLKIEN'S MIDDLE-earth, it is certainly appropriate to consider the question: *What is a plant?* Plants are often defined broadly as any living organism that is eukaryotic (i.e., whose cells contain a membrane-bound nucleus in which the DNA is linked with proteins, thus forming chromosomes), has usually cellulose-containing cell walls, and is autotrophic (i.e., capable of converting light into chemical energy through the process of photosynthesis, thus producing carbohydrates from carbon dioxide and water). *Autotrophic* organisms can make their own food, in contrast to *heterotrophic* life forms such as animals (which must ingest, or eat, their food and have cells lacking stiffened walls) or fungi (which must absorb their food and have cells surrounded by chitin-containing walls). The chemical reactions of photosynthesis occur within specialized intracellular organelles called *chloroplasts,* which are ultimately derived from a cyanobacterial ancestor. Chloroplasts have their own DNA and should be considered *endosymbionts*—organisms distinct from the cells in which they live, but which exist in a manner that is closely and obligately linked with that of their host. The chemical reactions of photosynthesis were invented by this group of bacteria: the *cyanobacteria* (or *blue-green bacteria*—so-called because they contain the green molecule chlorophyll a, along with particular accessory pigments). To recap, all photosynthetic eukaryotic organisms are often considered as plants. The logical next question—or pair of related questions is: *How are these various photosynthetic organisms related to each other, and how many times did photosynthetic endosymbiosis (i.e., the incorporation of a photosynthetic organism into a eukaryotic host cell) occur?*

These questions involve the history of life on Earth and thus must be approached phylogenetically—in the context of their evolutionary history—which can be visually represented as a branching diagram representing the phylogenetic relationships (or evolutionary history) of a group of organisms (i.e., a tree of life). Figure 3.1 is a

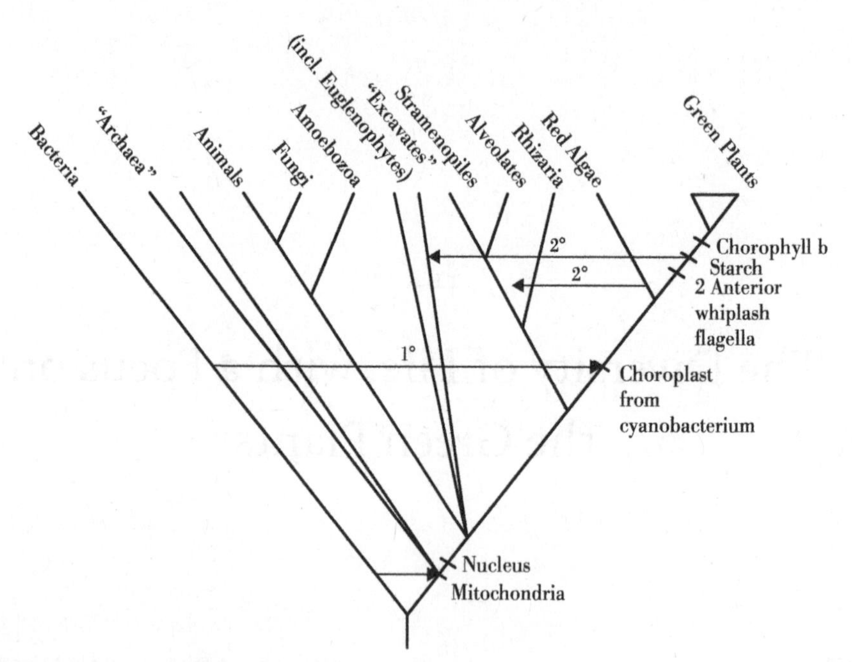

FIGURE 3.1 *Simplified evolutionary tree of living organisms.*

simplified evolutionary tree of living organisms that shows our most strongly supported hypothesis of phylogenetic relationships among living organisms. In such evolutionary trees, the lines represent *lineages* (i.e., ancestral/descendent sequences of populations, considered as moving through time), and the branching points indicate that the divergent organisms, represented by each branch, share a unique common ancestor. Thus, in this figure, the green plants (i.e., those organisms with chlorophyll a and b, starch as a carbohydrate storage product, and motile cells with two anterior whiplash flagella [modified or lost in some]), are hypothesized to share a unique common ancestor with the red algae, a mainly marine aquatic group lacking both starch and flagella and possessing only chlorophyll a. A group such as red algae + green plants, which is hypothesized to have evolved from a unique common ancestor and to include all the descendants of that ancestor, is called a *clade* or *monophyletic group*. A detailed presentation of the methodology by which evolutionary trees are constructed is beyond the scope of this chapter (interested readers should consult an introductory plant systematics textbook, such as Judd et al., 2016, *Plant Systematics*; see chapters 1, 2, and 6 from which the two figures in this chapter are derived). However, such trees are based on considerations such as minimizing the acquisition of evolutionary novelties (known as the *principle of parsimony*). This is seen in the fact that red algae and green plants both have chloroplasts surrounded by two membranes and have detailed internal features that indicate that they have been derived from free-living cyanobacterial ancestors (and thus are the result of primary endosymbiosis); thus, they are shown as arising from a common ancestor. In this case, it is simpler to assume that the incorporation of a free-living cyanobacterium within the eukaryotic cell occurred just once (instead of twice or even numerous times) in the history of life on earth, and thus a single common ancestor of these two plant groups is hypothesized. Likewise, all the groups, from animals through green plants (i.e., the eukaryotes; see Figure 3.1), have cells with a membrane-bound nucleus and *mitochondria* (i.e., membrane-bound intracellular organelles in which aerobic respiration occurs), and it is thus assumed that this large and diverse group of organisms has a single common ancestor in which a nucleus evolved and which engulfed an alpha-proteobacterium (which became the mitochondrial endosymbiont). Again, it is simpler to assume that the evolution of a complex structure, such as a nucleus containing chromosomes, or an unlikely event, such as the incorporation of an alpha-proteobacterium into a cell as an internal symbiont, each occurred just once in the history of life than to assume that several such events occurred. The principle of parsimony is, therefore, a common sense idea: choose the simplest hypothesis that explains the data—which in this case is represented by the shortest evolutionary tree explaining the array of characteristics of the life forms being considered. Using this approach, we see that related organisms share particular derived characteristics (e.g., the eukaryotes all share the derived characteristics of a membrane-bound nucleus and the presence of mitochondria in each of their cells, and the members of the group comprising the green plants + red algae share chloroplasts surrounded by only two membranes). However, evolutionary trees can also be constructed using various models of evolution; for example, in a *maximum likelihood approach*, the best evolutionary tree is the one that maximizes the likelihood of the data (array of characteristics of a group of organisms) given a particular branching pattern of the tree, the lengths of the various branches, and a model of evolutionary change.

Using a *phylogenetic approach* (and noting the branching patterns seen in the figure under discussion), we can answer the two questions posed earlier. *How are the various photosynthetic eukaryotes related to each other*? And, *how many times has endosymbiosis occurred*? As mentioned earlier, the green plants and the red algae are shown as closely related, both having chloroplasts surrounded by two membranes, resulting from a primary endosymbiosis event. However, it can be seen in Figure 3.1 that *alveolates* (including both photosynthetic organisms such as the dinoflagellates and nonphotosynthetic organisms such as the ciliates [e.g., the paramecium and its relatives] and apicomplexans [e.g., the malarial parasite and its relatives]) and *stramenopiles* (including photosynthetic organisms such as the brown algae, golden algae, and diatoms and nonphotosynthetic organisms such as the water molds) have chloroplasts derived from a red algal endosymbiont. These are surrounded by three or four membranes resulting from secondary endosymbiosis (i.e., incorporation of a eukaryotic cell, in this case a red alga, as an endosymbiont). The brown algae, the diatoms, and the dinoflagellates acquired their chloroplasts (and photosynthetic abilities) independently from that of the green plants (through a separate endosymbiotic event) and are more closely related to some nonphotosynthetic groups (e.g., the rhizaria) than they are to the green plants. Similarly, the unicellular, photosynthetic organisms called euglenas acquired their chloroplasts by incorporating an endosymbiotic green alga as their chloroplast. They are more closely related to various nonphotosynthetic groups (e.g., the kinetoplastids, an "excavate" organismal group) than they are to the green plants. Therefore, the various photosynthetic eukaryotes are not each other's closest relatives; instead, photosynthetic ability has been acquired several times (through primary as well as secondary endosymbiosis).

Only monophyletic groups should receive names in scientific classifications, and thus the group "plants," when defined as all photosynthetic eukaryotes, is not a phylogenetically meaningful or biologically predictive grouping. A more restrictive definition is needed. Thus, we consider plants to be the monophyletic group comprised of only the green plants, the red algae, and a third, more obscure group, the *glaucophytes.* Plants, so defined, constitute all those photosynthetic eukaryotes that have chloroplasts surrounded by two membranes (and which are endosymbiotic cyanobacteria). We note that this is the most inclusive grouping of photosynthetic organisms to which the name "plants" can be applied. For more information on phylogenetic relationships within photosynthetic organisms, including the evolutionary history of their chloroplasts, see Keeling (2004), Palmer et al. (2004), and Judd et al. (2016).

The major monophyletic subgroups of the plants are shown in Figure 3.2, which highlights the most important phylogenetic relationships within the green plants. In this tree, we see that, within the green plants, there is a large monophyletic subgroup, the *embryophytes,* which consists of all of those plants that have a multicellular sporophyte (= spore-producing) body, an embryo in their life cycle, a waxy layer or cuticle that protects the plant from desiccation, specialized gamete-protecting structures, and thick-walled spores. These characteristics all relate to the problems faced by plants living in terrestrial environments (the ancestors of the embryophytes, by contrast, were freshwater or marine aquatics). Embryophytes (i.e., the land plants) include small organisms such as liverworts and mosses, but in this book we mainly focus on the

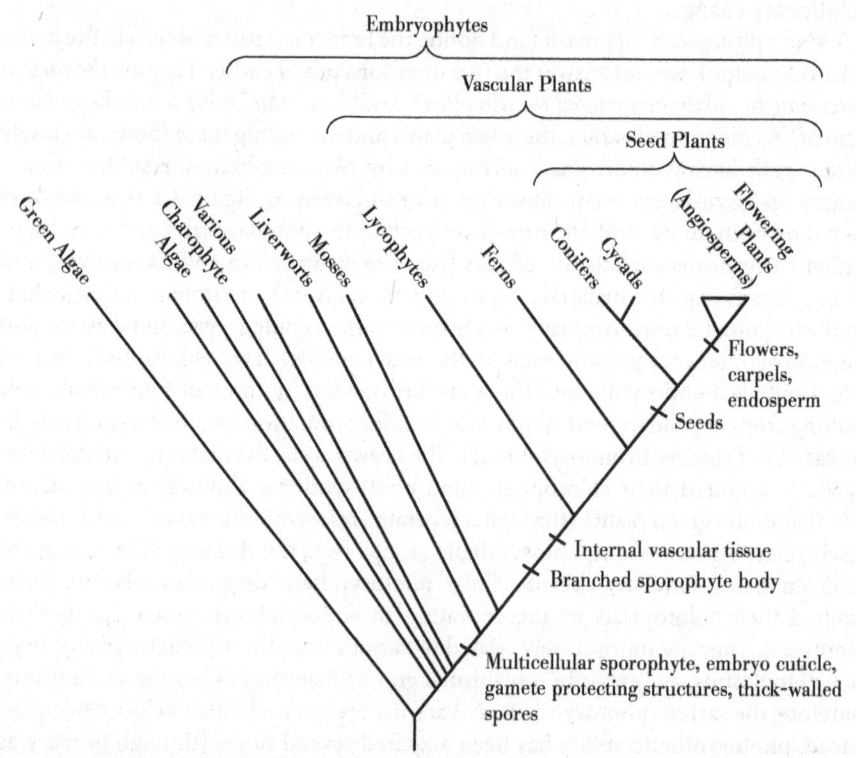

FIGURE 3.2 *Evolutionary tree of green plants.*

dominant groups of land plants—the various species of vascular plants—which are much larger in stature, have branched spore-producing bodies, and contain internal water-conducting tissue (called *xylem*) and sucrose-conducting tissue (called *phloem*). The vascular plants include both free-sporing groups, such as lycophytes and ferns, and also the plants that reproduce by seeds. The major groups of seed plants are the conifers (which produce their seeds in cones and have resinous, simple leaves), the cycads (which bear seeds on modified leaves often clustered into cone-like structures and have mucilage-containing, pinnately compound leaves), and the flowering plants (which produce their seeds in fruits developed from the carpels of the flowers and have leaves that may be simple or compound but that usually have a well-developed network of veins lacking in conifers or cycads); see Table 3.1 for a more detailed comparison of ferns, cycads, conifers, and flowering plants. Existing today are about 1,200 species of lycophytes, 12,000 species of ferns, 130 species of cycads, 600 species of conifers, and 260,000 species of flowering plants. Clearly, the seed plants, and especially the flowering plants, are the dominant group of terrestrial plant life (see Judd et al., 2016; Soltis et al., 2005, 2017). The flowering plants can be divided into several major subclades—most notably the *magnoliids* (magnolias, avocados, black pepper, and their relatives), *monocots* (grasses, lilies, orchids, and relatives usually with parallel-veined leaves, flowers with six perianth parts, and pollen grains with a single aperture), and *eudicots* (an especially diverse group including legumes, roses, mustards, citrus trees, cacti, blueberries, mints, umbellifers, and asters usually having net-veined leaves, flowers with four or five petals, and pollen grains with three apertures or related conditions) (Judd et al., 2016). In Chapter 7, "The Plants of Middle-earth," we consider species representing (in order of the number of occurrences in J. R. R. Tolkien's writings) flowering plants, conifers, ferns, and mosses among the embryophytes. However, because plants were traditionally treated quite broadly, including all photosynthetic eukaryotes and sometimes even the fungi (the latter actually closely related to animals), other non-animal living organisms are also considered in Chapter 7, such as seaweeds (various brown algae, in the stramenopile clade), seafire (dinoflagellates, in the alveolate clade), mushrooms (basidiomycete fungi), beard lichens (ascomycete fungi, more or less living symbiotically with a green alga), and even the cyanobacteria. Numerous animals occur within Tolkien's legendarium; however, they are not the subject of this book.

When the evolutionary tree of living organisms is considered (see Figure 3.1), we immediately realize that these organisms represent several major evolutionary lineages and cannot be simplistically divided into plants versus animals. However, since both plants and animals (and, to a lesser extent, fungi and many brown algae that, as noted earlier, are sometimes incorrectly considered "plants") are large organisms, these two groups are visually dominant—attracting our attention much more than the diverse array of microscopic life—and also have loomed large in traditional human classifications. The same was true of elven classifications. Indeed, corresponding elven words are provided in the *Silmarillion* for both groups: *olvar* (defined in the index of *The Silmarillion* as "growing things with roots in the earth") and *kelvar* ("animals, living things that move"). In the context of Tolkien's mythology, *olvar* and *kelvar*, and secondarily all the other groups of living organisms, are the special concern of Yavanna (see SILM 2).

In conclusion, one may perhaps view skeptically our evolutionary approach to the consideration of the diversity of living organisms—especially when contrasted

TABLE 3.1 Comparison of the major clades of seed plants: ferns, cycads, conifers, and flowering plants[a]

Ferns	Leaves simple to compound, and veins usually not forming a reticulum	Spores usually all alike, usually borne in sporangia on lower leaf surface, released and dispersed by wind	Seeds absent (instead, the plant free-sporing)	Life cycle with free-living gamete-producing plant	Flowers absent	Sperm flagellated, usually swimming to or splashed to the egg-holding structure
Cycads	Leaves pinnately compound, with each leaflet usually with a single vein or numerous ± parallel veins	Megaspores and microspores, the former retained in integumented sporangium (ovule)	Seeds present (and borne on modified leaves, these sometimes densely clustered into conelike structures)	Life cycle with gamete-producing plant reduced (the egg-producing plant in ovule and sperm-producing plant in pollen grain)	Flowers absent	Sperm flagellated, inside pollen grains; the latter is transported to ovule, usually by insects (beetles)
Conifers	Leaves simple, usually with only a single vein	Megaspores and microspores, the former retained in integumented sporangium (ovule)	Seeds present (and borne on cone scales of the seed-cones)	Life cycle with gamete-producing plant reduced (the egg-producing plant in ovule and sperm-producing plant in pollen grain)	Flowers absent	Sperm nonflagellated, inside pollen grains, which are transported to ovule by wind
Flowering plants	Leaves simple to variously compound, usually with venation forming a reticulate network	Megaspores and microspores, the former retained in integumented sporangium (ovule)	Seeds present (in fruits, which develop from the carpel or fused carpels of a flower)	Life cycle with gamete-producing plant reduced (the egg-producing plant in ovule and sperm-producing plant in pollen grain)	Flowers present, consisting of perianth, stamens, and carpels (enclosing the ovules)	Sperm nonflagellated, inside pollen grains, which are transported to stigma (of carpel) by wind, water, or various animals and then brought to ovule by growth of pollen tube

[a] *For more information on each of these groups, see Judd et al. (2016) and Soltis et al. (2005, 2017).*

with J. R. R. Tolkien's creation myth, which tells how in the time of the two lamps, "Yavanna planted . . . the seeds that she had long devised," and these quickly began "to sprout and to burgeon, and there arose a multitude of growing things great and small, mosses and grasses and great ferns, and trees whose tops were crowned with cloud as they were living mountains" (SILM 1). Yet, from his letters, it is clear that Middle-earth is to be equated with our own world, and it is also evident that Tolkien thought that the pattern of diversity among plants and their grouping into "families was sound, and that in general this grouping did point to actual physical kinship in descent" (Letters: No. 312). The wording used by Tolkien here is significant because evolution is often defined as descent with modification from a common ancestor. This is exactly the concept expressed by Tolkien in his letter. In fact, he even commented on finding in a botanical garden a morphological "link" between two of the evolutionary branches of the plant family Scrophulariaceae. We think that there is, in fact, no conceptual conflict between the mythological statements concerning the creation of plants provided in *The Silmarillion*—such as the fashioning and planting of seeds by Yavanna (reflecting her part in the song of creation as related in the *Ainulindalë*)—and his view that the pattern of plant diversity in our world reflects genealogical descent with modification. The former statements are of religious significance and represent concepts that can be best expressed through myth: that the world of plants is beautiful and good, and the pattern of the natural world expresses the plan of creation, reflective of the musical theme of Ilúvatar (God). However, this plan (and pattern) was only slowly achieved through ongoing physical and biological processes taking place across the depths of time. Certainly, Tolkien would have no objection to the view of many in the Church that there is no essential conflict between Christianity and the theory of evolution. He likely held a theistic evolutionary viewpoint. This view likely was first expressed by Asa Gray (1880), the well-known Harvard plant systematist and early supporter of Darwin.

FOUR

Introduction to Plant Morphology

Learning the Language of Plant Descriptions

MORPHOLOGICAL CHARACTERISTICS ARE FEATURES OF EXTERNAL FORM or appearance. They provide most of the characters used for plant identification; such characters are easily observed and find practical use in both identification keys (see Chapter 5) and descriptions (see Chapters 6 and 7). Characteristics useful in plant identification are found in all parts of the plant, both vegetative (relating to the form of the roots, stems, and leaves) and reproductive (relating to the form of the flowers and fruits of the flowering plants, of the cones of conifers, or of the spore clusters of ferns). We emphasize characteristics (or characters) relating to the *tracheophytes* (i.e., vascular plants, such as ferns, conifers, and flowering plants) because most of the organisms treated in this book belong to this group. In our descriptions of the plants of Middle-earth, we have tried to keep the number of technical terms used to a minimum but some are nonetheless needed to achieve some efficiency of communication and to open our eyes to the natural variation in plant form. Many if not all of the terms used should be considered merely as convenient points along a continuum of variation in form. Although these terms are useful in communication of information regarding variation in plant form, intermediate conditions will be encountered (and should not be a source of concern!).

Humans suffer from a condition referred to as *plant blindness* (i.e., the "inability to see or notice the plants in our own environment"), which then leads to the "inability to recognize the importance of plants in the biosphere and in human affairs" (see J. Wandersee and E. Schussler, 2001, and W. Allen, 2003). Humans mentally process only a tiny fraction of the visual information taken in at any particular moment, and generally we emphasize movement, conspicuous colors or patterns, and objects that are known and/or are the source of potential danger. Because plants are static, blend into the background, and are generally not dangerous, they typically do not get our attention. Thus many walk around seeing plants only as vague green blobs without definite form, and any variation between one plant and another goes unnoticed. Thus, if we want to appreciate plants in our surroundings—including those plants characteristic of J. R. R. Tolkien's legendarium—we must begin to overcome this problem. Our goal should be to mimic the awareness of the elf Voronwë who wandered in Nan-Tathren "naming new flowers" (UT 1: I) or at least of Túrin, a human being who, in his more reflective moments, walked far in the forests of Middle-earth observing his surroundings—including the wildflowers—and hearing "the names of the flowers of Doriath as echoes of an old tongue almost forgotten" (CoH VI). We cannot really understand plant variation without having at least a few descriptive terms that can be used to describe such variation—and the goal of this chapter is to provide these basic terms. For a more detailed introduction to morphological terminology, the reader should consult any plant systematics textbook (e.g., Judd et al., 2016; Murrell, 2010; Simpson, 2010; Spichiger et al., 2002). And once we really begin to see plants—and understand their variation—we can then, like Túrin, remember their names (even those in Latin, certainly an "old tongue")!

How do we begin the process of actually seeing plants—paying attention to the variation in plant form that we see around us? It is helpful to break the observational process down into several steps through a categorization of the kinds of variation that we are encountering. Thus, we will present our useful identification terms around the following foci: (1) **Duration and Habit**, (2) **Roots**, (3) **Stems**, (4) **Leaves**, (5) **Hairs**, and, for the flowering plants, (6) **Inflorescences**, (7) **Flowers**, and (8) **Fruits** and **Seeds**; or, for the conifers, (9) **Pollen** and **Seed Cones**. Occasionally, we will also mention anatomical characteristics (i.e., those related to the internal structure of plants), although these are generally less useful in practical identification because they are more difficult to observe. Finally, in identification, we must not neglect chemical characters—those we experience through taste or smell—and such characteristics are occasionally used in our identification keys and descriptions. In the following paragraphs, important descriptive terms are indicated in **bold type**, and these terms are also included in the Glossary at the end of the book.

DURATION AND HABIT

Duration is the life span of an individual plant. An **annual** plant lives for a single growing season, whereas a **biennial** plant lives two seasons, growing vegetatively in the first and flowering and fruiting in the second. **Perennial** plants live for three or more years, and they usually flower and fruit (or produce cones) repeatedly. Perennials may be herbaceous (lacking woody tissue), with only the underground portions living for several years, or they may be woody, with persistent above-ground stems.

The general appearance, or *habit*, of plants varies greatly. Woody tissue is present in trees and shrubs but is lacking in **herbs**. **Trees** produce one main trunk, while **shrubs** are usually shorter and produce several trunks. Climbing plants may be woody (such plants are called **lianas**) or herbaceous (and are then called **vines**). Herbs or shrubs that grow on another plant that is used for support are called **epiphytes**. It is interesting that Sindarin-speaking elves divide trees into two groups: those that are spreading and have thick trunks with dense branches are *galadh*, while those more slender trunks and more upright branches are *orn*. In contrast, in Quenya, the name for all trees is *alda*. The Sindarin word for vine/liana is *gwîn*, while the Quenya word is *vinë*.

ROOTS

Roots are usually underground and branch irregularly; they have the primary function of holding the plant in place and absorbing water and minerals. Some are thickened and also store water and carbohydrates. A few have other functions, such as the **aerial roots** of epiphytes or the specialized roots of parasitic plants, which penetrate the tissues of a host species. Some plants have **fibrous roots** in which all portions of the root system are of more or less equal thickness and well-branched, while others have a **taproot**, a major root that usually is enlarged, often fleshy, and grows downward. Roots frequently are quite uniform in appearance, and a plant usually cannot be identified without its above-ground parts (especially leaves and reproductive structures).The Sindarin word for root is *thond* (and Quenya, *sulca*).

STEMS

Stems are the above-ground axes of plants and are frequently useful in identification. *Olf* is Sindarin for stem or branch, and the corresponding Quenya word is *olba*. Stems bear leaves with axillary buds; the points at which the leaves (and buds) are attached are called **nodes**, and the portions of the stem between the nodes are the **internodes** (see Figure 4.1). All stems on a plant may be the same, which is the typical condition, but some plants (e.g., pines and larches) have two kinds of stems, called short shoots and long shoots. Stems with elongated internodes are **long shoots**; others that produce only a few leaves each year, all with only very short internodes, are **short shoots**.

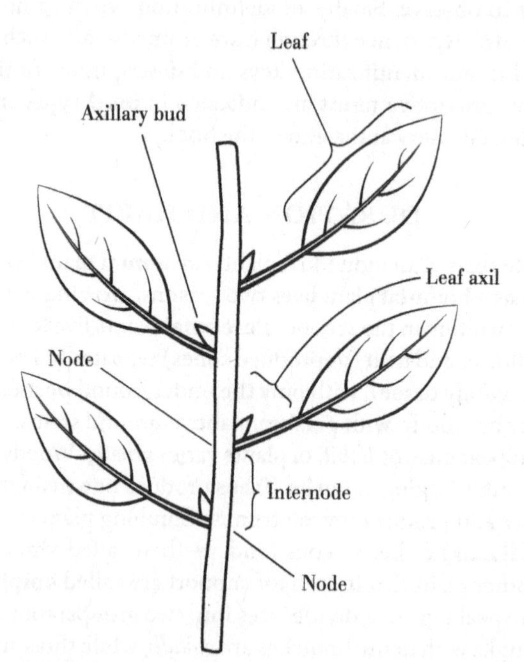

FIGURE 4.1 *Stem of a flowering plant.*

Other plants, such as coffee, also show differentiated stems, with the central axis growing erectly and thus distinguished from the lateral branches, which grow horizontally. Stems are usually elongated and function in exposing the leaves to sunlight, flowers to pollinating agents, and fruits to dispersal agents. A few, however, are succulent and store water or carbohydrates, and others are modified for climbing (e.g., the twining stems of vines and lianas). Finally, some are protective, forming sharp-pointed structures (**thorns**). **Buds** (Sindarin, *tui*; Quenya, *tuima*) are short embryonic stems, and they are often protected by scales, sticky secretions, and/or a dense covering of hairs. They are positioned at the end of the stems and/or in the **leaf axils** (in each angle formed by the stem and a leaf). In the temperate zone, they are dormant during the winter and may grow out in the spring, producing new vegetative branches, clusters of flowers, or both. A **bulb** is a short, erect, underground stem surrounded by thick, fleshy leaves or leaf bases. A **corm** is a short, erect, underground, more or less fleshy stem covered with thin, dry leaves or leaf bases. A **rhizome** is a horizontal stem, usually underground, bearing scale-like leaves. A **tuber** is a swollen, fleshy portion of a rhizome. **Lenticels** are wart-like protuberances on the stem surface that are involved in gas exchange; they can be dot-like or elongated horizontally. As stems increase in girth, the **bark** develops; it is a variously colored protective covering that may be smooth or roughened, blocky or vertically fissured, and sometimes peeling. *Rîf* is the Sindarin word for bark, and the Quenya word is *parma* (which also means "book" and refers to use of skins or bark as primitive writing materials).

LEAVES

Leaves are the major photosynthetic parts of most plants. The Sindarin word for leaves is *lais* (singlular: *las* or *lass*), while the Quenya word is *lassi* (singular: *lassë*). Leaves are borne at the nodes of a stem, usually just below a bud (see Figure 4.2). In contrast to stems, they do not continue to grow year after year. They are usually flat and have one

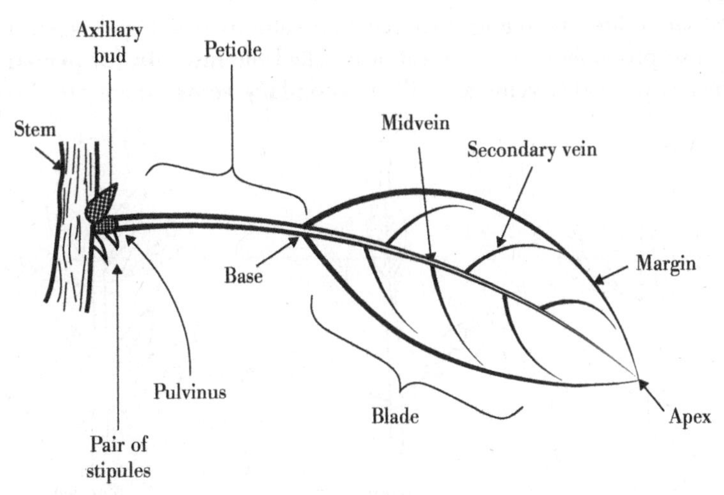

FIGURE 4.2 *Typical leaf of a flowering plant.*

surface facing upward and another facing downward. The expanded, flat portion of the leaf is the **blade**, while the stalk of the leaf is the **petiole**. In grasses, lilies, irises, and their relatives (the monocots), the leaf base is usually sheathing. In some plants (e.g., grasses and gingers) there is a flap of tissue (**ligule**) at the junction of the sheath and the blade. Some leaves are associated with **stipules** (i.e., paired appendages located on either side of or on the petiole base). Stipules may be leaflike, scalelike, spinelike, glandular, twining, or much reduced. Leaf form is especially important in identifying plants, and thus leaves are discussed here in some detail. Leaves may function for only a few days to many years, but most function for only one or two growing seasons. **Deciduous** leaves fall at the end of the growing season, whereas **evergreen** plants are leafy throughout the year. Leaves are occasionally modified for protection, forming sharply pointed structures (**spines**). They can also store water and/or carbohydrates, and are then "succulent." In some vines or lianas, a portion of the leaf may be modified as a climbing structure (**tendril**).

Leaves may be arranged in one of three major patterns (see Figure 4.3). **Alternate leaves** are borne singly and are usually arranged in a **spiral** pattern along the stem. However, alternate leaves can also be placed along just two sides of the stem (a condition called **two-ranked**) or just three sides of the stem (**three-ranked**).

Opposite leaves are borne in pairs, the members of which are positioned on opposing sides of the stem. Opposite leaves are usually positioned so that the leaves of adjacent nodes are rotated 90 degrees (a pattern called **decussate**). Finally, when three or more leaves are positioned at each node, the arrangement is called **whorled.**

A leaf with a single blade is termed **simple** (Figure 4.4); a leaf with two or more blades (or **leaflets**) is said to be **compound.** The distinction between simple and compound leaves can be confusing but is resolved by locating the axillary bud. An axillary bud is subtended by the entire leaf and never by individual leaflets. Leaflets can be arranged in various ways, the two major being **pinnate** (with leaflets along the sides of leaf axis, borne like the teeth on a comb) and **palmate** (with the leaflets radiating like the fingers from a palm) (see Figure 4.4). A leaf with only three leaflets is called **trifoliolate.**

Leaves are quite variable in their venation patterns (see Figure 4.5). If there is a single most-prominent vein in a leaf, it is called the **midvein** (or **primary vein**), and branches from this veins are called **secondary veins.** There are three major

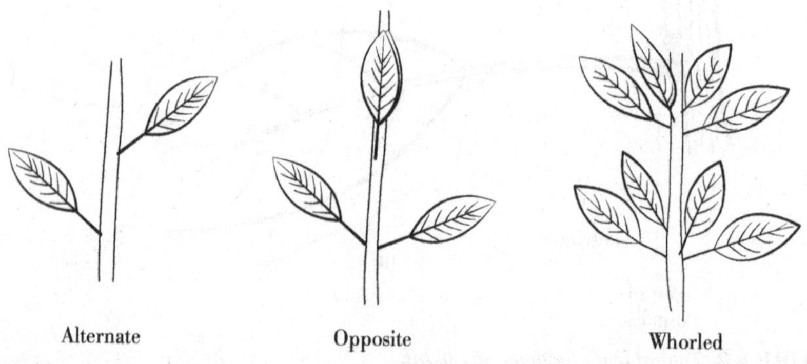

FIGURE 4.3 *Three leaf arrangements (i.e., alternate, opposite, and whorled leaves).*

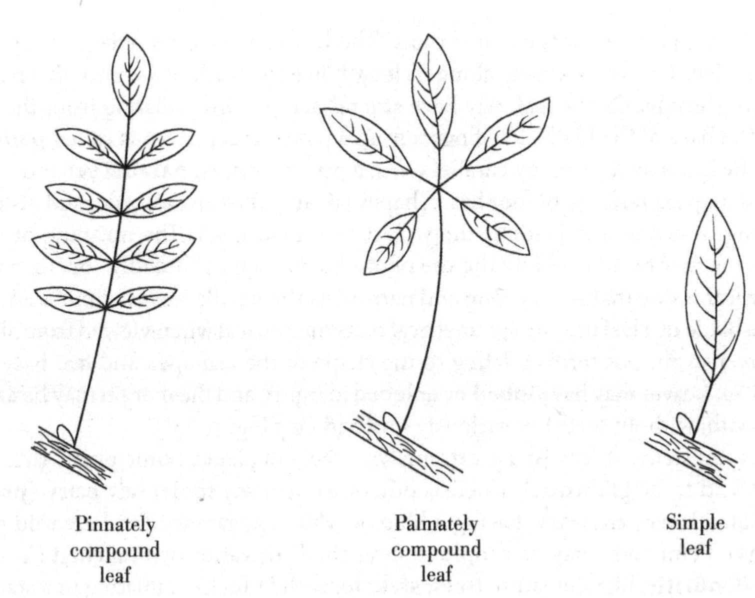

FIGURE 4.4 *Simple and compound leaves.*

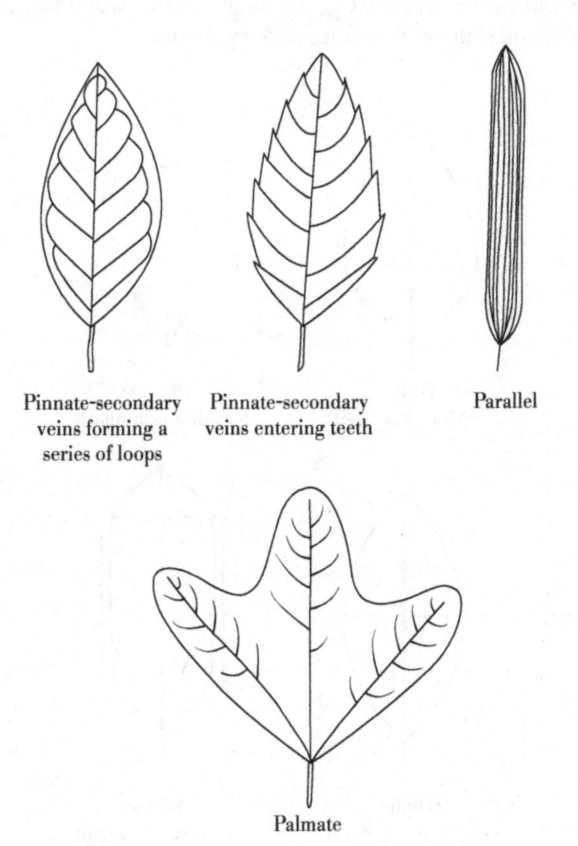

FIGURE 4.5 *Leaf venation patterns.*

organizational patterns of the major veins. The leaf may have a single primary vein, with the secondary veins arising along its length like the teeth of a comb; this pattern is termed **pinnate**. Or the leaf may have several major veins radiating from the base or near the base of the blade, like fingers from a palm. This pattern is called **palmate**. Finally, the leaf may have many parallel veins, a pattern termed **parallel** venation.

A leaf may express one of four basic shapes (**ovate, obovate, elliptic**, and **oblong**) depending on where the blade is the widest (see Figure 4.6).The meaning of these shape terms may be adjusted by the use of modifiers such as "broadly" or "narrowly." A **linear** leaf is one that is very long and narrow, as the needle leaves of pines and firs. Also, the blade of a leaf may be symmetrical or asymmetrical when viewed from above. There are also various terms relating to the shape of the leaf apex and leaf base (see Figure 4.7). Leaves may have **lobed** or unlobed margins, and the margin may be **entire** (i.e., smooth, without teeth) or variously **toothed** (see Figure 4.8).

Hairs (on stems or leaves) are extremely variable in plants. Some plants lack hairs (and are said to be **glabrous**), whereas others are sparsely to densely hairy (**pubescent**). Some leaves are waxy, having a blue or white appearance (and are said to be **glaucous**). Plant hairs may be simple (unbranched) or variously branched (Y- or T-shaped; **dendritic**, like miniature trees; **stellate**, with branches radiating in a star-like pattern), unicellular or multicellular, nonglandular or with a glandular head, and **terete** (cylindrical) or variously flattened, e.g., forming a shield-shaped structure (**peltate**). Hairs that are stiff and sharp-pointed are called **prickles**.

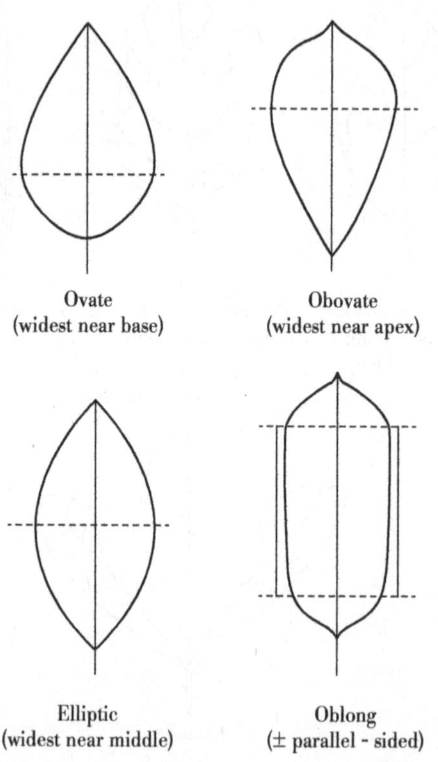

FIGURE 4.6 *Leaf shapes.*

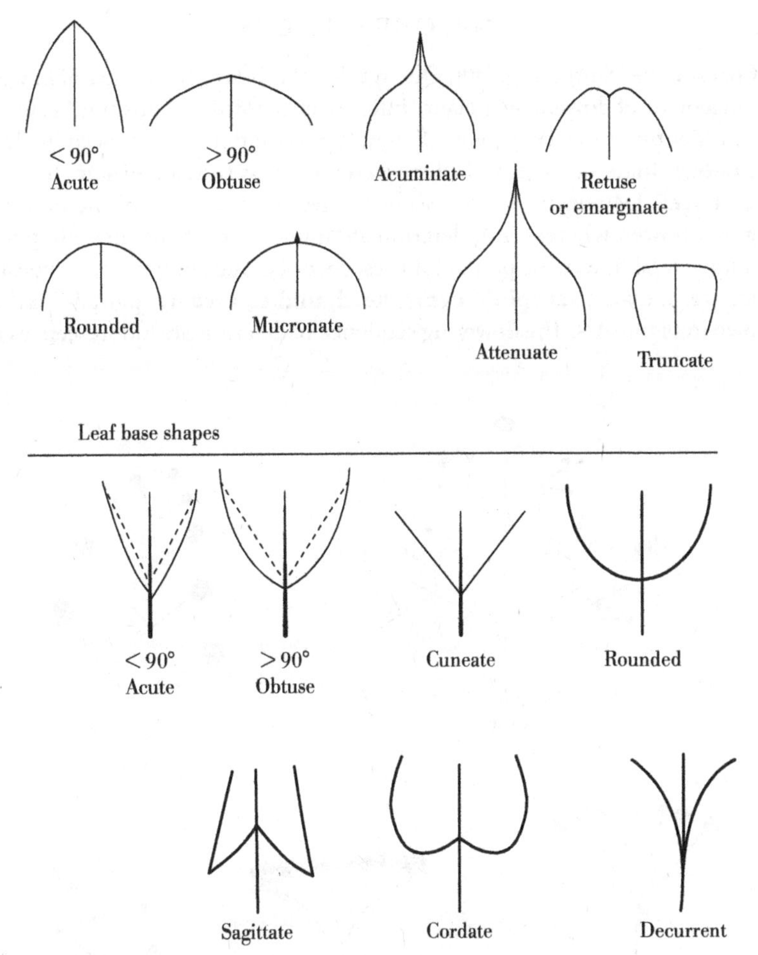

FIGURE 4.7 *Leaf apex and base shapes.*

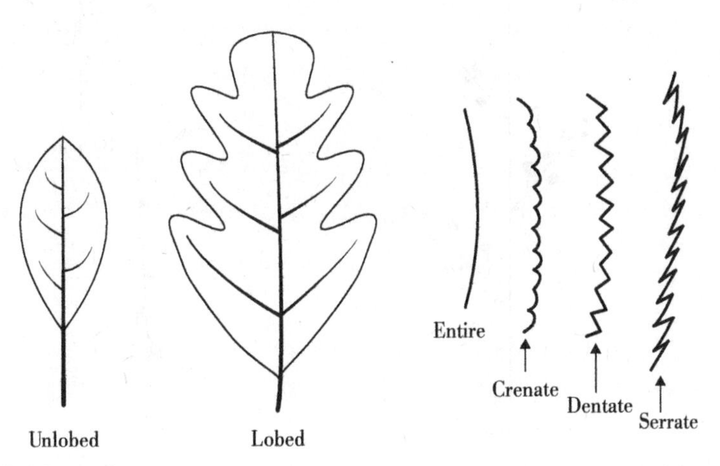

FIGURE 4.8 *Leaf margins.*

INFLORESCENCES

An **inflorescence** (Sindarin, *goloth*; Quenya, *lossë*) is basically a cluster of flowers, and the arrangement of flowers on a plant (inflorescence form and position) is important in routine identification. Two quite different types of inflorescences occur in flowering plants: **determinate** or **cymose inflorescences** versus **indeterminate** or **racemose inflorescences**. Determinate inflorescences have the main axis of the inflorescence ending in a flower, whereas in indeterminate inflorescences the growing point produces only lateral flowers or partial inflorescences (groups of flowers). Typical inflorescences (e.g., **cyme, scorpioid cyme, head, umbel, raceme, panicle**, and **spike**) are shown in Figure 4.9. The flowering sequence of determinate inflorescences usually

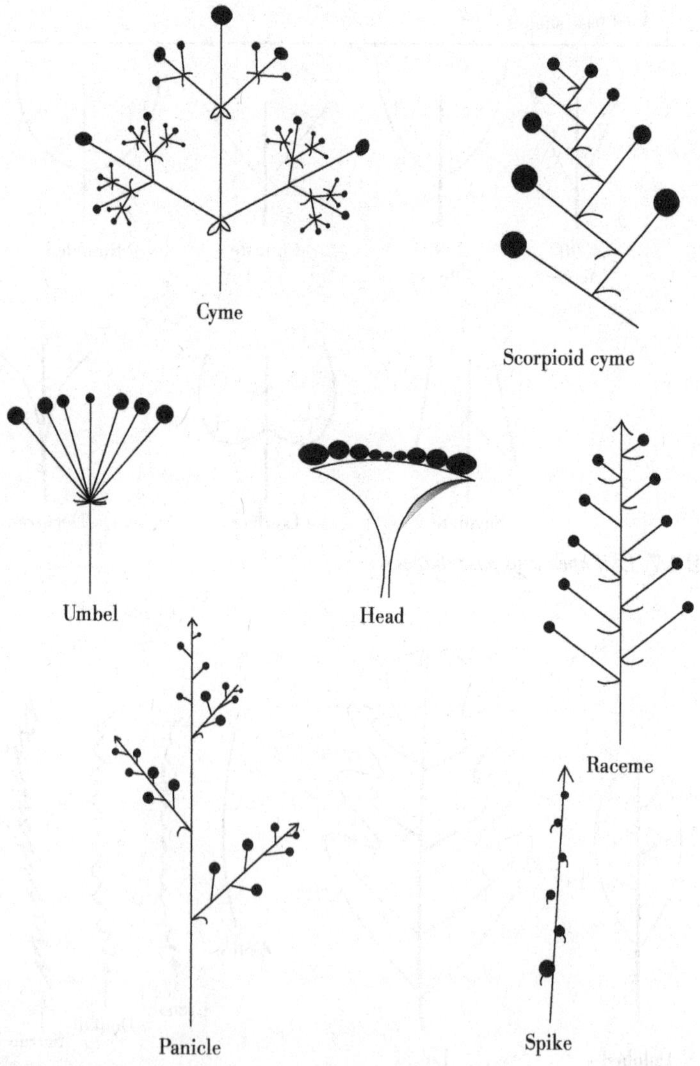

FIGURE 4.9 *Inflorescences.*

begins with the terminal flower, which is at the top or center of the floral cluster. In contrast, indeterminate inflorescences start blooming at the base (or outside) of the cluster. The term **catkin** is used for any elongated inflorescence composed of numerous inconspicuous (usually wind-pollinated) flowers, as in birches and oaks. Occasionally inflorescences are modified, forming climbing structures (**tendrils**).

FLOWERS

A **flower** (Sindarin, *alf* or *loth*; Quenya, *alma, lós,* or *lótë*) is a highly modified shoot bearing specialized appendages (highly modified leaves), as shown in Figure 4.10. The floral axis is called the **receptacle**; the floral stalk is referred to as the **pedicel**, and flowers are often associated with modified leaves (**bracts**). Flowers have up to three major kinds of parts: **perianth** (outer protective and/or colorful structures), **androecium** (plural **androecia**; pollen-producing structures, the stamens), and **gynoecium** (plural **gynoecia**; ovule-producing structures, the carpels). If one or more of these structures is absent, the flower is said to be incomplete. If at least the androecium and gynoecium are present, the flower is termed **bisexual**. If either is lacking, the flower is **unisexual**—and then it is either **staminate**, if only the stamens (or a stamen) are present, or **carpellate**, if only the carpels (or a single carpel) are present. Unisexual flowers can be distributed variously: two common patterns have (1) the staminate and carpellate flowers on separate plants or (2) together on the same plant. Sometimes, carpellate flowers have nonfunctional stamens (**staminodes**). Likewise, a sterile carpel (in a staminate flower) is called a **carpellode**.

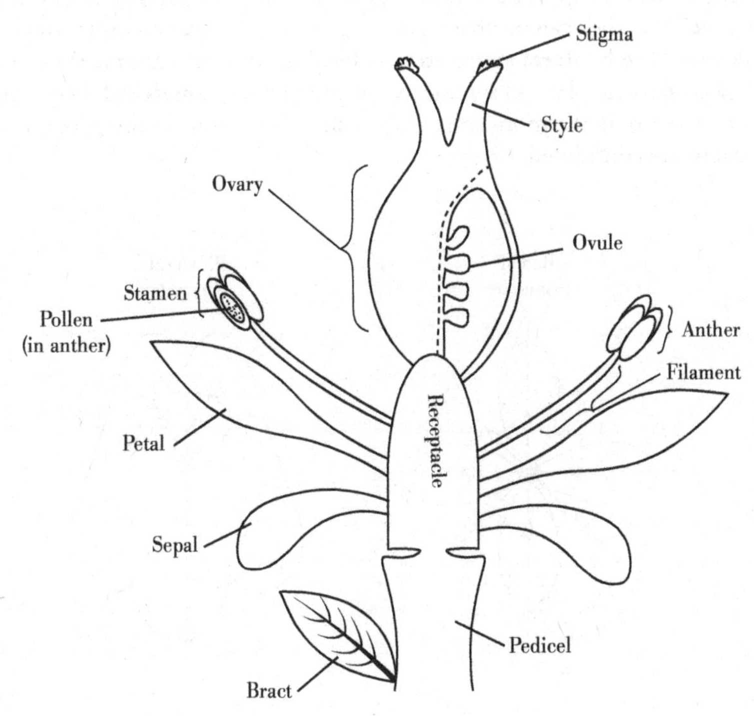

FIGURE 4.10 *A generalized flower.*

The **perianth** is always the outermost series of parts in the flower, followed by the **androecium**, with the **gynoecium** in the center of the flower. The perianth parts may be undifferentiated (i.e., all alike), in which case the parts are called **tepals**, as seen in the flowers of lilies. Alternatively, the perianth may be differentiated into two kinds of parts (as in most flowers)—an outer whorl of **sepals**, collectively called the **calyx**, and an inner whorl of **petals**, collectively called the **corolla**. The calyx is usually green and typically protects the inner flower parts in the bud, and the corolla is usually colorful and assists in attracting pollinators. The androecium comprises all the **stamens** of the flower. Stamens are usually differentiated into an **anther** (with its pollen-producing sacs) and a **filament** (stalk). The pollen sacs are joined to each other and to the filament by the **connective**. The gynoecium comprises all of the **carpels** of the flower. The carpel is the site of **pollination** and **fertilization**. Carpels are typically composed of a **stigma** (which collects and facilitates the germination of pollen carried to it by wind, water, or various animals, especially insects and birds), a **style** (slender region specialized for pollen tube growth), and an **ovary** (an enlarged basal portion that surrounds and protects the ovules). Each **ovule** contains an egg, and, after fertilization, the ovary grows into the fruit, while the ovule becomes the seed. Ovules can be arranged within the ovary in various ways. Finally, many flowers contain **nectaries** (nectar-producing glands) because nectar is a common floral reward provided to pollinators.

Flowers, like many biological objects, have symmetry (see Figure 4.11). Some flowers have their parts arranged so that two or more planes bisecting the flower through the center will produce symmetrical halves. Such flowers have **radial symmetry**, as seen in those of apples, anemones, or stonecrops. The parts of other flowers are arranged such that they can be divided into symmetrical halves on only a single plane. These flowers have **bilateral symmetry**, as seen in those of snapdragons or orchids. A few flowers have no plane of symmetry and are thus **asymmetrical**. In determining floral symmetry, only the more conspicuous structures (usually the perianth and/or androecium) are considered.

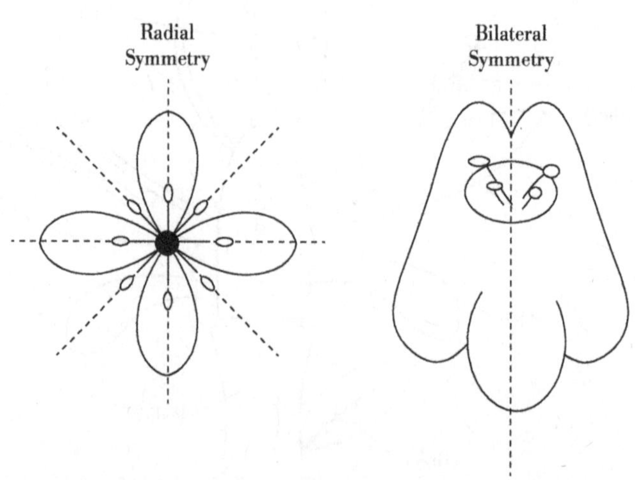

FIGURE 4.11 *Floral symmetry (radial symmetry vs bilateral symmetry).*

The floral parts may be fused together in various ways. Like parts (e.g., all the petals of a flower) may be **fused**, or they may be **distinct**. Also, unlike parts (e.g., the stamens to the petals) may be fused, or they may be **free**. Flowers also differ in the number of sepals, petals, stamens, and carpels. The numbers of parts are usually easily determined by counting, but extreme fusion, especially of the carpels, may cause difficulties. Almost every flower is based on a particular numerical plan; that is, on a pattern of three, four, five, or a multiple of one of those numbers.

Floral parts may be attached to the receptacle (or floral axis) in various ways. Three major insertion types are recognized: hypogynous, perigynous, and epigynous (see Figure 4.12). The position of the ovary in relation to the attachment of floral parts also varies, from superior to inferior. Flowers in which the perianth and androecium are inserted below the gynoecium are called **hypogynous**; the ovary of such flowers is said to be **superior**. Flowers with a cuplike or tubular structure surrounding the gynoecium, but without being fused to it, are called **perigynous**. In such flowers the perianth and androecium are attached to the rim of this structure, which is called the **floral-cup**, **floral-tube**, or **hypanthium**. Such flowers also have a superior ovary (because the floral-cup or hypanthium attaches to the floral axis below the ovary). Flowers in which the perianth and stamens appear to be attached to the upper part of

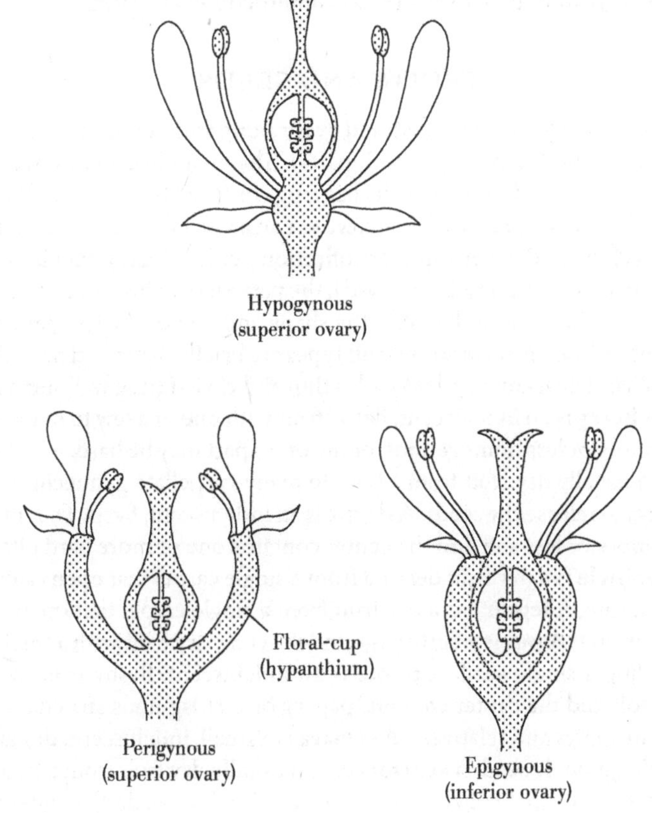

FIGURE 4.12 *Flower insertion types.*

the ovary (usually because of fusion of the hypanthium to the ovary) are called **epigynous**. The ovary of such flowers is said to be **inferior**.

Plants are stationary and thus depend on external forced to bring their gametes together. We've mentioned already that each ovule (tiny structures inside the ovary of a flower) contains an egg. The sperm of seed plants (e.g., flowering plants, conifers) is packaged inside pollen grains (with two sperm inside each pollen grain). The transport of pollen (from the site of its production, the anther, to the stigma[s] of a carpel or a group of fused carpels) is referred to as **pollination**. In contrast, the movement of the sperm and its fusion with the egg is called **fertilization**. Conifers are pollinated exclusively by wind, while flowering plants are pollinated by animals (e.g., usually various insects, birds, bats) as well as by wind or water. Most plants are predominantly out-crossing, with the pollen moving between plants, but some are selfing (with the pollen from one flower landing on the stigma of the same flower and germinating, and eventually the sperm fusing with the eggs of that flower). Flowers pollinated by various agents are often quite different in form. For example, moth-pollinated flowers are usually pale in color; have a strong, sweet odor; and bloom at night. Bird-pollinated flowers are brightly colored (often red), have no odor, and bloom during the day. Bee-pollinated flowers are often blue, yellow, or purple; have a strong scent; are day-blooming; and frequently have bilateral symmetry. Fly-pollinated flowers often are purplish brown and produce a strong carrion odor. For more details regarding pollination syndromes, see Faegri and van der Pijl (1979) and Proctor et al. (1996).

FRUITS AND SEEDS

A **fruit** (Quenya, *yávien, yávë*) is a matured ovary along with fused accessory structures (most often the hypanthium or perianth parts). The great diversity of size, form, texture, and means of opening among fruits has long confused botanists, and many different fruit types have been proposed. Here, we use an artificial system of descriptive fruit terms that is widely used in keys and identification guides. This system is based on the texture of the fruit wall (fleshy, dry, or hard), the pattern of dehiscence or indehiscence (type of fruit opening or lack thereof), the shape and size of the fruit, and the carpel and seed number. The most common fruit types are briefly described here. An **achene** is a fairly small, indehiscent, dry fruit with a thin and close-fitting wall surrounding the single seed. A **berry** is an indehiscent, fleshy fruit with one or a few to many seeds. The flesh may be more or less homogenous, or the outer part may be hard, firm, or leathery. A **capsule** is a usually dry fruit from a two- to many-carpellate gynoecium that opens in various ways to release the seeds. A **drupe** is an indehiscent, fleshy fruit in which the outer part is more or less soft and the center contains one or more hard pits enclosing the seeds. A **follicle** is a dry fruit derived from a single carpel that opens along a single longitudinal suture. A **legume** is a dry fruit from a single carpel that opens along two longitudinal sutures. A **nut** is a fairly large, indehiscent, dry fruit with a thick and bony wall surrounding a single seed. A **pome** is an indehiscent, fleshy fruit in which the outer part is soft and the center contains papery or cartilaginous structures enclosing the seeds (as in apples and relatives). A **samara** is a small, indehiscent, dry and winged fruit, containing a single seed. A **schizocarp** is a usually dry fruit from a two- to many-carpellate gynoecium that splits into one-seeded (or few-seeded) segments. Finally, a **silique** is a fruit derived from a two-carpellate gynoecium in which the two halves of the fruit split away from a persistent partition (around the rim of which the seeds are attached). The shape of these three-dimensional objects can be described by similar terms as were introduced for leaf blades (which, of course are usually two-dimensional): that is, ovoid, obovoid, ellipsoid. Ball-shaped fruits are called globose.

A **seed** (Sindarin, *eredh*; Quenya, *erdë*) is a matured ovule that contains an embryo and often its nutritive tissues (the female gamete-producing plant in conifers and the endosperm in flowering plants). The seed is surrounded by a protective seed coat, which develops from the integument or integuments, the outer layer or layers of the ovule. Some seeds are associated with a hard to soft, oily to fleshy, and often brightly colored structure called an **aril**. It is usually an outgrowth of the stalk of the ovule.

Either the fruit or the seed may function in dispersal, and clearly the diversity of dispersal mechanisms is responsible for the phenomenal variation in fruit and seed characteristics seen in the flowering plants. Most fruit types can be dispersed by a variety of agents, and different parts of the fruit, seed, or associated structures (especially perianth parts or hypanthium) may be modified for particular dispersal-related functions. For example, plants that are wind dispersed may have a tuft of hairs or wings on their seeds, winged fruits, wing-like perianth parts, or bracts (associated with the fruit or fruits). Likewise, plants that are dispersed by birds may have fleshy and bright red seeds, red berries or drupes, or fruits associated with fleshy, red accessory structures (flower stalks, perianth parts, etc.). Thus, very different fruit types (and morphological structures) often function in similar ways and vice versa. For more information on dispersal mechanisms in plants, see van der Pijl (1972).

POLLEN AND SEED CONES

Conifers (i.e., pines, firs, junipers, cedars, and their relatives) do not have flowers or fruits: instead, they have **cones** as their reproductive structures. The so-called pollen cone (or strobilus) is a modified branch bearing reduced scale-leaves, each bearing sporangia (that produce pollen). Their ovules are borne on **cone scales** (or **ovulate scales**) in globular to elongated cones. A cone is a modified branch, bearing **bracts**, in which each bract is associated with a short shoot (the cone scale or ovulate scale). Characteristics of the cones are critical in conifer identification, especially the shape and size of the cone, whether or not the bracts are fused to the ovulate scales, the size of the bract in relation to the ovulate scale, the number of ovules (and seeds) per ovulate or cone scale, and the form of the cone scale itself. Seed cones may be held erectly or may be pendulous. The seed cones may open to release the seeds, or they may fragment at maturity (releasing the seeds in the process of fragmentation). A few conifers, such as the yews, have, in the course of evolution, lost their cones. In addition, some conifers have variously winged seeds while others do not. Pollination in conifers is exclusively by wind, which carries the pollen grains from the pollen cones to the pollination droplets of the very young ovulate cones. The seeds of conifers usually are dispersed by wind or water, but some, such as junipers and podocarps, have reduced fleshy cones and are dispersed by birds. The seeds of yews are surrounded by a large, fleshy aril and are also bird dispersed.

FIVE

Identification of the Plants of Middle-earth

BEFORE ATTEMPTING TO IDENTIFY A PLANT, YOU SHOULD CAREFULLY observe its morphological characteristics, and a 10× hand lens will be useful in this process. The metric system is used in the following keys (and in the descriptions in Chapters 6 and 7), and thus a millimeter rule also is essential for measuring things like leaf or petal length. Important observations include (1) the habit of the plant, i.e., whether woody or herbaceous, and whether climbing or not; (2) leaf arrangement (e.g., alternate or opposite), whether simple or compound, and leaf shape, apex, base, margin, and venation condition; (3) hair characteristics, especially hairs on the stems and/or leaves; (4) whether or not the plant has an odor; (5) whether the plant has a milky or colored latex (i.e., sap); (6) floral features, such as symmetry, whether the perianth is differentiated into a calyx (green) and corolla (colorful parts) or undifferentiated (consisting only of tepals); whether or not the floral parts are distinct from each other or fused together; the number of perianth parts, stamens, and carpels; and the ovary position (i.e., superior, with the perianth parts attached below the ovary, or inferior, with the perianth parts attached upon the ovary); and (7) fruit type (e.g., capsule, berry, drupe, etc.).

Systematic botanists use several tools in identifying plants, but the most important of these are *dichotomous keys*. A dichotomous key presents the user with a series of choices between two mutually exclusive and parallel statements (i.e., the leads of a couplet). If the user chooses correctly, he or she will be led to the name of the unknown plant. Dichotomous keys always have a flowchart structure, and they often are written in a bracketed format (as in the two keys presented in this chapter). Bracketed keys present the two leads of a couplet side by side, and numbers direct the user to the subsequent couplets; for ease of usage, the two leads of a couplet usually start with the same word (a noun, followed by adjectives). The two leads are parallel; that is, if flower color is given in the first, a contrasting condition is presented in the second (e.g., petals red vs. petals white). When using a dichotomous key, always read both choices, do not guess on measurements, and look up any terms that you do not understand. Remember that living things are variable, so be sure to look at several leaves, flowers, or fruits in the process of observing a plant.

Another kind of key is the *multiaccess key* (or *polyclave*), as experienced in computer interactive identification—you may encounter such keys at some websites (e.g., an interactive key to the plants of the British Isles can be found at www.botanicalkeys.co.uk/flora, and an interactive key to the plants of New England can be found at https://gobotany.newenglandwild.org). In such keys, characteristics to be used in identification may be selected in any sequence, and characteristics may be repeatedly provided to the interactive system until only one possible plant remains that has the characteristics matching those provided.

Keys are often presented in books together with plant descriptions, illustrations, and other biological information (as in the following two chapters). It is important to read the appropriate plant description after using a key because the description should match the named plant in hand. The order of information in plant descriptions has been standardized as follows:

1. Habit (the plant as a whole)
2. Underground parts (roots, tubers, bulbs, etc.)
3. Stems
4. Leaves (including arrangement, structure, shape of blade, venation, base of blade, apex of blade, margin of blade)
5. Inflorescences (or cone position in conifers)
6. Flowers (including symmetry, sexual condition, sepals, the calyx, petals, the corolla, stamens, the androecium, and carpels, the gynoecium, and nectaries, or the form of the cones in conifers)
7. Fruits (including whether or not the fruit opens, how it opens, and characteristics of the fruit wall; e.g., hard, fleshy, in flowering plants, but fruits not present in conifers)
8. Seeds

Illustrations, of course, are also extremely useful in identification, and these are provided here in conjunction with the descriptions. A *flora* is an account of the plants occurring in a particular area, including keys, descriptions, and illustrations. Floras may be local, applying to a relatively small region, or continental in scale, such as *Flora Europaea* (Tutin et al., 1964–80) or the *Flora of North America* (Flora of North America Editorial Committee, 1993–2016). Most floras have been published as books; however, there are many advantages of online or "electronic" floras because these can be more easily searched, accessed, and updated. You may want to become familiar with the plants of your region or state, and a good way to start learning is to purchase a flora or field guide to the trees or wildflowers of your region; many such books are available. A listing of major US state floras is provided in the textbook *Plant Systematics: A Phylogenetic Approach*, 4th ed. (Judd et al., 2016). This reference also contains additional information on descriptive terms, plant relationships (including a presentation of the major families of vascular plants), and plant collecting and identification, as do several other introductory texts (e.g., Murrell, 2010; Simpson, 2010). A listing of published floras for any location in the world can be found in Frodin (2001).

The Internet is an ever-growing resource for plant identification, and photos of many species can be viewed often merely by typing the scientific name of the plant into Google. (Keep in mind, however, that the Internet has no editorial oversight, so you are sure to encounter misidentified photos as well as properly identified ones.) In addition, many websites present keys, descriptions, and/or photos of the plants of particular geographical regions (e.g., the *Flora of China* at www.efloras.org/flora_page.aspx?flora_id=2, *Flora Europaea* at rbg-web2.rbge.org.uk/FE/fe.html, and the *Flora of North America* at www.efloras.org/flora_page.aspx?flora_id=1).

Provided here are two dichotomous keys that will allow identification of the plants of Middle-earth. Begin with Key 1, and if the plant is a herbaceous flowering plant you will be led to Key 2. In addition to being useful in identification, the following two keys will allow you to focus in on the salient characteristics differentiating the species found in J. R. R. Tolkien's legendarium. Since the flora of Middle-earth is predominantly that of temperate regions of the Northern Hemisphere, the following key will be most useful in temperate Europe, North America, and Asia. If you live in any of these regions, you will surely encounter many Middle-earth plants in the meadows or forests near your home—and perhaps even in your own backyard. You will also find species that are not part of Tolkien's imaginative world; for example, among the conifers, trees such as spruces (*Picea*), the red and white cedars (*Thuja, Chamaecyparis*), hemlocks (*Tsuga*), Douglas-fir (*Pseudotsuga*), and bald-cypress (*Taxodium*), and, among the flowering trees, maples (*Acer*), hickories (*Carya*), sycamores (*Platanus*), hackberries or sugarberries (*Celtis*), redbuds (*Cercis*), manzanitas (*Arctostaphylos*), magnolias (*Magnolia*), mulberries (*Morus*), soapberries (*Sapindus*), sweetgums (*Liquidambar*), and tupelos (*Nyssa*), among many others—and these will become apparent as you compare the plants included in this book with those listed in your local floras or field guides. Most of these are plants that do not occur natively in Europe; among those just listed, the only genera occurring in Europe yet not mentioned in Tolkien's writings are the spruces (*Picea*), maples (*Acer*), sycamores (*Platanus*), hackberries (*Celtis*), and mulberries (*Morus*), and, of these five, only *Acer* is native to England (where it is represented by *A. campestre*). The absence of maples from Middle-earth is thus surprising, and we like to think that perhaps maples were seen but not recognized by Frodo and Sam when they were in Ithilien! Finally, if, as you attempt an identification using the keys, neither lead seems to match your plant or the description associated with the name you arrive at in the key doesn't match the plant in your hand, then it is likely that the plant is not one of those of Tolkien's Middle-earth.

KEY 1: KEY TO PLANTS OF MIDDLE-EARTH

1. Organisms microscopic, their bodies unicellular or of cell-colonies consisting of filaments, flat sheets, or globular masses 2
1. Organisms macroscopic, their bodies multicellular and easily visible 3
2. Cells without membrane-bound nucleus, and flagella absent; organism not bioluminescent . . . . **Cyanobacteria** (or blue-green bacteria).

2. Cells with a membrane-bound nucleus, with two flagella, one in an equatorial groove around the cell, and the second attached near the first but passing down a longitudinal groove and extending behind the cell like a tail; organism bioluminescent . . . . **Seafire**.
3. Organisms growing submerged in marine habitats or floating in ocean currents . . . . **Seaweeds**.
3. Organisms terrestrial or fresh-water aquatics, or growing in salt-marshes (and thus daily flooded by tidal action), but not submerged in marine waters 4
4. Leaves absent; organisms fungi or lichens, variously colored, white to brown, yellow-green or gray-green 5
4. Leaves present; organisms plants, and green in color 6
5. Organisms terrestrial, growing from the ground or on dead wood or organic material; with branched filamentous mycelium producing mushroom-shaped reproductive bodies, each with a stalk bearing a fleshy, white to brown cap, with radiating gills on the lower surface, each gill bearing numerous spores . . . . **Mushrooms**.
5. Organisms epiphytic, on tree branches, and well-branched, erect to pendent, yellow-green to gray-green (because the fungal component of the lichen is associated with microscopic green algae), eventually producing cup-shaped or disk-shaped reproductive structures, each with numerous spores . . . . **Beard lichens**.
6. Plant reproducing by dustlike spores, and lacking flowers, fruits, and cones 7
6. Plant reproducing by seeds, and these borne either in cones or in fruits (developing from flowers) 10
7. Plants tiny, usually only to 5 cm tall; stems lacking vascular tissues (i.e., without internal water- and carbohydrate-conducting cells); sporangia atop tiny, leafless stalks, each sporangium usually with a apical rim of minute teeth . . . . **Mosses**.
7. Plants larger; stems with vascular tissues (i.e., with internal water-conducting cells and carbohydrate-conducting cells); sporangia in clusters (sori) on underside of leaf, and each sporangium ripped open due to action of thick-walled annulus cells 8
8. Leaves simple and unlobed, with cordate base, with prominent midvein and obscure, pinnate, and dichotomously branched secondary veins; with sori linear, borne along the veins, so more or less perpendicular to midvein, and protected by a lateral indusium . . . . **Hart's-tongue**.
8. Leaves compound or simple and pinnately lobed, or if simple and unlobed, then without the above combination of characters. . . . . .9.
9. Leaves widely spaced along long-creeping rhizomes, each leaf three-times pinnately compound, and the leaflets also deeply pinnately lobed, the petiole deeply grooved on upper surface, and sporangia in elongate sori, continuous along leaf margins, and protected by inrolled leaf margin on one side and an elongated indusium on the other. . . . . **Bracken**.
9. Leaves without the above combination of characters . . . . **Ferns** (other genera).

10. Plants without flowers and producing their seeds in cones (or terminal, naked seed surrounded by a red aril); leaves needle-like to scale-like. 11
10. Plants with flowers and producing their seeds in various fruits; leaves various, but often broad and with an evident network of veins. 17
11. Plant with seeds in cones, borne on cone scales, not associated with arils. . 12
11. Plant with seeds solitary, each surrounded by a scarlet to orange-scarlet aril. . . . . **Yews.**
12. Plants with long and short shoots, so needlelike leaves appearing to be fascicled. 13
12. Plants with all shoots equivalent, so scalelike; awl-shaped or needlelike leaves individually borne along stems 15
13. Photosynthetic leaves in clusters of (1–)2–5 on short shoots, the short shoots produced in a single year and eventually dropping, and thus with a set number of leaves; leaves of two forms, needle-like and photosynthetic on short shoots (along with sheathing, scale leaves) and all scalelike and nonphotosynthetic on the long shoots. . . . . **Pines.**
13. Photosynthetic leaves in clusters of 10, numerous on short shoots, the short shoots continuing to grow for several years, producing additional leaves; leaves all needlelike and alike, photosynthetic on all shoots 14
14. Cones 1–5 cm long, with persistent scales (so cone opening to release the seeds); trees deciduous and wood not fragrant . . . . **Larches.**
14. Cones 5–12.5 cm long, with deciduous scales (so cone fragmenting at maturity, thus releasing the seeds); trees evergreen and wood fragrant . . . . . . **Cedar.**
15. Leaves needlelike, alternate, and spirally arranged; seed cones dry at maturity, 3.5–20 cm long, with deciduous scales (so cone fragmenting when mature); the bract evident and free from the ovulate scale (= cone scale); seeds each with a conspicuous, single, and elongated wing (with wing tissue derived from cone scale) . . . . **Firs.**
15. Leaves awl-shaped or scale-like, opposite and decussate or three-whorled; seed cones dry to fleshy at maturity, 0.4–4 cm long, with persistent scales (so remaining intact when mature); the bract not evident, fused to the ovulate scale (= cone scale); seeds unwinged or narrowly two-winged (with wing tissue, if present, derived from seed coat) 16
16. Cones dry and woody, not berrylike, eventually opening (sometimes only after fire). . . . . **Cypress.**
16. Cones more or less fleshy, berrylike, remaining closed at maturity, and the seeds thus retained.. . . . . **Junipers.**
17. Plants climbing (vines or lianas) 18
17. Plants not climbing (trees, shrubs, herbs) 28
18. Plant climbing by means of aerial roots that emerge from along the stems; stems with secretory canals, with an unpleasant resinous odor . . . . **Ivy.**
18. Plants without aerial roots and climbing by other means; stems without secretory canals, without a resinous odor. 19
19. Leaves opposite 20

19. Leaves alternate 21
20. Leaves often compound, occasionally simple; flowers radially symmetrical, with perianth parts all alike, of distinct, petaloid tepals, numerous stamens and numerous distinct carpels; fruits achenes . . . .**Clematis.**
20. Leaves simple; flowers bilaterally symmetrical, with a calyx and corolla, the petals fused, five stamens and two fused carpels; fruits berries . . . .**Goatleaf.**
21. Leaves pinnately compound or trifoliolate; flowers pealike; fruits legumes. 22
21. Leaves simple (or rarely palmately compound); flowers various, but not pealike; fruits capsules, achenes, drupes, or berries. 23
22. Leaves pinnately compound, with four or six leaflets, ending in a terminal, branched tendril; stipules large, foliaceous; seeds more or less globose . . . . **Peas.**
22. Leaves trifoliolate (i.e., with only three leaflets), not ending in a tendril; stipules very small; seeds bean-shaped . . . . **Pole beans.**
23. Petiole attachment peltate (i.e., away from the leaf margin); flowers bilaterally symmetrical, with sepals forming a nectar spur. . . . . **Nasturtians, nasturtiums, Indian cresses.**
23. Petiole attachment not peltate (i.e., attached at the leaf margin); flowers radially symmetrical, with sepals not forming a nectar spur 24
24. Vine climbing by means of tendrils arising from the nodes. 25
24. Vine climbing by means of a twining stem. 26
25. Tendrils arising from the stem at a point somewhat deflexed from the leaf axils; flowers conspicuous, opening to reveal bright yellow, fused petals; the stamens fused by their filaments; berry developing from inferior ovary . . . . **Cucumbers.**
25. Tendrils arising from the stem at a point opposite the leaves; flowers inconspicuous, the greenish or cream-colored petals falling as a cap as flower opens; the stamens distinct; berry developing from superior ovary . . . . . **Grape vines.**
26. Leaves usually lobed, with margin serrate; fruits achenes. . . . . **Hops.**
26. Leaves unlobed, with margin entire; fruits drupes or triangular and three-winged capsules 27
27. Flowers with six tepals; fruits triangular and three-winged capsules (and not hot tasting); leaves lacking pellucid dots; plants producing aerial and/or underground tubers . . . .**Roots, Yams** (*Dioscorea*).
27. Flowers lacking perianth parts; fruits globose drupes (and hot to the taste); leaves with pellucid dots; plants not producing tubers. . . . .**Pepper.**
28. Plants woody (i.e., shrubs to trees), forming hard woody stems. 29
28. Plants herbaceous, without woody tissue . . . . . See Key 2 (flowering herbs)
29. Leaves opposite or whorled. 30
29. Leaves alternate 41
30. Leaves compound. 31
30. Leaves simple 32
31. Leaves usually pinnately compound; flowers with only two stamens; fruits samaras (i.e., each winged, with a single seed). . . . .**Ashes.**

31. Leaves palmately compound; flowers with five stamens; fruits thick-valved capsules (with 1–3 large seeds). **Horse-chestnuts.**
32. Leaves whorled . . . . **Heath.**
32. Leaves opposite 33
33. Flowers with inferior ovary 34
33. Flowers with superior ovary. 36
34. Node with stipules, and these interpetiolar (i.e., triangular stipule placed between the petiole bases); flowers with conspicuously fused petals . . . . **Coffee.**
34. Node lacking stipules; flowers with distinct petals 35
35. Leaves evergreen, with pellucid dots (aromatic); flowers with five petals, numerous stamens; fruit a blue-black berry . . . . **Myrtle.**
35. Leaves deciduous, without pellucid dots; flowers with four petals, four stamens; fruit a red drupe . . . . **Cornel.**
36. Flowers with distinct tepals or petals.. 37
36. Flowers with fused petals 38
37. Perianth of sepals and petals, the sepals dimorphic (two small and three large), the petals five, conspicuous, usually yellow . . . . **Rockroses.**
37. Perianth of tepals, greenish, inconspicuous, and without the above combination of characters . . . . **Box, boxwood.**
38. Flowers bilaterally symmetrical, the corolla of five petals, two-lipped, with two petals in an upper lip and three petals in a lower lip; ovary deeply four-lobed; fruits schizocarps of four nutlets 39
38. Flowers radially symmetrical, the corolla of four petals, the lobes all the same size; ovary more or less globose, unlobed; fruits capsules or drupes 40
39. Inflorescences densely spicate (due to clustering of the pseudo-whorls of flowers), raised above the leafy part of the plant; flowers purple or blue . . . . **Lavender.**
39. Inflorescences not densely spicate and flowers in obvious and well-separated pseudo-whorls on distal portion of branches; flowers white to rose or purple. . . . . **Thyme.**
40. Low shrub; leaves tiny, with obscure venation and margins strongly revolute (so abaxial surface is hidden); flowers pink to rose or purple; stamens eight, the anthers each with a pair of appendages; fruits capsules . . . . . **Heather, ling.**
40. Tree; leaves with evident, pinnate venation and margins not strongly revolute (so abaxial surface is clearly visible); flowers white; stamens two, the anthers without appendages; fruits drupes. . . . . **Olive.**
41. Leaves compound, with three to several leaflets arranged pinnately or twice-pinnately 42
41. Leaves simple 46
42. Leaves twice-pinnately compound . . . . **Mimosa.**
42. Leaves once-pinnately compound or trifoliolate (i.e., with three leaflets). 43
43. Leaves aromatic, with resin canals; flowers unisexual (and staminate and carpellate flowers on different plants); fruits drupes . . . . **Terebinth.**

43. Leaves not aromatic, without resin canals; flowers bisexual; fruits pomes or legumes 44
44. Leaves pinnately compound; flowers radially symmetrical; petals white; stamens numerous; ovary inferior, of 2–5 fused carpels; fruit a colorful pome . . . . **Rowan.**
44. Leaves trifoliolate (i.e., with three leaflets); flowers bilaterally symmetrical, like pea-flowers; petals yellow; stamens 10; ovary superior, a single carpel; fruit a legume. . . . . 45.
45. Flowers in pendulous racemes; leaves larger than the flowers . . . . . **Laburnums.**
45. Flowers solitary in leaf axils, on erect branches; leaves smaller than the flowers . . . . **Broom.**
46. Leaves minute, scalelike, with salt-excreting glands . . . . **Tamarisk.**
46. Leaves larger, with an obvious blade, without salt-excreting glands 47
47. Leaves with entire margins 48
47. Leaves with serrate, dentate, or crenate margins 56
48. Bark smooth and snow-white. . . . . **White Tree of Gondor.**
48. Bark smooth to variously roughened and furrowed, brown to gray, but not white 49
49. Stipules (i.e., minute to conspicuous leaflike structures at the nodes, paired at the petiole bases) present 50
49. Stipules absent. 54
50. Leaves usually palmately lobed and veined, with mucilage canals; flowers associated with enlarged bracts . . . . **Cotton.**
50. Leaves unlobed, pinnately veined, without mucilage canals; flowers not associated with enlarged bracts 51
51. Flowers showy, with perianth of green sepals and 4–6 white to dark pink petals; fruits fleshy, colorful 52
51. Flowers inconspicuous, with perianth absent or of very inconspicuous tepals; fruits dry, greenish to brown 53
52. Flowers unisexual, with number of stamens (or staminodes) equaling the petals; style very short or absent; fruit a drupe with several pits. . . . . . . **Holly.**
52. Flowers bisexual, with the stamens more numerous than the petals; style elongate; fruit a drupe with a single pit. . . . . **Sloe, plums, cherries.**
53. Flowers associated with nectar glands; both staminate and carpellate flowers in catkins; fused carpels with numerous ovules; fruits capsules, opening to release hairy seeds . . . . **Willows.**
53. Flowers without nectar glands; only staminate flowers in catkins; fused carpels with only six ovules and only one of these functional; fruits nuts, with each nut surrounded by a scaly cuplike structure (acorn) . . . . **Oaks.**
54. Flowers with perianth of 3–7 sepals and 3–7 fused petals, the corolla more or less urn-shaped and pendulous; carpels 3–8, fused; fruits berries . . . . . . **Billberry, whortleberry.**
54. Flowers with a perianth of four or six distinct tepals; fruits drupes 55

55. Flowers with perianth of four tepals; carpel one ; drupes black . . . . **Bay.**
55. Flowers with perianth of six tepals; drupes red . . . . . **Oiolairë (one of the Fragrant Trees of Eressëa).**
56. Flowers inconspicuous, in catkins or catkinlike clusters (or at least the staminate flowers in such clusters) and usually wind pollinated (except insect pollinated in **Willows**); petals absent or very inconspicuous 57
56. Flowers showy, not in catkins, insect-pollinated; petals evident, conspicuous 64
57. Leaf teeth with globose, glandular apices; staminate and carpellate flowers on different plants; seeds with tuft of hairs 58
57. Leaf teeth without globose, glandular apices; flowers bisexual or, if unisexual, then the staminate and carpellate flowers on the same plant; seeds lacking hair tufts 59
58. Buds with several bud scales; flowers not associated with nectar glands, with disk- or cuplike perianth . . . . **Poplars.**
58. Buds with a single bud scale; flowers associated with nectar glands, without a perianth . . . . **Willows.**
59. Leaf base asymmetrical; flowers bisexual . . . . **Elms.**
59. Leaf base symmetrical; flowers unisexual 60
60. Only the staminate flowers in catkins; flowers with three fused carpels; fruits borne in a spiny or scaly cuplike structure (= cupule) 61
60. Both staminate and carpellate flowers in catkins; flowers with two fused carpels; fruits associated with bracts, but not in a cuplike structure 62
61. Fruit an acorn (i.e., a solitary nut), circular in cross-section, associated with a scaly cuplike structure, without sutures or valves . . . . **Oaks.**
61. Fruit a pair of nuts, each triangular in cross-section, associated with a four-valved, spiny cuplike structure . . . . **Beeches.**
62. Fruits nuts (i.e., globose, unwinged, and more than 1 cm in diameter); fruits associated with foliaceous bracts . . . . **Hazels and filberts.**
62. Fruits samaras (i.e., flattened, with two-winged or at least two-ridged, and much smaller than above); fruits associated with small bracts 63
63. Carpellate flowers two per bract complex, and bract complex with five lobes, thickened, woody, and persisting long after release of fruits, thus old conelike carpellate infructescences present on the plant; twigs with winter buds stalked . . . . **Alders.**
63. Carpellate flowers usually three per bract complex, and bract complex with three lobes, thinner and not woody, usually deciduous with release of fruits, so old carpellate infructescences not present on the plant; twigs with winter buds not stalked . . . . **Birches.**
64. Flowers bilaterally symmetrical, with four colorful and showy tepals; carpel one . . . . **Unnamed hollylike tree** (*Banksia*).
64. Flowers radially symmetrical, with green sepals and showy petals; carpels one to several (and then fused) 65
65. Petals fused 66

65. Petals distinct 68
66. Flowers pendulous; petals strongly fused, forming a cylindrical, globose, urn- or bell-shaped corolla . . . . **Billberry, whortleberry.**
66. Flowers not pendulous; petals only slightly fused, the corolla more or less spreading 67
67. Flowers unisexual, the staminate ones with 4–6 stamens, equaling the number of petals . . . . **Holly.**
67. Flowers bisexual, with numerous stamens, much exceeding the number of petals. . . . . **Tea (and Camellia).**
68. Petiole more or less winged; leaves with pellucid dots (and aromatic citrus odor). . . . . **Oranges.**
68. Petiole not winged; leaves without pellucid dots (and not aromatic) 69
69. Flowers yellow-gold; lower leaf surface silver (due to reflective, silky hairs), and petiole flattened, thus leaves strongly fluttering . . . . **Mallorn.**
69. Flowers usually white, pale green, to pink or red, occasionally pale yellowish; lower leaf surface usually pale green, but occasionally white (due to dense stellate hairs) and petiole not flattened, the leaves not strongly fluttering. 70
70. Inflorescence stalk fused to an enlarged and elongated papery bract; base of the leaves asymmetrical . . . . **Lindens.**
70. Inflorescence stalk not fused to an enlarged bract; base of the leaves symmetrical 71
71. Flowers without a floral cup (= hypanthium) . . . . **Tea (and Camellia).**
71. Flowers with a floral cup (= hypanthium) that is nectiferous on its inner surface 72
72. Fruit a drupe; flowers with a single carpel, with the ovary superior. . . . . **Sloe, plums, cherries.**
72. Fruit a pome; flowers usually with 2–5 fused carpels, with the ovary inferior 73
73. Core of pome woody or bony; fruits 6–20 mm in diameter . . . . **Hawthorns, thorns.**
73. Core of pome cartilaginous; fruits 20–60 mm in diameter. . . . . **Apples, wild apples.**

KEY 2: FLOWERING HERBS OF MIDDLE-EARTH

1. Leaves with parallel venation 2
1. Leaves with a network of veins, and with the major veins typically pinnate (i.e., with a single midvein and smaller veins branching from it) or palmate (i.e., with several prominent and diverging veins) 15
2. Flowers showy, with a perianth of petaloid tepals, and attracting various insects as pollinators 3
2. Flowers inconspicuous, the perianth parts absent or very small, and pollinated by wind 10

3. Flowers with three stamens; leaves equitant (i.e., in two ranks and with blades flattened in the plane of leaf insertion); ovary inferior. . . . **Flag-lily, yellow iris.**
3. Flowers with six stamens, plants with leaves variously arranged, but not equitant; ovary inferior or superior 4
4. Ovary inferior 5
4. Ovary superior. 6
5. Flowers with a corona (i.e., cup- or trumpetlike, petaloid outgrowth of the perianth); perianth parts yellow or white; plants blooming in the spring. . . . . **Daffodils.**
5. Flowers without a corona; perianth parts white, but the inner ones each with a green blotch; plants blooming in winter . . . . **Niphredil.**
6. Plants with onion odor (due to sulfur-containing compounds); flowers in an umbel atop a long scape . . . . **Onions.**
6. Plants lacking an onion odor; flowers in various inflorescences or solitary, but if umbellate then not atop a scape 7
7. Plants from rhizomes . . . . **Asphodel.**
7. Plants from bulbs 8
8. Flowers numerous, in racemes; perianth usually blue or purple, occasionally pink, or white, not spotted; nectar produced in the septa of the ovary; capsules with black seeds . . . . **Hyacinth.**
8. Flowers solitary or few and in racemes or umbels; perianth orange, red, yellow, pink, or white, spotted or without markings; nectar produced at the bases of the tepals; capsules with yellowish, tan, or brown seeds 9
9. Inflorescences reduced to a solitary flower; tepals not reflexed, yellow . . . **Alfirin and Mallos.**
9. Inflorescences racemose, umbellate, or paniculate, sometimes reduced to a solitary flower; tepals recurved or reflexed, and orange, red, yellow, pink, or white, often spotted . . . . **Lilies.**
10. Leaves in three ranks; stem in cross-section circular to triangular. 11
10. Leaves in two ranks; stem in cross-section circular to elliptic (i.e., the grasses: four species included in the key, although J. R. R. Tolkien's writings contain numerous other references to unidentified grasses, especially those genera dominating steppes or prairies) 12
11. Stems round; the leaf sheath wrapping around the stem, but the margins not fused, thus sheath said to be "open"; flowers with six small tepals, the bracts various but never flask-shaped; fruit a capsule . . . . **Rushes.**
11. Stems more or less triangular; the leaf sheath fused into a tube around the stem, thus sheath "closed"; flowers lacking tepals, but the carpellate flowers surrounded by a flask-shaped bract; fruit an achene . . . . **Sedges.**
12. Inflorescence a panicle 13
12. Inflorescence a raceme or spike 14
13. Spikelets with elongated hairs; leaf blades 2–4 cm wide; plants to 4 m tall . . . . **Reeds.**

13. Spikelets without elongated hairs; leaf blades 0.3–2 cm wide; plants shorter, usually to only 1.5–3 m tall . . . . **Salt-marsh grasses, cordgrasses.**
14. Spikelets solitary at each node of the inflorescence axis. . . . . **Wheat.**
14. Spikelets in clusters of three at each node of the inflorescence axis . . . **Barley.**
15. Inflorescence a thick, fleshy axis, associated with a demented white bract . . . . **White Flowers of Morgul Vale.**
15. Inflorescence not as above. 16
16. Leaves two-ranked, composed of a blade and sheath, with flap of tissue (ligule) at their junction . . . . **Cardamon.**
16. Leaves various, but not as above, without a ligule 17
17. Plants with flowers in dense heads, each of which is surrounded by an involucre of bracts 18
17. Plants with flowers in various inflorescences, but not in involucrate heads 25
18. Tissues with milky sap 19
18. Tissues without milky sap 20
19. Heads solitary, each on a scape; leaves oblong to obovate in a basal rosette . . . . **Dandelion.**
19. Heads clustered and not scapose; leaves ovate to orbicular, basal and borne along stem . . . . **Lettuce.**
20. Flowers of head all alike (and radially symmetrical) 21
20. Flowers of head differentiated, with peripheral ray flowers (that are bilaterally symmetrical) and central disk flowers (that are radially symmetrical) 22
21. Leaves with bristle-teeth to spine-teeth; flowers white, pink, red, purple, or yellow . . . . **Thistles.**
21. Leaves without spines or bristles; flowers bright yellow. . . . . **Mugwort.**
22. Ray flowers yellow to orange 23
22. Ray flowers white, pink, to purple 24
23. Fruits curved; plants short, to ca. 80 cm . . . .**Marigolds.**
23. Fruits compressed and more or less straight; plants tall, 50–400 cm. . . . .**Sunflowers.**
24. Heads clustered in elongated inflorescences; disk flowers white, pink, to purple . . . . **Butterbur.**
24. Heads solitary, each atop a scapelike stem; disk flowers yellow . . . . . **Daisy.**
25. Plants with flowers in compound umbels; foliage with a resinous odor and petioles sheathing the stems 26
25. Plants with flowers in various inflorescences, but not in compound umbels; foliage with odor or not, but not resinous, and petioles not sheathing the stems 30
26. Fruits covered in rows of hooked or straight bristles; roots white, yellow, orange, or red . . . . **Carrots.**
26. Fruits lacking hooked or straight bristles; roots more or less white 27

27. Fruits flattened and strongly winged; leaf sheaths extremely expanded and cuplike . . . . **Angelica.**
27. Fruits terete or only slightly flattened, not winged, but ridged or unridged; leaf sheaths more or less expanded but never cuplike 28
28. Fruits more than twice as long as wide, terminating in a narrowed beak; the two segments not ridged . . . . **Wood-parsley.**
28. Fruits less than twice as long as wide, without a beak; the two segments moderately to strongly ridged 29
29. Flowers white; stems reddish spotted; fruit usually with undulated ribs . . . . **Hemlock.**
29. Flowers greenish-yellow or yellow; stems not spotted; fruit ribs not undulated . . . . **Parsley.**
30. Plants aquatic, with leaves more or less circular, with a dissection in region where petiole is attached, the blade floating on the water's surface, with elongated petiole in water column; each flower on an elongate peduncle (stalk), with numerous petals and stamens . . . . **Waterlilies.**
30. Plants terrestrial, or if aquatic then not as above 31
31. Flowers bilaterally symmetrical 32
31. Flowers radially symmetrical 44
32. Leaves compound, with three leaflets; flowers pealike (i.e., with uppermost petal a banner, the two lateral petals winged, and two lowermost petals fused and keel-like; stamens fused by their filaments 33
32. Leaves simple; flowers not as above; stamens not fused 35
33. Leaflets with entire margins; flowers with keel petals spirally coiled; fruits elongated . . . . **Beans.**
33. Leaflets with clearly to minutely serrate margins; flowers with keel petals straight; fruits short 34
34. Corolla withering but persistent after pollination; inflorescences headlike or more or less short . . . . **Clovers.**
34. Corolla deciduous after pollination; inflorescences slender and elongated . . . . **Melilot.**
35. Petals distinct 36
35. Petals fused into a tube, with two petals usually forming an upper lobe and three petals forming a lower lobe of the corolla 38
36. Petals four; ovary inferior, comprised of four fused carpels; seeds with conspicuous hair tufts . . . . **Fireweed.**
36. Petals five; ovary superior, composed of three fused carpels; seeds not hair-tufted 37
37. Leaves peltate (i.e., petiole attached near center of blade); plants pungent, with mustard oils; flowers with a nectar spur formed by the sepals, with eight stamens; fruit a schizocarp, separating into three drupe- or nutlike segments . . . . **Nasturtians.**

37. Leaves not peltate (i.e., petiole attached at base of blade); plants not pungent, lacking mustard oils; flowers with a nectar spur formed by petals and stamens, with five stamens; fruit a capsule . . . . **Pansy.**
38. Plant with milky sap; flowers with a dorsal slit in the corolla; ovary inferior . . . . **Lobelia.**
38. Plant without milky sap; flowers without a dorsal slit in the corolla; ovary superior 39
39. Ovary globose, with numerous ovules, with style arising from the apex of the ovary; fruits capsules, opening by pores; inflorescences racemes . . . . **Snapdragons.**
39. Ovary four-lobed, with only four ovules, with style arising out of a depression in the middle of the four-lobed ovary; fruits four nutlets; inflorescences are pairs of reduced cymes, forming pseudo-whorls to spicate clusters 40
40. Corolla strongly two-lipped, the upper lip more or less concave; stamens two, each anther with a much-expanded connective, so the two halves of the anther are widely separated . . . . **Sages.**
40. Corolla two-lipped to nearly radial, the lips only weakly differentiated, the upper lip more or less flat; stamens four, each anther with connective not or only slightly expanded 41
41. Corolla nearly radially symmetrical, seemingly four-lobed, but the wider upper lobe actually two fused petals . . . . **Mentha.**
41. Corolla distinctly two-lipped 42
42. Inflorescences densely spicate (due to clustering of the pseudo-whorls of flowers), raised above the leafy part of the plant; flowers purple or blue . . . . **Lavender.**
42. Inflorescences not densely spicate and flowers in obvious and well-separated pseudo-whorls on distal portion of branches; flowers white to rose or purple 43
43. Calyx radial; flowers bisexual . . . . **Marjoram.**
43. Calyx two-lipped; flowers bisexual and carpellate . . . . **Thyme.**
44. Petals or tepals fused 45
44. Petals or tepals distinct 54
45. Leaves pinnately or twice pinnately compound 46
45. Leaves simple 47
46. Leaves twice-pinnately compound, all leaflets the same size (i.e., all minute); flowers with stamens much more conspicuous than the corolla, and the white, pink, or lavender stamens not contrasting in color with the petals; stipules present; fruits legumes; plants without tubers . . . . **Mimosa.**
46. Leaves pinnately compound with leaflets very divergent in size (i.e., some large and other small); flowers with stamens and corolla equally conspicuous, and the stamens yellow, contrasting with the white petals; stipules absent; fruits berries; plants producing tubers . . . . **Potatoes.**

47. Leaves opposite and decussate 48
47. Leaves alternate and spirally arranged. 49
48. Petals yellow or occasionally silver (and sometimes having both colors) . . . . **Elanor.**
48. Petals white, pink, red, orange, or blue . . . . **Pimpernel.**
49. Flowers with four-lobed ovary, the style arising out of depression in center of lobed ovary; inflorescences one-sided cymes 50
49. Flowers with more or less globose ovary, the style arising from the ovary apex; inflorescences various, but not one-sided cymes 51
50. Plant aromatic; leaves large, hoary (densely hairy); flowers nodding, without five short scales . . . . **Kingsfoil.**
50. Plant not aromatic; leaves smaller, not hoary; flowers more or less erect, with throat obstructed by five short scales . . . . **Forget-me-nots.**
51. Carpels four or five 52
51. Carpels two. 53
52. Ovary, in cross-section with a single basal ovule; leaves with only midvein evident. . . . . **Thrift.**
52. Ovary in cross-section with numerous ovules on a central axis; leaves with obvious pinnate venation . . . . **Primeroles, primroses.**
53. Inflorescences terminal; corolla with tube white and the lobes pink to rose; fruits brown capsules; leaves can be smoked . . . . **Pipe-weed, tobacco.**
53. Inflorescences axillary; corolla purple; fruits black berries; leaves are not smoked . . . . **Belladonna.**
54. Flowers with distinct carpels (and thus with several ovule-bearing structures in the center of the flower) 55
54. Flowers with fused carpels (thus only a single ovule-bearing structure in the center of the flower) or only a single carpel 63
55. Leaves succulent. . . . . **Seregon and stonecrops.**
55. Leaves thin and herbaceous 56
56. Leaves and stems with sharp-pointed prickles 57
56. Leaves and stems without prickles. 58
57. Flower with floral cup (= hypanthium) urn-shaped; fruits dry and hard (achenes), surrounded by orange to red fleshy cuplike structure (i.e., forming a rose hip). . . . . **Roses.**
57. Flower with floral cup flat; fruits fleshy, red or black (druplets), exposed . . . . **Blackberries, raspberries, brambles.**
58. Stipules present; flowers with a floral cup (= hypanthium); floral receptacle expanding as fruits develop, forming a fleshy, red structure on which the fruits (achenes) are borne. . . . . **Strawberries.**
58. Stipules absent; flowers without a hypanthium; floral receptacle neither expanding nor becoming fleshy; fruits achenes or follicles 59
59. Leaves opposite . . . . **Clematis.**

59. Leaves alternate (but pair of bracts below inflorescence may be opposite) . 60
60. Flowers with 3–5 sepals and five to numerous petals 61
60. Flowers with four to numerous tepals. 62
61. Petals lacking nectar glands, white to pink or red, purple, or yellow; fruits follicles . . . . **Peony.**
61. Petals with basal nectar glands, usually yellow; fruits achenes . . . **Buttercups.**
62. Herb blooming throughout the year; tepals white . . . . **Evermind.**
62. Herb blooming only in the spring; tepals white, red, blue, purple, green, or yellow . . . . **Anemones.**
63. Leaves opposite 64
63. Leaves alternate 68
64. Flowers wind-pollinated; perianth of tepals, inconspicuous 65
64. Flowers insect-pollinated; perianth of sepals and petals, the petals showy, colorful 66
65. Leaves palmately compound; carpellate flowers with two styles and stigmas (so two carpellate) . . . . **Hemp.**
65. Leaves simple; carpellate flowers with a single brush-like stigma (and seeming to be a single carpel, but actually two) . . . . **Nettles.**
66. Calyx with distinct sepals, with two small sepals and three larger ones; petals usually yellow, the margin entire; stamens numerous . . . . **Rockroses.**
66. Calyx with fused sepals, all the same size; petals white to red or purple, apically notched or fringed; stamens 10. 67
67. Styles three; petals with a pair of appendages at the junction of the narrow base and the spreading blade . . . . **Campion.**
67. Styles two; petals lacking a pair of appendages . . . . **Gilly, gillyflower.**
68. Plant with white, yellow, orange, or red sap (latex); flowers with two quickly deciduous sepals and four wrinkled petals 69
68. Plant without latex (or sap merely clear); flowers not as above, with four or five sepals and the same number of petals, or with six tepals (and then the inner larger than the outer), the petals not wrinkled 70
69. Carpels two; petals yellow. . . . . **Celandine.**
69. Carpels three or more; petals pink to red or lilac, orange, or yellow . . . . **Poppy.**
70. Each node of the stem with ocrea (i.e., cylindrical fused stipules) surrounding the stem; flowers with six tepals, the inner three larger than the outer ones . . . . **Sorrel.**
70. Each node of the stem with or without stipules, but if present, then not as above; flowers usually with four or five sepals and petals (but 3–5 tepals in **Amaranth**) 71
71. Leaves twice-pinnately compound . . . . **Mimosa.**
71. Leaves simple to once-pinnately compound. 72
72. Flowers with a perianth of 3–5 tepals . . . . **Amaranth.**
72. Flowers with a perianth of four or five sepals and four or five petals 73

73. Flowers with five petals 74
73. Flowers with four petals 76
74. Stamens 5 or 10, with each anther opening by two slits; tissues without mucilage canals; sepals distinct. 75
74. Stamens numerous, with each anther opening by a single slit; tissues with mucilage canals (slime); sepals connate . . . . **Malva.**
75. Flowers with five stamens; leaves narrow, with only three visible veins . . . **Flax.**
75. Flowers with 10 stamens; leaves broader, with numerous visible veins in a pinnate or palmate arrangement . . . . **Saxifrages.**
76. Flowers white; basal leaves more or less pinnately compound. . . . . . . **Cress, watercress.**
76. Flowers yellow; basal leaves simple to pinnately compound 77
77. Petals usually 18–25 mm long; terminal segments of fruits 4–11 mm long; cultivated form without fleshy roots, but with apical, globose head formed by overlapping leaves . . . . **Cabbage.**
77. Petals usually 6–11 mm long; terminal segments of fruits 8–22 mm long; cultivated form with fleshy taproot, but stem lacking globose head formed by overlapping leaves . . . . **Turnips.**

SIX

Telperion and Laurelin

The Two Trees of Valinor

Under her song the saplings grew and became fair and tall, and came to flower; and thus there awoke in the world the Two Trees of Valinor. Of all things which Yavanna made they have most renown, and about their fate all the tales of the Elder Days are woven. The one had leaves of dark green that beneath were as shining silver, and from each of his countless flowers a dew of silver light was ever falling, and the earth beneath was dappled with the shadows of his fluttering leaves. The other bore leaves of a young green like the new-opened beech; their edges were of glittering gold. Flowers swung upon her branches in clusters of yellow flame, formed each to a glowing horn that spilled a golden rain upon the ground; and from the blossom of that tree there came forth warmth and a great light. Telperion the one was called in Valinor . . . but Laurelin the other was . . . and many names in song beside. (SILM 1).

In the depths of time, the Two Trees of Valinor, Telperion and Laurelin, were sung into existence by Yavanna, one of the Queens of the Valar and a lover of all growing things (Figure 6.1). The trees provided light to the realm of Valinor—the land of these Powers. Telperion flowered alone for the first six hours of each day, and during the seventh hour his silver light mingled with the golden light of Laurelin. She blossomed during the latter half of the 12-hour Valien day, and, at the end of her blossoming period, her light again mingled with the regenerating light of Telperion. Thus, twice each day, there was a precious period of mingled gold and silver light. During the time of the Two Trees, Middle-earth lay in near darkness, lit only by starlight, as the light of the Two Trees was restricted to the vicinity of Valinor. This light was produced by the trees' flowers—almost as if it were nectar—and was gathered in shining lakes, illuminating all of Valinor. After the awakening of the Elves, the Valar summoned them to Valinor, "there to be gathered at the knees of the Powers in the light of the Trees," although Ulmo spoke against this decision, stating that they should be left free "to walk as they would in Middle-earth, and with their gifts of skill to order all the lands and heal their hurts"

FIGURE 6.1 *Fëanor and the Two Trees of Valinor.*

(SILM 3). Thus, Ingwë, Finwë, and Elwë were brought as ambassadors to Valinor. They were filled with awe at the sight of the Valar and "greatly desired the light and splendor of the Trees" (SILM 3). Returning to Cuiviénen, they spoke to their people, urging them to heed the summons—and thus began the long migration of the three hosts into the west. Eventually, the people of Ingwë, Finwë, and Elwë became, respectively, the Vanyar (the Fair Elves), Noldor (the Deep Elves, in the sense of possessing great knowledge), and Teleri (the Last-comers, as this host arrived last in Valinor, and, in fact, many of the Teleri [e.g., the Sindar and Nandor] never left Middle-earth). The Calaquendi (i.e., the elves who reached Valinor and had seen the holy light of the Two Trees) much surpassed the Dark Elves (i.e., those who had remained behind in Middle-earth) in wisdom, skill, and beauty. The bliss of Valinor lasted while the trees lived. Tragically, however, they were killed by Morgoth and the spider-like Ungoliant, and, after their death, their holy light remained only in the light of the Silmarils (which had been fashioned by the elf Fëanor, see Figure 6.1) and in the light of some of the brighter stars. Yavanna, through her singing, and Nienna, through her tears, sought to heal the trees, but they did not succeed. Yet their efforts caused Telperion, at last, to bear one great, silver flower, and Laurelin to produce a single golden fruit—and (as related in the myths of the Noldor) from these the Valar fashioned the moon and sun in order to bring light to all of Middle-earth. It is important to remember that during the Third Age the unsullied light of the Two Trees (i.e., light produced before they were attacked by Morgoth) only survived in

Eärendil's star—arising from the Silmaril in his possession—which had been recovered from Morgoth by Beren and Lúthien. A small portion of this light was caught in the phial of Galadriel, which she gave to Frodo, to be a light in dark places—"when all other lights go out" (LotR 2: VIII). Thus, we see that the light of the Two Trees is more than a memory, but is an ongoing force within Arda—a holy light, which contributed to the success of Frodo's quest and that can still be seen in the morning star.

How shall we imagine these two holy trees—the greatest of the works of Yavanna, the Giver of Fruits? Telperion was the most beloved by the elves, and Yavanna made an image or copy of the Elder Tree for the elves of Tirion; this image of Telperion was called Galathilion. It grew in the courtyard beside the Tower of Ingwë and was identical to Telperion in every way—except that it did not produce its own light. Galathilion was the ancestor of the long lineage of White Trees, ending in the various White Trees of Gondor. Thus, a living image of Telperion survived into the Third and Fourth Ages (see the White Tree of Gondor, in Chapter 7, for a detailed description of this tree). No comparable image of Laurelin survived the attack by Morgoth and Ungoliant. The description and illustration of Laurelin (see Figure 6.2) presented here is based on J. R. R. Tolkien's description (see initial quote) and thus assumes some artistic and botanical license. It is clear that Telperion and Laurelin are not based, in any simple way, on any currently existing flowering tree. However, the white flowers of Telperion are said to resemble those of a cherry (*Prunus,* a genus of Rosaceae; see HoM-E Vol. 5, Part 2: VI, p. 209), while the hanging clusters of yellow flowers of Laurelin are clearly reminiscent of those of the golden-chain tree (*Laburnum,* a genus of legumes, the Fabaceae or Leguminosae). In fact, in the original *Quenta Silmarillion* (see HoM-E Vol. 5, Part 2: VI, p. 209), it is stated that the hanging

FIGURE 6.2 *Laurelin.*

clusters of yellow flowers of Laurelin were like the hanging blossoms of "those trees Men now call Golden-rain"—a common name usually applied to *Koelreuteria* (especially the species *K. paniculata* and *K. elegans*, of the Sapindaceae). However, the yellow-flowered inflorescences of *Koelreuteria* are erect, not hanging. It is thus more likely that Tolkien was thinking of *Laburnum*, and these species are occasionally also called golden-rain trees (instead of the more commonly used golden-chain). Yet it would be a mistake to think that Telperion was a merely species of *Prunus* and Laurelin a species of *Laburnum*. The flowers of Galathilion (i.e., Sindarin for "Radiant-Holy-Moon") were moonlike in shape, implying that their petals were at least slightly fused—not distinct from each other as in cherries. Likewise, the flowers of Laurelin are described as horn-shaped, spilling "golden rain upon the ground" (SILM 1). This, likewise, implies that their petals were fused, allowing the horn-like corollas to hold light as if it were nectar, as shown in the beautiful illustration by Roger Garland (http://www.theonering.com/galleries/professional-artists/the-silmarillion/two-trees-of-valilnor-roger-garland). The form of the flowers of both Telperion and Laurelin, therefore, are more in line with that of members of the asterid clade (i.e., Asteridae or Sympetalae), a group of flowering plants that is characterized by flowers with fused petals, forming various tubular or bell-shaped corollas. Thus, we have shown the flowers of both trees as having fused petals. Those of Telperion are reminiscent of some Ericales, while those of Laurelin are similar to many Lamiales, and especially some Bignoniaceae (e.g., *Tecoma stans*). It would have been overly simplistic to visualize the flowers of Laurelin as having the form of pea-flowers (as would be appropriate for *Laburnum*, a member of the Fabaceae, a group with separate petals) or the flowers of Telperion merely as five-petaled and with a floral cup (as would be appropriate for *Prunus*, of the Rosaceae, another group with distinct petals). The leaves of Telperion and Laurelin seem to draw upon yet other trees. Those of Telperion are said to be silver beneath, and such leaves, for example, are characteristic of some Elaeagnaceae (e.g., *Shepherdia* or some species of *Elaeagnus*, which have silver peltate-scales covering the lower surface) or Sapotaceae (e.g., species of *Chrysophyllum*, which have dense, T-shaped, silver or gold hairs covering the lower surface). They are also said to tremble, as is characteristic of the leaves of many species of *Populus* (of the Salicaceae) or the leaflets of *Schefflera tremula* (Araliaceae), among others. The leaves of Laurelin are compared with newly opened beech leaves (i.e., *Fagus*, Fagaceae), and thus we assume that they were simple (i.e., each leaf with a single blade), not pinnately compound (i.e., with several leaflets) as in *Laburnum* or *Koelreuteria*. It is clear that, in Laurelin and Telperion, we see elements of several species pulled together in unconventional, yet lovely combinations. Each of the Two Trees is unique, existing only in Tolkien's legendarium and representing beautiful elements drawn from a diverse array of flowering plants.

The Two Trees are central in the mythology of Middle-earth, as stated in the *Silmarillion:* "about their fate all the tales of the Elder Days are woven" (SILM 1). It is not really surprising that the fate of Telperion and Laurelin is at the center of Tolkien's legendarium because trees are frequently of great religious and/or mythological significance. Many examples could be cited, but we will mention only a few here. Two trees are an essential part of the story of Adam and Eve (see Genesis: chapters 2 and 3): the Tree of the Knowledge of Good and Evil and the Tree of Life. Another, quite different mythological vision of two trees is that found in the fourth chapter of Zechariah, where the prophet saw two olive trees (*Olea europaea*), one on each side of a golden lampstand, representing Joshua (the high priest) and Zerubbabel (the political ruler) and the power and grace needed to accomplish the rebuilding of the temple. In Egyptian mythology, two sycamore-figs (*Ficus sycomorus*) stood at the eastern gate of the heavens through which Ra emerged every morning, and these represented the Tree of Life. Another example is the World Tree, Yagdrasil, possibly identified with an ash (*Fraxinus*; see also Watts, 2007), of Norse mythology, which extended from the heavens to the underworld. Trees have been considered sacred in many religions, often symbolizing strength, splendor, immortality, and/or growth and fertility, and worship frequently occurred in sacred groves (e.g., in the ancient religions of Greece and Rome, the Middle East, or of the Celts, Druids, and other peoples of northern Europe). This connection of trees (and especially oaks, i.e., *Quercus* spp.) with the divine is seen, for example, in the Hebrew words for oak (i.e., *allon* and *elon*) both of which are associated with the word for "god"—*el*. It is perhaps significant that Joshua erected a stone beneath an oak when he and his people renewed their covenant with the Lord. Oaks were sacred in the religions of the ancient peoples of Greece (where they were associated with Zeus), the Baltic region (associated with Perkons, a god of thunder), the Scandinavian region (where they were associated with Thor, another thunder god), and various Celtic groups (where oaks, again, were associated with thunder gods). This connection may result from the fact that long-lived oaks are frequently hit by, but survive, lightning strikes. Finally, we have in Tolkien's mythology the connection of trees and light. Light is almost universally of religious significance and is symbolically connected with the divine. Thus, we see in the *Silmarillion* that the silver and golden light of the Two Trees is holy, and those who live in their light are transformed—the Calaquendi, for example, have gained skills and knowledge far above those of the Dark elves who had refused the summons. Even the "opening eyes of Men" (SILM 12) were first turned toward the rising of the sun—second-hand light, derived from the Two Trees—in the West.[1] The first humans to enter Beleriand did so because they had heard that there was "Light in the West" (see SILM 17), but unfortunately, instead they found before them Morgoth, the Lord of the Dark and killer of the Two Trees.

ETYMOLOGY

The Quenya name *Telperion* is derived from the Telerin *telpë* (meaning silver); the meaning of *-rion* is unclear but may be a masculine name suffix. An English translation of the name is unclear, but perhaps Silver Tree is the most appropriate (as glossed in the index of the 50th anniversary edition of the *Lord of the Rings*). In Hammond and Scull's *The Lord of the Rings: A Reader's Companion* (see p. 637), the suggested translation is "?silver-white." Telperion is also given the name *Silpion* (from the Quenya

[1] According to J. R. R. Tolkien's mythology (as related in Chapter 11 of *The Silmarillion*) both the moon and the sun, after they were fashioned in Valinor, first arose in the west, although later Varda commanded that they rise in the east. She also established their current movements, allowing a time of shadow and half-light in Middle-earth.

element *sil* meaning white, and thus White Tree) and *Ninquelótë* (from the Quenya *ninquë*, meaning white, and *lótë*, meaning flower). The Sindarin equivalent of the latter is *Nimloth*. The Quenya name *Laurelin* may mean either singing-gold (if based on the stem *laurelind-*) or hanging-gold (if based on the stem *laureling-*). Alternative names for this tree are the Quenya *Culúrien*, based on the stem *kul-* (meaning golden-red, so perhaps meaning Tree of Gold), and *Malinalda*, based on *malina* (yellow or golden) and *alda* (tree).

DESCRIPTION OF TELPERION

See description of the White Tree of Gondor (Chapter 7), which was similar to Telperion in every way except that the flowers of the White Trees did not produce their own light.

DESCRIPTION OF LAURELIN

Tree with erect branches; the leaves alternate and spirally arranged, simple, ovate to elliptic, with pinnate venation, the base and apex acute, the margin entire; stipules absent. The inflorescences axillary racemes; flowers bisexual, bilaterally symmetrical, with five sepals, five connate petals, forming a hornlike, yellow corolla, with two upper lobes and three lower lobes, perhaps four stamens, and two fused carpels, with a superior ovary. Fruits berries, globose and yellow.

SEVEN

The Plants of Middle-earth

All the plants recorded from J. R. R. Tolkien's legendarium are considered in this chapter (along with a few fungi and other nonanimal living organisms). They are listed alphabetically (arranged by common name) since many grew in more than one geographical region and occurred from the First through the Fourth Ages, thus making any arrangement based on geographical locality (e.g., plants of the Shire, Rivendell, or Lothlórien) or reference (i.e., plants of *The Silmarillion, The Hobbit,* or *The Lord of the Rings*) arbitrary and impractical. Most users of this chapter likely are more familiar with the plant names as given in Tolkien's writings than they are with the detailed characteristics on which the plant families (e.g., rose family, Rosaceae; pea family, Fabaceae or Leguminosae; or heath family, Ericaceae) are based, thus we have used an alphabetical arrangement instead of grouping these plants by their family and arranging these families in a system reflecting their phylogenetic relationships (but see Chapter 4).

For each of the important plants in Tolkien's legendarium, the treatment includes (1) common and scientific name, along with an indication of the family to which the plant belongs; (2) a brief quote from one of Tolkien's works in which the plant is referenced; (3) a discussion of the significance of the plant in the context of Tolkien's legendarium; (4) the etymology, relating to both the English common name and the scientific name derived from Latin or Greek and, where relevant, the name in one or more of the languages of Middle-earth; (5) a brief description of the plant's geographical distribution and ecology; (6) economic uses, both traditional and current; and (7) a brief description of the plant (whether a species or group of related species actually existing today or one unique to Tolkien's created world). Most of these major species are also provided with a woodcut-style illustration (as an aid to identification) along with an inset vignette illustrating one of the events in the history of Middle-earth in which the plant played a role. Less important species have been clustered into one of four categories and considered in the illustrated treatments: Plants of Ithilien, Food Plants of Middle-earth, Bree names, and Hobbit Names. For additional information on the families of vascular plants, readers should consult one or more introductory plant systematics texts (e.g., Judd et al., 2016; Murrell, 2010; Simpson, 2010; Spichiger et al., 2004). Additional information relating to the genera of vascular plants, their economic uses, and geographical distributions can be found in Mabberley (2008). Additional etymological information is at the Online Etymology Dictionary (Harper, 2001–16) and more information regarding elven names can be found in Tolkien's (1987) *The Etymologies* and his *Nomenclature of the Lord of the Rings* (published in Hammond and Scull, 2005), as well as Giraudeau's (2011) *Sindarin Corpus*, Fauskanger's (2008) *Quenya–English* dictionary, Carpenter's (2014) *Sindarin–English* dictionary, and Sallo's (2016) website, *A Qenya Botany*. The website *Parf Edhellen* (https://www.elfdict.com) is also very useful.

We hope that the information and illustrations provided here will add to the reader's understanding and appreciation of Tolkien's works and the importance of plants in his legendarium. A list of the 141 plants (and nonanimal life forms) of Middle-earth is provided here, followed by the detailed treatments.

LIST OF PLANTS OF J. R. R. TOLKIEN'S LEGENDARIUM

1. **Aeglos,** and **gorse** or **whin** (*Ulex europaeus*)
2. **Alders** (*Alnus* spp.)
3. **Alfirin** and **mallos** (*Fritillaria* spp., yellow-flowered)
4. **Anemones** (*Anemone* spp.)
5. **Apples** (*Malus domestica, M. sylvestris,* and related species)
6. **Ashes** (*Fraxinus* spp.)
7. **Barley** (*Hordeum vulgare*)
8. **Bay** (*Laurus nobilis*)
9. **Beard lichens** (*Usnea* spp.)
10. **Beeches** (*Fagus* spp.)
11. **Billberry** or **whortleberry** (*Vaccinium myrtillus, V. uliginosum,* and related species)
12. **Birches** (*Betula* spp.)
13. **Blackberries, raspberries, brambles** (*Rubus* spp.)
14. **Bracken** (*Pteridium aquilinum*)
15. **Buttercups** (*Ranunculus* spp.)
16. **Campion** (*Silene* spp., especially *S. uniflora*)
17. **Cedar** (*Cedrus libani*)
18. **Clovers** (*Trifolium* spp.)
19. **Coffee** (*Coffea arabica* and *C. canephora*)
20. **Cress, watercress** (*Nasturtium officinale,* and related species)
21. **Cyanobacteria, blue-green bacteria** (various genera of Cyanobacteria, a subgroup of the Bacteria)
22. **Daffodils** (*Narcissus* spp.)
23. **Daisy** (*Bellis perennis*)
24. **Dandelions** (*Taraxacum officinale, T. erythrospermum,* and relatives)

25. **Ebony** (*Diospyros* spp.)
26. **Elanor** (a species of *Anagallis*, unique to J. R. R. Tolkien's legendarium)
27. **Elms** (*Ulmus* spp.)
28. **Evermind, simbelmynë, uilos** (a species of *Anemone*, unique to J. R. R. Tolkien's legendarium)
29. **Ferns** (*Adiantum, Asplenium, Athryium, Blechnum, Dryopteris, Polypodium, Polystichum, Thelypteris, Woodsia*, and others)
30. **Fireweed** (*Chamerion angustifolium*)
31. **Firs** (*Abies* spp.)
32. **Flag-lilies** or **yellow iris** (*Iris pseudacorus*)
33. **Flax** (*Linum usitatissimum*)
34. **Forget-me-nots** (*Myosotis* spp.)
35. **Fragrant trees from Eressëa** (various species) [oiolairë, lairelossë, nessamelda, vardarianna, taniquelassë, yavannamirë, lavaralda]
36. **Grape vines, wine** (*Vitis vinifera*)
37. **Grass** (various genera dominating temperate prairies or steppes, e.g., *Agropyron, Andropogon, Bouteloua, Bromus, Festuca, Koeleria, Panicum, Poa, Schizachyrium, Sorghastrum, Stipa*)
38. **Hart's-tongue** (*Asplenium scolopendrium*)
39. **Hawthorns, thorns** (*Crataegus* spp.)
40. **Hazels** and **filberts** (*Corylus avellana, C. maxima*, and related species)
41. **Heath** (*Erica* spp.)
42. **Heather, ling** (*Calluna vulgaris*)
43. **Hemlock, hemlock-umbels** (*Conium maculatum*, or perhaps *Anthriscus sylvestris*)
44. **Hemp, gallow-grass** (*Cannabis sativa*)
45. **Holly** (*Ilex aquifolium*, and related species)
46. **Horse-chestnuts, chestnuts** (*Aesculus* spp.)
47. **Ivy** (*Hedera helix*)
48. **Kingsfoil, athelias** (based, at least in part, on *Symphytum* spp.)
49. **Laburnums, Field of Cormallen, culumalda** (*Laburnum anagyroides, L. alpinum*) [see also laurinquë]
50. **Larches** (*Larix* spp.)
51. **Lavender** (*Lavandula angustifolia* and relatives)
52. **Lilies** (*Lilium* spp.)
53. **Lindens** (*Tilia* spp.)
54. **Lissuin** (an unknown species, unique to J. R. R. Tolkien's legendarium)
55. **Mallorn-trees** (a species unique to J. R. R. Tolkien's legendarium, based in part upon *Fagus, Prunus*)
56. **Moss** (various genera)
57. **Mushrooms** (*Agaricus bisporus*, and related species)

58. **Nasturtians, nasturtiums, Indian cresses** (*Tropaeolum majus*, and related species)
59. **Nettles** (*Urtica dioica*, and related species)
60. **Niphredil** (a species unique to J. R. R. Tolkien's legendarium, but based upon *Galanthus nivalis*)
61. **Oaks** (*Quercus* spp.)
62. **Olive** (*Olea europaea*)
63. **Pines** (*Pinus* spp.)
64. **Pipe-weed, tobacco** (*Nicotiana tabacum*)
65. **Pole beans, beans** (*Phaseolus* spp.)
66. **Poplars** (*Populus* spp.)
67. **Potatoes, taters** (*Solanum tuberosum*)
68. **Reeds** (*Phragmites australis*)
69. **Rockroses** (*Helianthemum nummularium*, and relatives)
70. **Roots** (probably *Dioscorea* spp.)
71. **Roses, wild roses, eglantine** (*Rosa* spp., especially *R. rubiginosa* and *R. arvensis*)
72. **Rowan** (*Sorbus aucuparia*, and related species)
73. **Rushes** (*Juncus* spp.)
74. **Sages** (*Salvia* spp.)
75. **Salt marsh grasses, cordgrasses** (*Spartina* spp.)
76. **Saxifrages** (*Saxifraga* spp.)
77. **Seafire** (various bioluminescent dinoflagellates)
78. **Seaweeds** (various large marine algae, such as *Macrocystus, Fucus, Laminaria, Saccharina, Alaria, Undaria*, and *Sargassum*)
79. **Sedges** (*Carex* spp.)
80. **Seregon** and **stonecrops** (*Sedum* spp., and the former a species unique to J. R. R. Tolkien's legendarium)
81. **Sloe, plums, and cherries** (*Prunus spinosa, P. domestica, P. cerasus, P. avium*, and relatives)
82. **Snapdragons** (*Antirrhinum majus*, and related species)
83. **Sorrel** (*Rumex acetosa*, and related species of *Rumex* with sour leaves; i.e., *Rumex* subgenus *Acetosa*)
84. **Sunflower** (*Helianthus annuus*)
85. **Tea** (*Camellia sinensis*)
86. **Thistles** (*Cirsium* spp.)
87. **Thrift** (*Armeria maritima*)
88. **Turnips** (*Brassica rapa*)
89. **Unnamed hollylike tree** (*Banksia* sp., perhaps related to *B. integrifolia*)
90. **Waterlilies** (*Nymphaea alba*, and related species)
91. **Wheat** (*Triticum aestivum*, and close relatives)

92. **White flowers of the Morgul Vale** (a species unique to J. R. R. Tolkien's legendarium, but based on *Arum* spp.)
93. **White tree of Gondor** (a species unique to J. R. R. Tolkien's legendarium)
94. **Willows** (*Salix* spp.)
95. **Wood-parsley** (*Anthricus sylvestris*)
96. **Yews** (*Taxus* spp.)
97. **Plants of Ithilien** (including the following, in part)
    - 97a. Asphodel (*Asphodelus* spp.)
    - 97b. Box or boxwood (*Buxus sempervirens*, and relatives)
    - 97c. Broom (*Cytisus scoparius*)
    - 97d. Celandine (probably *Chelidonium majus*)
    - 97e. Clematis (*Clematis* spp.)
    - 97f. Cornel (*Cornus mas*)
    - 97g. Cypress (*Cypressus sempervirens*, and relatives)
    - 97h. Hyacinth (*Hyacinthus orientalis*)
    - 97i. Junipers (*Juniperus* spp.)
    - 97j. Marjoram (*Origanum majorana*)
    - 97k. Myrtle (*Myrtus communis*)
    - 97l. Primeroles, primroses (*Primula* spp.)
    - 97m. Tamarisks (*Tamarix* spp.)
    - 97n. Terebinth (*Pistacia terebinthus*)
    - 97o. Thyme (*Thymus vulgaris*, and relatives)
    - 97p. Parsley (*Petroselinum crispum*)
98. **Food Plants of Middle-earth** (including the following, in part)
    - 98a. Cabbage (*Brassica oleracea*)
    - 98b. Cardamon (*Elettaria cardamomum*)
    - 98c. Carrots (*Daucus carota*)
    - 98d. Lettuce (*Lactuca sativa*)
    - 98e. Hops (*Humulus lupulus*)
    - 98f. Onions (*Allium cepa*)
    - 98g. Oranges (*Citrus* × *aurantium*, Sweet Orange Group)
    - 98h. Peas (*Psium sativum*)
    - 98i. Pepper (*Piper nigrum*)
    - 98j. Pickles, cucumbers (*Cucumis sativa*)
    - 98k. Strawberries (*Fragaria vesca*, and close relatives)
99. **Hobbit Names** (including the following, in part)
    - 99a. Amaranth (*Amaranthus* sp.)
    - 99b. Angelica (*Angelica sylvestris*)
    - 99c. Belladonna (*Atropa belladonna*)
    - 99d. Camellia (*Camellia* spp.)
    - 99e. Cotton (*Gossypium* spp.)
    - 99f. Gilly, gillyflower (*Dianthus caryophyllus*)
    - 99g. Lobelia (*Lobelia* spp.)
    - 99h. Malva (*Malva* spp.)
    - 99i. Marigold (*Calendula officinalis*)

AEGLOS, AND GORSE OR WHIN

(The former a species unique to Tolkien's legendarium, but based upon gorse or whin, [Ulex europaeus]—in the Legume family [Fabaceae or Leguminosae].)

> *They passed over the tumbled stones, and began to climb; for Amon Rûdh stood upon the eastern edge of the high moorlands that rose between the vales of Sirion and Narog, and even above the stony heath at its base its crown was reared up a thousand feet and more. Upon the eastern side a broken land climbed slowly up to the high ridges among knots of birch and rowan, and ancient thorn-trees rooted in rock. Beyond, upon the moors and about the lower slopes of Amon Rûdh there grew thickets of aeglos; but its steep grey head was bare, save for the red seregon that mantled the stone. (CoH VII)*

As the petty-dwarf, Mîm, led Túrin and his outlaw band to his dwelling on Amon Rûdh (the vegetation of which is described in the preceding quote from the *Children of Húrin*), they came to steeper ground (Figure 7.1) and "passed under the shadows of ancient rowan-trees, into aisles of long-legged *aeglos*: a gloom filled with a sweet scent" (CoH VII). Then they encountered a rock wall, but, turning to the right and following a narrow path, Túrin saw a cleft in the wall, shrouded by trailing plants: the hidden opening to Mîm's home (which was soon to become Túrin's). The description of the plants of Amon Rûdh is extremely detailed, and aeglos is one of the more intriguing. It is not a pleasant plant, this relative of gorse (*Ulex europaeus*; see UT 1: II, note 14); the gloom of its shade, along with its numerous, branched, and sharply pointed thorns, is off-putting, perhaps even ominous, and is counterbalanced only by its fragrant, white flowers. Aeglos (meaning snow-thorn) was later the name of the spear of Gil-galad, so perhaps we see in this name a foreshadowing of Túrin's bravery in his soon-to-intensify struggle against Morgoth. The mountain is capped by another, more mysterious plant, seregon, with its blood-red flowers adding to the sense of foreboding (see seregon, *Sedum* sp.). Amon Rûdh is a focal point in the story of Turin, and it is clear that Tolkien thought that it needed to be carefully described—and its plants are an important part of this description. Their mention is far from mere descriptive detail, and although the placement of these plants on this mountain is ecologically appropriate, neither do they

FIGURE 7.1 *Aeglos, gorse, or whin.*

simply contribute to the veracity of Middle-earth. Plants such as aeglos and seregon contribute to the mood and give the reader a clearer understanding of the unfolding narrative theme. When the reader understands aeglos and seregon, we better understand the tragic life of Túrin. We don't expect, in the end, much good to come from Túrin's new home.

Gorse (or whin, *Ulex europaeus*) is a very similar plant to aeglos but has yellow flowers instead of white ones. It is mentioned only twice in *The Lord of the Rings*, the first encounter (using the name whin) takes place along the road from the Gates of Moria, shortly before the Company saw Durin's Stone, and the second (using the more commonly applied name gorse) is when Frodo and Sam, led by Gollum, climb up a ridge in northern Ithilien, and, at this occurrence, the plant is given a surprisingly detailed description: "The gorse-bushes became more frequent as they got nearer the top; very old and tall they were, gaunt and leggy below but thick above, and already putting out yellow flowers that glimmered in the gloom and gave a faint sweet scent. So tall were the spiny thickets that the hobbits could walk upright under them, passing through long dry aisles carpeted with a deep prickly mould" (LotR 4: VII). It grew among thorn trees (*Crataegus* spp.), whortleberry (*Vaccinium myrtillus* and/or *V. uliginosum*), and brambles (*Rubus* spp., especially those of the *R. fruticosus* complex). It is in this spot that the hobbits found a sheltered hollow where they spent the night—not the most pleasant location since both gorse and thorn trees have sharp-pointed branches (thorns) and brambles have sharp-pointed, straight to curved hairs (prickles).

Etymology: According to the index of *Unfinished Tales, aeglos* is Sindarin for "snow-thorn," and in *The Silmarillion* the name is glossed as "snow-point." The meaning of both names clearly is the same since thorns are modified, sharply pointed branches. The English common name "gorse" is derived from Old English *gors*, derived through Proto-Germanic **gorst*-[1] from Proto-Indo-European **ghers*- (meaning prickly or bristly). The common name "whin" is less commonly used and is derived from the Middle English *whynne*, which comes from the Old Norse *hvein* (meaning gorse). The scientific name *Ulex* is derived from the ancient Latin name for the plant (or perhaps some other similar species).

Distribution and Ecology: Aeglos, it seems, only grew in the vicinity of Amon Rûdh in western Beleriand, although it is undoubtedly related to furze or gorse (*Ulex europeaus*), from which it differed in its larger stature and white (not yellow) flowers. Gorse is widespread in Europe and also grows in northwestern Africa; it has been introduced (and is an invasive, weedy shrub) in parts of western North America, southern South America, Jamaica, and Australia. Gorse was probably also widely distributed in the Middle-earth of the Third Age since we know that it grew just west of the Misty Mountains near the Drimrill Dale and also in Ithilien. Both gorse and aeglos are thorny shrubs, with fragrant, bee-pollinated flowers and explosively dehiscent fruits. Gorse

[1] The asterisk before this word indicates that it is not recorded from texts, but instead is a hypothetically reconstructed word, based on hypotheses of phylogenetic relationship among extant languages and the form of the words considered to be descended from this ancestral word in those languages. Remember, extant languages are related to each other by evolutionary histories, just as extant plant species show such phylogenetic relationships (see Chapter 3).

and presumably also aeglos occur in open grassy areas, moors, heaths, open rocky and grassy slopes, and montane thickets.

Economic Uses: The flowers of gorse are edible (used in salads), and the leaves provide fodder for livestock. They are attractive shrubs and are thus also used horticulturally (but in some places have become problematic ecologically due to their ability to naturalize). The wood of gorse is very flammable and can be used to start fires.

Description: Erect shrub, the branches, at least when young, green, ridged, and sharply pointed at the apices (i.e., forming thorns). Leaves alternate, spirally arranged, simple (but compound, with three leaflets in seedlings), very narrowly ovate to linear, with a single midvein, other veins obscure; the apex acute and spinose, the base broad, the margin entire, both surfaces with simple hairs; stipules absent. Flowers solitary, stalked and in the axils of small leaves, each associated with two small bracts, bilaterally symmetrical, bisexual, and very fragrant; the sepals five, fused into two parts, the upper minutely two-toothed, the lower minutely three-toothed, these pale green, hairy; the petals white (in aeglos; yellow in gorse), modified, the uppermost forming the standard and positioned on the outside in bud; the two lateral forming the wings and the two lowermost fused, forming the keel; the corolla thus like a pea-flower; the stamens 10, fused into a tube and the single carpel elongated, with a curved style; ovary superior. Fruit an explosively dehiscent, hairy legume (Figure 7.1)

ALDERS (*ALNUS* SPP.)

(In the Birch family [Betulaceae].)

> *After some time they crossed the Water, west of Hobbiton, by a narrow plank-bridge. The stream was there no more than a winding black ribbon bordered with leaning alder-trees. (LotR 1: III)*

Alders nearly always grow near water, and this close association is stressed in nearly every one of J. R. R. Tolkien's references to these common trees or shrubs (as in the preceding quote) and is also shown in Tolkien's watercolor painting *Alder by a Stream* (see fig. 7 in W. G. Hammond and C. Scull's, 1995, *J. R. R. Tolkien: Artist and Illustrator*). Furthermore, a person can easily get lost in alder thickets, which are unpleasant because one almost always has wet feet after hiking through them. Thus, it is not surprising that alder habitats early acquired a reputation as hideouts or places of danger. This sinister reputation is echoed in *The Silmarillion*, in the scene in which Beren reaches the site of his father's death and "as he drew near the carrion-birds rose from the ground and sat in the alder-trees beside Tarn Aeluin, and croaked in mockery" (SILM 19) (see Figure 7.2). *Alnus glutinosa*, the only species to occur natively in Great Britain, has the common name of black alder and is clearly referenced by Tom Bombadil when he tells the four hobbits to "Fear no alder black!" (LOTR 1: VI). And certainly many of Tom's adventures had taken place along the Withywindle, among reeds (*Phragmites australis*), willows (*Salix alba*, and relatives), and "under leaning alder" (TATB, "Bombadil Goes Boating," line 22). Other common species of the Northern Hemisphere include green alder (*A. alnobetula*), grey or speckled alder (*A. incana*), red alder (*A. rubra*), smooth alder (*A. serrulata*), and white alder (*A. rhombifolia*).

FIGURE 7.2 *Alders.*

Etymology: The English name "alder" is derived from the Old English *alor*. The related Latin name, *Alnus*, is the classical name for this plant, which may have been derived from the verb *alo* (to nourish), an allusion to the habitats of these shrubs or trees, which are closely associated with water. Alternatively, the name could go back to the Proto-Indo-European **el*- meaning red or brown, a possible reference to the wood, which, when cut, is initially white and then turns red. The name in Quenya may be *ulwë* or *uluswë*.

Distribution and Ecology: Alders, a group of some 30 species, are widely distributed in the Northern Hemisphere and also extend into South America along the Andes. China and Japan have the highest diversity (ca. 17 species). They are sun-loving plants and almost always grow in wet habitats, occurring, for example, in swamps, along streams, on wet floodplains, and in bogs or muskegs. They are ecologically significant in that the root nodules contain symbiotic bacteria (*Frankia*) that convert atmospheric nitrogen into a form that can be used by plants. This nitrogen is transported to the leaves, which, when dropped, enrich the soil. The species are differentiated by characters such as the form of the winter buds, leaf shape, form of the apex, and marginal condition, and whether the carpellate catkins are clustered or solitary. The plants are wind-pollinated, and the fruits are dispersed by wind or water.

Economic Uses: The wood, when submerged, is rot resistant and thus has been used in pilings (e.g., much of Venice was originally built on alder pilings). The wood also is used in barrels, furniture, is prized in electric guitars, and makes good charcoal. The cone-like fruiting clusters occasionally are used in dried floral arrangements.

Description: Deciduous trees and shrubs, the bark smooth to scaly, with circular to horizontally elliptic lenticels; the roots with nitrogen-fixing nodules. Leaves alternate, simple, ovate to ± elliptic, pinnately veined, with secondary veins running into the teeth, the apex acute or obtuse to slightly acuminate, the base cuneate or acute to slightly cordate, the margin usually double serrate (i.e., with small and large teeth), with simple, straight to crisped hairs and minute, gland-headed hairs, petiolate, with well-developed stipules. Flowers inconspicuous, unisexual (and both staminate and carpellate flowers on each plant), usually blooming in spring before or as the leaves emerge; the staminate flowers in elongate, dangling catkins, with usually four tiny perianth parts, these slightly fused, and usually four stamens; the carpellate flowers in erect catkins that become ovoid to ellipsoid and cone-like at maturity due to their persistent bracts, with perianth lacking and two fused carpels; the styles two, reddish or purple, and ovary nude (inferior). Fruit small, slightly flattened, a nutlet or samara (Figure 7.2).

ALFIRIN AND MALLOS

(Fritillaria, especially yellow-flowered species such as F. aurea, F. carica, F. conica, F. euboeica, and F. pudica; the Lily family [Liliaceae].)

> *And the golden bells are shaken of mallos and alfirin*
> *In the green fields of Lebennin.* (LotR 5: IX)

This species has long been a source of confusion to readers of *The Lord of the Rings*. It is sometimes considered to be the same plant as *simbelmynë* (or *evermind*, *uilos*), which grew on the burial mounds of the kings of Rohan and also at Amon Anwar, where

FIGURE 7.3 *Alfirin and mallos.*

Elendil was once buried. The name *alfirin* is used for both species—for *simbelmynë*, which has more or less erect, starlike, white flowers (see entry for Evermind), and also the plant considered here, which has nodding, long-lasting, bell-shaped, golden-yellow flowers, and is mentioned in Legolas's poem describing the "green fields of Lebennin" that he recited to Merry and Pippin in Minas Tirith (see preceding quote and Figure 7.3). These descriptions are so different, however, that it seems unlikely that they can refer to the same plant, as noted by both Christopher Tolkien (see *Unfinished Tales*, in footnote 38 to "Cirion and Eorl and the Friendship of Gondor and Rohan") and W. G. Hammond and C. Scull (*The Lord of the Rings: A Reader's Companion*, 2005, pp. 397–398). We agree and consider alfirin, as described in Legolas's poem, to be a species distinct from simbelmynë. Yet, the question remains, what sort of plant is this alfirin? Among the European wildflowers, the most likely possibility, in our opinion, is that it is a yellow-flowered species of fritillary (i.e., a member of the genus *Fritillaria*, a group of lilylike plants that arise from bulbs and occur commonly in open habitats in Europe). These plants have nodding, bell-like flowers, and they often grow in grass-dominated habitats, fitting the ecological setting of Legolas's poem—in "green fields" (LotR 5: IX). Lebennin had a southern European climate, and it is interesting that several bright yellow species of *Fritillaria* grow in Greece (e.g., *F. conica, F. euboeica*) and Turkey (e.g., *F. aurea, F. carica*). These plants perfectly fit the description and ecological setting provided in Legolas's poem. A final, and related, problem concerns the identity of mallos. This name also is Sindarin, and means "golden-flower." Is this merely another name for alfirin, or is it a different plant? The use of both names in the poem supports the latter, and we suggest that perhaps *mallos* represents yet another yellow-flowered species of *Fritillaria*. This is a distinct possibility since several yellow-flowered species occur in this genus and in addition to those just listed (e.g., *Fritillaria pudica*, a western North American species; *F. crassicaulis*, a Chinese species; and the yellow variants of *F. imperialis*, a widely cultivated species that is native from Turkey and Iraq eastward across Iran and extending into the western Himalayas). Another genus of Liliaceae that potentially matches the brief description of alfirin is *Erythronium*, a group of 27 species of woodlands, montane meadows and prairies species that grows in North America and Eurasia. They have more strongly reflexed petals than the flowers of *Fritillaria* (as shown in the lower flower in the illustration) and may have white, yellow, pink, or violet tepals.

Etymology: The Sindarin name *alfirin* is derived from *al-* (not) and *firin* (mortal) and thus means immortal (see *Parma Eldalamberon* 17, by David Giraudeau), although why the plant receives this name is not entirely clear. In a letter to Amy Ronald (Letters: No. 312) J. R. R. Tolkien compared it with an immortelle (i.e., members of the genus *Helichrysum*, in the plant family Asteraceae, which have dry, papery, and thus long-lasting flowers when picked). Thus, the name may relate to the fact that *alfirin* was envisioned by Tolkien as a plant with very long-lasting flowers. He specifically noted that the flowers of *alfirin* should not be thought of as "dry and papery," and his description of the plant as having "golden bells" also does not at all agree with those of *Helichrysum*, which has tiny flowers in involucrate heads. It thus seems that only the concept of long-lasting flowers and the name immortelle are taken from *Helichrysum*, while the morphology of the plant itself agrees much more with that of a petaloid monocot, such as *Fritillaria* or *Erythronium*.

The scientific name *Fritillaria* comes from the Latin *fritillus*, meaning dice box, and is a reference to the checkered pattern of the perianth parts (i.e., tepals) of many species of this genus.

Distribution and Ecology: Fritillaries or *mission bells* (members of the large genus *Fritillaria,* with ca. 130 species) are widely distributed in the Northern Hemisphere but are most diverse in the Mediterranean region, southwestern and central Asia, and western North America. Twenty-three species occur in Europe, 31 in Turkey (and eastern Aegean Islands), 24 species in China, and 20 in North America. They are characteristic of open habitats such as alpine and subalpine meadows, pastures, various grasslands, open woods and scrub, rocky places, and scree slopes. Species can be distinguished by characteristics such as form of the bulb scales; leaf size and arrangement; inflorescence structure; the shape, size, and orientation of the flowers; color of the tepals (yellow, orange, pink, red, pale to dark purple, brown, green, white, often with alternating squares of dark and light color); form of the nectar glands (on the tepals); floral fragrance (pleasant to quite unpleasant, or without an odor); length of the anther filaments; extent of style branching; and presence or absence of wings on the capsules. The showy flowers attract a variety of pollinators, such as birds, wasps, bumblebees, and flies. The flattened seeds may be wind dispersed.

Economic Uses: Some species of *Fritillaria* (fritillaries, mission-bells), such as *F. imperialis, F. meleagris,* and *F. persica* are popular garden ornamentals because of their large, colorful flowers, but many species, although beautiful, are difficult to grow (or are unknown in cultivation, having unknown horticultural characteristics). Many are quite poisonous, containing various alkaloids, and a few are of medicinal value.

Description: Perennial herbs, arising from scaly bulbs, these short and erect; the flowering stem erect, cylindrical; roots contractile, pulling the bulb deeper into the soil. Leaves alternate, opposite, or whorled, basal, or distributed along the stem, simple, linear to ovate, with parallel veins, the apex acute, the base broad, petiolate or not, the margin entire, glabrous to papillose-hairy; stipules absent. Inflorescences terminal, few to several flowered racemes, these occasionally umbel-like, with leaflike bracts, sometimes reduced to a solitary flower. Flowers radially symmetrical, usually pendent, fairly large, yellow (or in related species, orange, pink, red, pale to dark purple, brown, green, white, often with alternating squares of dark and light color), odorless to strongly fragrant (and then pleasant to very unpleasant), with a perianth of six tepals, these distinct, ovate to obovate, apically acute to obtuse, with three outer and three inner ones, arranged in bell-shaped manner, and with each tepal apically curved outward to more or less straight, and each with a basal nectar-producing region, but this better developed in the three inner ones. Stamens six, hidden within the perianth, with conspicuous, flexible anthers (attached to the filament at one end to near middle). Carpels three, fused, the style single or three-branched; ovary superior. Fruit a six-angled, sometimes six-winged capsule; seeds yellow to brown, flattened (Figure 7.3).

ANEMONES (*ANEMONE* SPP.)

(In the Buttercup family [Ranunculaceae].)

> *About them lay long launds of green grass dappled with celandine and anemones, white and blue, now folded for sleep. (LotR 4: VII)*

Anemones were seen by Sam and Frodo in Ithilien (Figure 7.4). They were first encountered shortly before they camped for the day by a small lake (and Sam stewed

FIGURE 7.4 *Anemones.*

rabbits), and at this locality they grew with primeroles (*Primula* spp.) in filbert brakes (thickets of *Corylus maxima*); later, they were seen on their journey to the cross-roads, where they occurred in open, grassy areas with celandine (possibly *Ranunculus ficaria*, but more likely *Chelidonium majus*). These spring wildflowers, among many others growing in Ithilien, lifted the spirits of the two hobbits. As J. R. R. Tolkien (in his role as narrator) comments, "Ithilien, the garden of Gondor now desolate kept still a disheveled dryad loveliness" (LotR 4: IV). The anemones are described as being white and blue (see quote), and it is likely, therefore, that this represents a mixed occurrence, possibly of the wood anemone (*A. nemorosa*, which usually has white flowers, flushed with pink or purple beneath, and is widely distributed in Europe) and the blue anemone (*A. apennina*, which has blue flowers and is widespread in southern Europe). The former grows natively in England, while the latter is commonly cultivated there and sometimes naturalizes.

Etymology: The English name "anemone" and the scientific generic name *Anemone* are derived from the classical Latin *anemone* and the Greek *anemone*, which means wind-flower (literally, "daughter of the wind"; from *anemos*, wind). The early American systematist, Asa Gray, suggested that this derivation is based on the mistaken idea that these flowers open only when the wind blows, but another possible link to wind is seen in their small, dry fruits, which are often plumose and thus wind-dispersed. However, many have doubted this etymological derivation, suggesting instead that the early Greeks borrowed the name from that of the Semitic deity *Naaman* (= Adonis), whose worship was associated with these flowers (said by Ovid in his *Metamorphoses* to have sprung from the blood of Adonis).

Distribution and Ecology: *Anemone* is a genus of about 150 species (of these, 25 occur in North America, north of Mexico; 17 grow in Europe; and more than 50 occur in China), and the group is broadly distributed, especially in temperate and arctic regions. Anemones occur in a variety of habitats, such as dry to wet meadows or prairies, lakeshores, streamsides, swamps, open to deep deciduous to mixed-coniferous woodlands, rocky slopes, and alpine meadows. Species distinctions are often difficult and based on characters such as whether or not the plant produces rhizomes or tubers; leaf form (simple or compound, extent of lobing or dissection); inflorescence structure and form of the associated bracts; number, shape, and color of the tepals; number of stamens; the length of the persistent style in fruit; whether or not it is persistently hairy; size and shape of the achene; length of associated hairs; and extent of wing development. The showy flowers are pollinated by various insects; the small, dry fruits are often associated with hairs or have a persistent hairy style and thus are dispersed by wind; some may be ant-dispersed.

Economic Uses: Several species of anemone are cultivated as ornamentals because of their beautiful flowers, which may be yellow, white to red, or blue. The plants are poisonous, as is typical of members of Ranunculaceae, containing the lactone glycoside ranunculin.

Description: Perennial herbs, sometimes producing rhizomes or tubers; hairs, when present, simple; stems erect or very short. Leaves usually alternate (but leaflike bracts below the inflorescence may be opposite), more or less basal, one to numerous, simple or compound, often lobed or dissected, with palmate venation; the apex acute or acuminate to obtuse; the margin entire, serrate or crenate, petiolate; stipules absent. Inflorescences terminal, few-flowered cymes, umbels, or reduced to a solitary flower and with leaflike to sepal-like, opposite or whorled bracts, these often appearing to be in tiers and positioned close to or distant from the flowers. Flowers radially symmetrical, bisexual, erect to nodding, usually with a perianth of tepals; these distinct, four to numerous, and white to red, blue, purple, green, or yellow; usually 10 to numerous stamens, these distinct; carpels numerous, distinct, the ovaries superior. Fruits a cluster of achenes, often associated with a tuft of elongate hairs, each with persistent, short to elongated and sometimes hairy style (Figure 7.4).

APPLES (*MALUS DOMESTICA*, *M. SYLVESTRIS*, AND RELATED SPECIES)

(The Rose family [Rosaceae].)

> *Sam was chewing an apple thoughtfully. He had a pocket full of them: a parting present from Nob and Bob. "Apples for walking, and a pipe for sitting," he said. "But I reckon I'll miss them both before long." (LotR 1: XI)*

Orchard apples (*Malus domestica*) were widely cultivated in the Middle-earth of the Third Age, as they are across the Northern Hemisphere today, although they must have been introduced there because they likely were originally restricted to a forested region east of the Sea of Rhûn (being native to Kazakhstan and surrounding regions, today). Apples are first mentioned (i.e., "Apple, thorn, and nut and sloe" LotR 1: III) in Bilbo's walking song, sung by the hobbits on their first full day of hiking through the Shire on their way to Crickhollow (Figure 7.5). This is likely a reference to the wild apple (or crabapple, *Malus sylvestris*), which was common in the hedgerows of the Shire. That evening, they encountered elves in the Woodyend, where they were served fruits "richer than the tended fruits of gardens" and were overwhelmed by the beauty of elven singing. Although he could never afterward fully describe it, that night remained etched in Sam's memory, and he would say "if I could grow apples like that, I would call myself a gardener. But it was the singing that went to my heart, if you know what I mean" (LotR 1: III). Clearly, the elves had shared their orchard apples, among other fruits, with the hobbits, and perhaps the elves, during their original migrations, had carried the species with them as they had moved westward across Middle-earth, starting from Cuiviénen on the shore of a great inland sea. Orchard apples, as an important cultivated fruit, were prized also by the entwives and are mentioned in the elvish song describing their estrangement from the ents. Although the ents loved the great trees of the wild woods, the entwives "gave their minds to the lesser trees . . . and they saw the sloe [*Prunus spinosa*] in the thicket, and the wild apple and the cherry [*Prunus* spp.] blossoming in spring" (LotR 3: IV). The "wild apple" referenced here, like the earlier one, is the European crabapple (*Malus sylvestris*), a species native to Europe that produces small, astringent fruits and has interbred with orchard apples, enriching their genome. The entwives ordered these trees—undoubtedly including both apples and crabapples—"to grow according to their wishes"—thus creating orchards and gardens, and, as a result, they slowly became estranged from the ents, who by this time lived far away, in pristine wilderness such as Fangorn. Finally, apples were appreciated by the human inhabitants of Gondor (and likely others regions as well). Beregond and Pippin were provided from the storehouse of the former's Company of the Guard

FIGURE 7.5 *Apples.*

"bread, and butter, and cheese and apples: the last of the winter store, wrinkled but sound and sweet" (LotR 5: I) shortly after Pippin had been sworn to the service of the Lord Denethor and the city of Gondor. According to Treebeard, humans learned many agricultural crafts from the entwives, so perhaps the entwives had passed on their knowledge of grafting, an essential skill in the cultivation and maintenance of desirable apple cultivars. Apple cultivars are not mentioned by name in *The Lord of the Rings*, but we are sure that the hobbits encountered a flavorful array of different kinds of apples as they traveled from the Shire to Gondor, just as there are many cultivated varieties of apples today. But apples could be put to other purposes—and as Strider was leading the hobbits out of Bree, Sam found one such use. Before leaving, he had filled his pocket with apples and was enjoying one when he (and companions) were insulted by Bill Ferny near the town's outskirts. Sam had had enough, especially when Bill Ferny implied that Sam would ill-treat their recently purchased pony. Sam turned quickly and, with a sudden flick, hit Bill right in the nose. As they walked on Sam noted—"Waste of a good apple" (LotR 1: XI)!

Etymology: The English word "apple" comes from Old English *æppel* (apple, any kind of fruit, except for berries), and the use of apple as a general fruit term existed as late as the seventeenth century. The Old English word may have been derived from the Proto-Germanic **ap(a)laz*, and from Proto-Indo-European **ab(e)l-* (original sense uncertain). The scientific generic name *Malus* is derived from the classical Latin *malum* (apple); note that when pronounced with a long "a" the word is the Roman name for apple, but when pronounced with a short "a," the word means bad or wicked. The Latin *malum* (apple) is derived from the ancient Greek *malon* or *melon* (fruit tree). The Sindarin name for apple is *cordof* and possibly also *orf*. *Orforn* has been suggested as meaning apple tree. The Quenya name may be *orva*.

Distribution and Ecology: Apples (members of the genus *Malus*) are distributed broadly in the Northern Hemisphere; the group is in need of systematic study, and species limits, especially, are unclear, with some scientists recognizing as many as 55 species. Ten species are native or naturalized in North America (the most widespread of which are the western crabapple, *M. ioensis*; the southern crabapple, *M. angustifolia*; and American or sweet crabapple, *M. coronaria*); six species occur in Europe, and the most widespread of these is the European crabapple (*M. sylvestris*). Eastern Asia is the region of highest diversity of apples, with about 28 species. Species are distinguished by characteristics relating to the leaf shape, pubescence, and margin (e.g., lobed or not, form of marginal teeth); shape, pubescence, and persistence of the sepals; color of the anthers; the number and pubescence of the styles; and size and coloration of the fruit. The orchard apple (*M. domestica*; also sometimes called *M. pumila*) originated in central Asia, in Kazakhstan, Kyrgyzstan, and adjacent China, where wild populations still occur (and are often identified as *M. sieversii*, although recent DNA evidence indicates that this should be considered conspecific with the orchard apple). Orchard apples likely were first domesticated in the Tian Shan Mountains and were then carried along the Silk Road, appearing in the Middle East about 4,000 years ago. They were later carried into Europe by the Greeks and Romans, and, by this time, grafting was already an important technique in their propagation (allowing maintenance of desirable cultivars since sexual reproduction in apples results in offspring that may have fruit characteristics very different from those of the female parent). While cultivated in Europe, orchard apples crossed with the native European crabapple (*M. sylvestris*), and evidence of these hybridization events remains in the apple genome (e.g., the 'Granny Smith' cultivar has the chloroplast genome of the European crabapple). The showy flowers open in the spring, before the emergence of the leaves, and they are pollinated by bees and flies; the small fruits characteristic of most species are dispersed by birds, but the large fruits of the wild populations of the orchard apple are eaten and dispersed by bears.

Economic Uses: Orchard apples (*M. domestica*) are the most important temperate fruit tree in the world and the fourth most important fruit in the world (after citrus, grapes, and bananas) in terms of economic production; more than 7,500 cultivars are known, and the large, green, yellow, or red fruits, quite variable in flavor, are eaten raw, cooked, or used to make cider (in the case of astringent cultivars, some of which is processed to form vinegar). Apples have moderate levels of a large number of nutrients. Apple wood (often from *M. sylvestris*, the European crab apple) is used for furniture, mallet heads, croquet balls, tool handles, and smoking meats. Finally, various species of *Malus* are used as ornamental trees because of their showy flowers and colorful fruits.

Description: Deciduous trees or shrubs, the bark brown to gray, eventually forming longitudinal and transverse fissures, thus forming plates or scales; the branches forming long and short shoots (and the latter often forming thorns), with or without simple hairs. Leaves alternate, spirally arranged, simple, ovate to obovate, pinnately veined, with secondary veins forming loops or running into the apices of the lobes; the apex acute or acuminate to rounded, the base cuneate or acute to slightly cordate, the margin clearly to obscurely serrate, double serrate, crenate, or occasionally nearly entire, sometimes also ± lobed, usually with simple, nonglandular hairs, petiolate; stipules present, usually deciduous. Inflorescences terminal, few to several-flowered cymes. Flowers radially symmetrical, bisexual, the sepals five, triangular; the petals five, basally narrowed, white to red; and the stamens numerous, with anthers globose, yellow or rose to purple, all borne on a nectar-producing cup (= hypanthium) with 3–5 fused carpels; the styles 3–5, the ovary inferior. Fruit small to large pomes; these globose or depressed-globose, obovoid, or oblong, green, yellow, or red to orange-red, with the flesh of most species lacking stone cells (Figure 7.5).

ASHES (*FRAXINUS* SPP.)

(In the Olive family [Oleaceae].)

> *Some recalled the ash: tall straight grey Ents with many-fingered hands and long legs. (LotR 3: IV)*

Ash trees are common in Middle-earth. They are frequently tall, with straight, gray-barked trunks, and their leaves are distinctive, being opposite and pinnately compound, with usually 5–11 leaflets arranged featherlike along the sides of the axis. These characteristics are reflected in the description (quoted here) of the ashlike ents seen by Merry and Pippin in Fangorn forest. It is thus not surprising that Aragorn and his companions observed the riders of Rohan carrying "tall spears of ash" (LotR 3: II; see also Figure 7.6) since its light yet strong wood was once the material of choice for such implements (and is still widely used in tool handles and sporting implements). The wood's elasticity makes it a good choice for walking sticks—and we know, thanks to Háma's observation, that Gandalf carried an ash staff as he entered, with Aragorn,

FIGURE 7.6 *Ashes.*

Legolas, and Gimli, the hall of Théoden in Edoras. Many times we see Gandalf's power reflected in his staff, as when (after his fall in Moria, revealing his white garments), he lifted up his staff and Gimli's axe "leaped from his grasp" and Aragorn's sword "blazed with a sudden fire" (LotR 3: V) during his encounter with Aragorn, Gimli, and Legolas in Fangorn. Wood must have great strength to channel such power, and this gives us a fresh appreciation for the ash.

Etymology: The English name, "ash," was derived through Old English, *æsc,* from the Germanic **askaz* or **askiz*. Interestingly *æsc* can mean ash tree or a spear made of ash wood since ash was the preferred source of wood for spear shafts. The Latin name, *Fraxinus*, is the classical name for this tree, and, interestingly, it can also mean a spear or javelin made from ash wood. The Sindarin name of these trees is *lith*.

Distribution and Ecology: Ashes belong to the genus *Fraxinus*, a group of 49 species distributed broadly across temperate regions of the Northern Hemisphere, although a few extend into the American tropics; 24 species occur in North America, while 4 occur in Europe, 16 in China, and 7 in Japan. The European ash (*Fraxinus excelsior*) is the most widespread and common species in Mediterranean region, although the narrow-leaved ash (*F. angustifolia*) is also common, while white ash (*F. americana*), pop ash (*F. caroliniana*), black ash (*F. nigra*), green ash (*F. pennsylvanica*), and blue ash (*F. quadrangulata*) are common in eastern North America. The most widespread western North American species include California ash (*F. dipetala*), Oregon ash (*F. latifolia*), and velvet ash (*F. velutina*). Numerous species also occur in eastern Asia. Floral form and fruit shape (especially the form of the wing) are very useful in distinguishing these species. Ashes occur in a wide variety of forested habitats, from swamps to arid uplands. Ashes are usually wind-pollinated, but a few species have showy white flowers (with two or four petals), attracting generalist insect pollinators, and their winged fruits are wind-dispersed. In North America, sadly, many ash trees are dying due to attack by a beetle, the emerald ash borer (*Agrilus planipennis*) recently introduced from eastern Asia; many European ashes also are under attack by the an ascomycete fungus (*Hymenoscyphus pseudoalbidus*), likely also introduced from Asia.

Economic Uses: Ash trees are important ornamentals, and they are thus valuable in urban forestry. Ashes provide a strong, fairly lightweight, yet elastic wood; they are therefore a popular source for tool handles, wooden sporting equipment, or weapons: bats, tennis racquets, billiard cues, hockey sticks, walking sticks, snowshoes, canoe paddles, bows, or spears. The wood is prized in making furniture and electric guitars; in thin strips, it is used in basket weaving. Ashes are also used as a source of firewood. The leaves and bark have been used medicinally, and in folklore ash was thought to protect from snakebite. Manna ash (*F. ornus*), a species of southern Europe, is the source of a sugary extract collected by making cuts in the bark and gathering the exuding sap when dry.

Description: Usually deciduous trees or shrubs, the bark rough, vertically furrowed. Leaves opposite, decussate, usually odd-pinnately compound, with leaflets (1–) 5–11, with the leaflets ovate to obovate, pinnately veined, with secondary veins forming obscure loops, the apex rounded to acuminate, the base slightly asymmetrical or not, decurrent to rounded, the margin entire to serrate, with simple hairs and/or glandular, peltate scales, petiolate or sessile; stipules absent. Inflorescences in axillary clusters, borne on shoot of the previous season, or terminal, borne on the shoot of the current season. Flowers radially symmetrical, bisexual or unisexual (and then plants usually with staminate flowers on one plant and carpellate flowers on another), usually inconspicuous and wind-pollinated, but occasionally showy, insect-pollinated; the sepals usually four, fused, small; the petals usually absent, but sometimes present, two or four, distinct or slightly fused, and white, with two anthers, and two fused carpels, style single, the ovary superior. Fruit a samara, with variably developed elongated wing (Figure 7.6).

BARLEY (*HORDEUM VULGARE*)

(The Grass family [Poaceae or Gramineae].)

> *The Northfarthing barley was so fine that the beer of 1420 malt was long remembered and became a byword. (LotR 6: IX)*

Barley was domesticated in the Middle East, in the region of the Fertile Crescent, at about the same time as wheat, about 10,000 years ago. Although barley is only rarely mentioned in J. R. R. Tolkien's legendarium, we can safely assume that this economically important grain was widely cultivated in the Middle-earth of the Third Age, as it is today across temperate regions of the Northern Hemisphere because of its importance in brewing beer (see "Economic Uses"), as is clear from the reference to "1420 malt" in the preceding quote. One can easily imagine the conversation between Sam Gamgee and Ted Sandyman occurring over a few pints as they sat in a corner of The Green Dragon pub at Bywater (Figure 7.7) discussing trouble on the borders of the Shire. And beer is frequently mentioned in *The Lord of the Rings*, as illustrated in the following few examples (listed in order of their occurrence). When Frodo and his three friends had dinner at the home of Farmer Maggot, there "was beer in plenty . . . beside much other solid farmhouse fare" (LotR 1: IV), and when they reached the Prancing Pony in Bree, Sam's misgivings relating to his being outside the Shire were "much relieved by the excellence of the beer" (LotR 1: IX). When Merry and Pippin dine with Aragorn, Legolas, and Gimli at the ruined entrance to Isengard, Merry asked if they would like "wine or beer" (LotR 3: IX). When Gandalf and the four hobbits stopped again in Bree on their return journey, they attempted to explain that Strider was now the king; Butterbur imagined the king in his great castle, "drinking wine out of a golden cup" and was surprised when Sam

FIGURE 7.7 *Barley.*

disagreed, telling him that "he says your beer is always good" (LotR 4: VII). Finally, when they returned to the Shire, they found that the Bridge Inn had been torn down and were dismayed by the changes they found, especially the changes in how hobbits interact with each other—following petty rules and distrusting others instead of providing hospitality. Sam, in frustration, said "All right, all right! . . . I don't want to hear no more. No welcome, no beer, no smoke, and a lot of rules and orc-talk instead" (LotR 4: VIII). Beer clearly is associated with hospitality, good conversation, and fellowship, reflecting Tolkien's own enjoyment in sharing a few beers with his friends.

Etymology: "Barley" is derived from the Old English *bærlic*, which was originally an adjective ("of barley"; with *bere* as the corresponding noun), from Proto-Germanic **bariz* (barley) + *lic* (like), and this probably derived from Proto-Indo-European **bhars-* (bristle, point), which also gave rise to the word "bristle." Interestingly, the word "barn" is derived from Old English *bereærn* (barley-house), as it is a place for the storage of grain. The scientific name *Hordeum*, the classical name for this plant, is related to the Latin *horrere* (to bristle), which is derived from Proto-Indo-European **ghers* (to bristle), both of which are references to the elongate awns on the inflorescences of this important grain.

Distribution and Ecology: Barley is native to western Asia and northeastern Africa, was domesticated in the Fertile Crescent region (in the area of Israel and Jordan) about 10,000 years ago, and was then spread widely as a cultivated grain. Wild populations have a disarticulating inflorescence (so disperse their own seeds), whereas in cultivated barley the grains are held on the plant (so they can be easily harvested and thus rely on humans for distribution). Cultivated plants also have wider leaves, shorter stems, shorter inflorescence awns (bristles), and larger grains. The small flowers are wind-pollinated (cultivated plants are largely selfing), and the grains of wild populations are dispersed by animals because their elongated awns (bristles) become entangled in fur or feathers. Barley is common in grasslands, open woodlands, or various disturbed habitats. It is quite tolerant of poor soils and saline conditions.

Economic Uses: Barley is economically important because the beer-brewing industry is mainly based on barley malt, which is made by soaking barley grains in water, allowing them to germinate (enzymatically converting their starch to sugars), and then rapidly kiln-drying them. In beer making, the malt is fermented to convert the sugars to ethyl alcohol (and then flavored with hops). The resulting alcoholic fluid (beer) can then be distilled, resulting in whiskey. However, barley is also used in making flat bread, and in soups and stews.

Description: Annual herbs, without rhizomes; stems erect, jointed, round in cross-section, hollow in internodal regions. Leaves alternate, two-ranked, simple and differentiated into a spreading blade, a sheath that closely encircles the stem, with its margins not fused, and with a membranaceous flap of tissue (the ligule) at the junction of the blade and sheath; the blade linear, flat, glabrous, but sometimes roughened. Inflorescences terminal, spikelike, with three spikelets per node, not breaking apart in cultivated plants, but falling apart (with spikelets falling in triplets) in wild populations. Flowers very tiny, arranged in spikelets, with only a single flower per spikelet; spikelets are arranged in two, four, or six rows along the axis, and each flower is surrounded by two bracts (a lemma and palea), some of these (the lemmas) often with elongated awns, with three sagittate anthers and two fused carpels; the ovary superior; the two stigmas plumose. Fruit a large plump grain, usually tightly enclosed by the bracts; nutritive tissue hard or mealy (Figure 7.7).

BAY (*LAURUS NOBILIS*)

(The Laurel family [Lauraceae].)

> *Beyond it was a thicket of dark-leaved bay-trees climbing up a steep bank that was crowned with old cedars. (LotR 4: IV)*

Sam had seen bay trees near the small lake by which they camped in Ithilien (see quote), and it was there he decided, in order to extend their food supply, that they should cook some fresh-caught game. Gollum cooperated, at least initially; he caught two small rabbits and fetched the water while Sam started a fire. Sam prepared to stew the rabbits in his camping pans. He had a precious bit of salt (from his pack), but he asked Gollum to bring him a "few bay-leaves, some thyme [*Thymus vulgaris*] and sage [*Salvia officinalis*]" before the water boiled. By that point, Gollum had had enough: he did not like the idea of cooking the rabbits, and he refused to look for herbs, stating that "Sméagol doesn't like smelly leaves" (LotR 4: IV). So Sam had to gather the fragrant herbs for the stew himself. He found them close by, "not out of sight of the place where his master lay, still sleeping." After Frodo awoke, the two ate the stew together, along with a bit of the Elvish waybread, sharing their single fork and spoon. After what they had experienced, it was like a feast. It is through such actions, even the little things like gathering bay leaves, that Sam expressed his love for his master and friend, Frodo. Even such small actions contributed to the success of the quest.

Etymology: The English common name "bay," for the tree *Laurus nobilis*, is derived from the late Middle English *baye* (or *baie*), originally referring only to the berrylike fruits of this tree, but from the 1520s extended to the tree itself. The Middle English word was derived from the Old English *beg* (berry), but was conflated with Old French, *baie* (berry), which was derived from Latin, *baca* (berry). The scientific name *Laurus* is the classical name for this tree, and from this is derived the English word "laurel," which is applied to *L. nobilis* (the true bay) and related species, such as the California bay (*Umbellularia californica*), red bay (*Persea borbonia*), and swamp bay (*P. palustris*). *Laurus* perhaps is etymologically related to the Greek *daphne* (meaning laurel). In Greek mythology, Daphne was a dryad (and daughter of the river god Peneus) who was amorously pursued by the god Apollo; she was turned into a bay tree to be safe from Apollo, and, as a result, he sadly wore a laurel wreath in her memory. Such wreaths became symbolic of what he embodied—especially victory and high achievement.

Distribution and Ecology: *Laurus*, a genus of only two species (*L. nobilis* and *L. azorica*), has a scattered distribution in the Mediterranean region and also occurs in the Azores, Madeira, and Canary Islands. Bays grow in remnant patches of evergreen forest and scrub. Their flowers produce nectar and are pollinated by insects (especially bees); their fleshy fruits (drupes) are bird dispersed.

Economic Uses: The major use of bay leaves is as a flavoring agent in food preparation, especially in soups and stocks. The leaves contain fragrant essential oils, especially 1,8-cineole, along with other terpenes and sesquiterpenes, that are stored in specialized spherical cells in the leaf tissue—thus, the leaves are minutely pellucid dotted when held up to the light. The leaves of related species of the laurel family (Lauraceae), such as California bay, red bay, and swamp bay, are used similarly. *Laurus nobilis* (the true bay, sweet bay, or bay laurel) is also used as an ornamental evergreen tree.

Description: Evergreen trees or shrubs, the branches with or without simple hairs, the bark more or less smooth, gray. Leaves alternate, spirally arranged, simple, ovate to slightly obovate, pinnately veined, the secondary veins forming loops, the apex acute to slightly acuminate, the base cuneate to acute, the margin entire, often undulate, with or without a few hairs beneath, shortly petiolate; stipules absent. Inflorescences axillary, in reduced cymes. Flowers radially symmetrical, unisexual (and the plants with the staminate and carpellate flowers on different plants), the perianth of four petaloid, white tepals, these distinct; the stamens in staminate flowers, 8–12, distinct, each stamen usually with two nectar-producing glands; its anther opening by two flaps, to which the yellow, sticky pollen adheres and, in carpellate flowers, reduced to 2–4 sterile structures (staminodes). The carpel one in the carpellate flowers, with a single style, the ovary superior. Fruit an ellipsoidal, black drupe, with a single pit.

BEARD LICHENS (*USNEA* SPP.)

(Parmeliaceae [a family of lichen-forming fungi].)

> *The entrance to the path was like a sort of arch leading into a gloomy tunnel made by two great trees leant together, too old and strangled with ivy and hung with lichen to bear more than a few blackened leaves. (Hobbit VIII)*

Usnea is the only lichen mentioned in J. R. R. Tolkien's writings, but it is included not only in his early works (*The Lay of the Children of Húrin* and *The Hobbit*) but also in *The Lord of the Rings* and contributes to the imagery surrounding several important events in the legendarium, as seen in the introductory quotation from *The Hobbit,* from the second sentence of the chapter "Flies and Spiders." Gandalf had just left, and Bilbo and the dwarves were about to enter Mirkwood, following the elven path. The

Forest Gate is described as archlike, formed by two gigantic trees leaning against each other, and these trees are "strangled with ivy and hung with lichen" and bear only a few old, damaged leaves. Here we see two distinctive characteristics of *Usnea*: first, its preference for sickly, dead, or dying trees that have fewer leaves and thus a more open canopy, allowing more sunlight to reach the lichens; and, second, its characteristic epiphytic and hanging habit—that is, it almost always grows on trees or shrubs on which it forms a much-branched system of often pendulous, pale gray to yellowish branchlets (Figure 7.8). This growth form is so characteristic of *Usnea*, a fruticose lichen (i.e., one that has a branched, miniature, shrubby or treelike form), that the species of this genus are called *beard lichens* (or *old man's beards*) because their hanging branches look like a graying beard. These common names are alluded to by Tolkien elsewhere, as when Merry and Pippin entered Fangorn forest (Figure 7.8) and saw "great trailing beards of lichen hung from" huge branches (LotR 3: III), and Pippin, picking up on the old feeling of the forest, exclaimed—"Look at all those weeping, trailing, beards and whiskers of lichen!" (LotR 3: IV). These descriptions perfectly match the appearance of many species of *Usnea*, which are widespread and diverse in Europe (with more than 30 species occurring there) and thus would have been very familiar to Tolkien. *Usnea* seems to have been as common in Middle-earth, and it adds to our mental image of—and gives a certain foreboding quality to—the great forests of Mirkwood and Fangorn. This expectation of evil is expressed most clearly in the very similar description of the forest gateway where the orc trail from Thangorodrim entered Taur-nu-Fuin: the Forest-Beneath-Night, so named because it was filled with terror and dark-enchantment by Morgoth. We read in *The Lay of the Children of Húrin* that Beleg and Gwindor saw

> [A]n archway opened. By ancient trunks
> it was framed darkly, that in far-off days
> the lightning felled, now leaning gaunt
> their lichen-leprous limbs uprooted. (Lays I: lines 936–939)

Again, we see the image of ancient dead trees covered with beard lichens. Their presence is described as "leprous" because of their gray-green to yellow-green color, but this term is also appropriate given that Taur-nu-Fuin itself is diseased and distorted by the evil actions of Morgoth. This forest, located in Dorthonion north of Beleriand in the First Age, was much more perilous than either Fangorn or Mirkwood. Yet it was here that Beleg found Gwindor and rescued Túrin (see SILM 21).

Etymology: The word "beard" is derived from Old English *beard*, and that from West Germanic **barthaz*, which was derived from Proto-Endo-European **bhardh-a-* (beard). The English word "lichen" is derived from Latin *lichen*, and Greek *leichen*, probably from *leichein* (to lick), and this word was used originally for liverworts. The scientific generic name *Usnea* is derived from the Arabic *usna*, meaning "moss."

Distribution and Ecology: The beard lichens (often also called *old man's beard* or *woman's long hair*) are members of the large (more than 500 species) and very widely distributed genus *Usnea*, which is common in Europe (with more than 30 species occurring there) and North America (ca. 80 species) as well as elsewhere, from the arctic to the tropics. Lichens are not a single organism; instead, each lichen is an integration of two very different species: a fungus and a photosynthetic organism (either

FIGURE 7.8 *Beard lichens.*

a cyanobacterium or a green alga) that interact in a more or less mutually beneficial relationship. In the case of *Usnea*, the fungal partner is an ascomycete (i.e., a sac fungus, a large group that, in reproduction, produce their sexually reproductive spores in an elongated sac, or ascus; the group also contains the cup fungi and morels), and the algal partner is a green alga (i.e., a chlorophyte, probably species of *Trebouxia*). Species of *Usnea* are distinctive in that they look like tiny erect to pendant, well-branched trees, and they commonly grow on the trunks and branches of trees, especially ones with a sparse crown, as in trees that are unhealthy or dying. *Usnea* is not a parasite and it does not harm the trees on which it grows, but its growth is favored by the loss of a leafy tree canopy because this allows more sunlight to penetrate into the crown, thus providing more light for the algal component of the lichen. *Usnea* is a good indicator of air quality because it is quite sensitive to air pollution. Their spores and vegetative propagules are dispersed by wind.

Economic Uses: The beard lichens are very useful medicinally because they contain usnic acid, which is an antibiotic. They are used in treating bacterial skin infections and are a component in some cosmetics and deodorants. They also have been used in making dyes and are an important wildlife food (and also provide nest materials for many birds).

Description: Fruticose (i.e., well-branched and shrubby) lichen, pendent to erect, yellow-green to gray-green, occasionally reddish, with the branches round to angular and having a central cartilaginous cord (of supporting tissue); the outer layer of each branch (i.e., the cortex) thin to thick, sometimes falling away, often with circular cracks around stem (so stem appearing segmented) and the branch surface smooth or roughened (because of tiny bumps, warts, or small, spiny side branches). Commonly with rounded mounds or depressed regions containing soredia (i.e., globose vegetative reproductive propagules consisting of one or a few algal cells surrounded by fungal filaments), and these often intermixed with isidia (i.e., small outgrowths of the lichen containing both algal cells and fungal filaments, which can break off and serve as propagules). The reproductive structures (producing the sacs, asci, bearing reproductive spores) cup- or disk-shaped; the spores colorless, ellipsoid, eight per ascus. Usnic acid present (Figure 7.8).

BEECHES (*FAGUS* SPP.)

(In the Beech or Oak family [Fagaceae].)

> *They bore back Beren Camlost son of Barahir upon a bier of branches with Huan the wolfhound at his side; and night fell ere they returned to Menegroth. At the feet of Hírilorn the great beech Lúthien met them walking slow, and some bore torches beside the bier. (SILM 19)*

Tolkien clearly loved beech trees; they are mentioned frequently in his works, occur commonly in the forests of Middle-earth, and are connected with central aspects of his legendarium. Beeches are mentioned several times in *The Hobbit*. They are common in parts of Mirkwood, and "great beeches came right down to the bank, till their feet were in the stream" (Hobbit IX) along the Forest River, by the bridge to Thranduil's halls, as seen in the illustration "The Elvenking's Gate" (see fig. 121 in

W. G. Hammond and C. Scull's, 1995, *J. R. R. Tolkien: Artist and Illustrator*; and also "Bilbo comes to the Huts of the Raftelves," fig. 124 in the same reference; these are also reproduced in Hammond and Scull's *The Art of the Hobbit by J. R. R. Tolkien*). Outside of poetry and song, in *The Lord of the Rings*, our first detailed description of beech comes when Sam awoke in Ithilien (after his and Frodo's rescue by the eagles Gwaihir, Landroval, and Meneldor) and found that "he was lying on some soft bed, but over him gently swayed wide beechen boughs, and through their young leaves sunlight glimmered, green and gold" (LotR 6: IV). In this passage, Tolkien beautifully captures the golden-green of their newly emerged, translucent leaves. We have to turn to *The Silmarillion* for a description of their striking trunks: stout, tall, and high-branched, with smooth, thin, and silver-gray bark. The forests of Neldoreth and Brethil, occupying the northern and western portions of Doriath where Thingol and Melian dwelt, were dominated by beeches intermixed with oaks (*Quercus* spp.), elms (*Ulmus* spp.), and horse-chestnuts (*Aesculus hippocastanum*), and, in their hidden halls, the cave-chambers were supported by pillars carved in the likeness of great beeches. Hírilorn, the greatest of all the beech trees of Neldoreth, grew near the gates of Menegroth and "it had three, trunks, equal in girth, smooth in rind, and exceeding tall" (SILM 19). According to *The Lays of Beleriand*, majestic beeches, with gray trunks and golden-russet leaves, also grew around the beautiful and hallowed pools of Ivrin, where Túrin, late in the year, was healed of his madness following the death of Beleg. However, they are most closely tied to the story of Lúthien and Beren, who first met in the beech-forest of Neldoreth and later walked there, "under mighty beeches silken-skinned, and sang of love that still shall be, though earth be foundered under sea" (Lays III: lines 3222–3226). Turning to the quote that introduces this section, under the great Hírilorn in which Lúthien had been imprisoned by her father, we see Beren, near death, but having recovered the Silmaril (Figure 7.9). It is here that Lúthien embraces and kisses him, asking him to await her beyond the Western Sea. In reading *The Silmarillion*, we begin to see beech trees, and the entire natural world, in a new way: the woods and meadows of our own world are transformed—and we hear the glorious music of the Ainur (see Bernthal, 2014).

Etymology: The English name "beech" was derived through Old English *béce*, from Proto-Germanic **bokjon*, and Proto-Indo-European **bhagos*, meaning "beech tree," and it is related to the Latin *fagus* (beech) and Greek *phagos* (edible oak). The word "book" possibly is derived from the same Proto-Indo-European word, which may relate to the early use by Germanic peoples of beechwood tablets as writing material. The Quenya name of beech is *feren* or *fernë* (pl. *ferni*), but this was replaced in Beleriand by *breth* (mast) and later *brethil* (beech, as seen in the name: Forest of Brethil)—which should not be confused with the Sindarin word *brethil*, which means "birch." Even more confusingly, in Thingol's kingdom, the name for beech was *neldor*, based on the prefix *nel-* (meaning three), and this is a reference to the great beech of Thingol with three trunks.

Distribution and Ecology: Beeches belong to the genus *Fagus*, a group of 10 species distributed across the Northern Hemisphere, mainly in mesic, broad-leaved, and temperate to subtropical forests. American beech (*Fagus grandifolia*) is widespread in eastern North America, while European beech (*F. sylvatica*) occurs widely across Europe, and oriental beech (*F. orientalis*) grows from Bulgaria across Turkey to northern Iran. Finally, seven species occur in eastern Asia. Species of beech are distinguished by characters of leaf shape, form of the margin, variation in pubescence, and in the form

FIGURE 7.9 *Beeches.*

of the scaly cupule (which surrounds the two triangular nuts). Their rather inconspicuous flowers are wind-pollinated, and their triangular nuts, within a spiny four-valved cupule, are eaten and dispersed by birds and mammals.

Economic Uses: Beech trees are important ornamentals; they provide shade trees in urban environments, and these trees are visually striking because of their thin, smooth, silver-gray bark and the beautiful yellow color of their leaves in autumn. They are important timber trees; the wood is heavy, hard, and strong, and is used in flooring, veneer, furniture, and musical instruments. The nuts are edible (at least in small quantities), and the trees provide food to a variety of wildlife. The smooth bark seems hard for people to resist using as a poster for writing, and in urban parks initials are frequently seen carved into beech trunks.

Description: Tardily deciduous trees, often high-branched, with the trunks stout and tall, covered with smooth, thin, and silver-gray bark. Leaves alternate, spirally arranged, simple, ovate to elliptic, pinnately veined, with secondary veins more or less parallel, running into the leaf margin; the apex acute to acuminate, the base acute to obtuse or cuneate, the margin crenate (without distinct teeth) to obscurely or clearly serrate, with or without unicellular hairs, occasionally also glandular, petiolate; stipules present. Flowers inconspicuous, radially symmetrical, unisexual (with the staminate and carpellate flowers borne on the same plant), with an inconspicuous perianth of usually six tepals; the staminate inflorescences axillary, dangling, globose clusters of numerous flowers, the carpellate inflorescences in leaf axils, two-flowered, surrounded by a four-valved cupule (= cuplike structure); staminate flowers each with six to many stamens; carpellate flowers with three fused carpels, with three styles, the ovary inferior. Fruits three-angled, paired nuts borne in a four-valved, prickly cupule (Figure 7.9).

BILBERRY OR WHORTLEBERRY (*VACCINIUM MYRTILLUS*, *V. ULIGINOSUM*, AND RELATED SPECIES)

(In the Heath family [Ericaceae].)

> *As quickly as they could they scrambled off the beaten way and up into the deep heather and bilberry brushwood on the slopes above, until they came to a small patch of thick-growing hazels. (LotR 1: XII)*

The names "bilberry" and "whortleberry" are applied to *Vaccinium myrtillus* but are also given to several related species, especially *V. uliginosum*. Both are low, rhizomatous shrubs (reaching only around a half meter in height) and occur across northern and central Europe in heaths, moors, thickets, and woods on acidic soils. J. R. R. Tolkien used both names and was most likely referring to *V. myrtillus*, which is the more widespread species, but *V. uliginosum* cannot be ruled out. *Vaccinium myrtillus* has leaves that are serrate (i.e., sharply saw-toothed) along the margin and acute at the apex and twigs that are angled, whereas *V. uliginosum* has leaves that are entire (i.e., smooth, without teeth) along the margin and obtuse at the apex and twigs that are terete. Perhaps he had both of these common European species in mind. The sole occurrence of the name "bilberry" in *The Lord of the Rings* occurs in the chapter "Flight to the Ford." Aragorn was leading the four hobbits through the Trollshaws, along the East-West road, attempting to escape the Black Riders and making toward Rivendell. After they had heard the sound of hoofs behind them, they scrambled off the road "up into the deep heather [*Calluna vulgaris*] and bilberry brushwood" (LotR 1: XII) to hide (Figure 7.10; see also entry for Hazels, and entry for Heather, Figure 7.33). Much later, near a stream that joined the Silverlode, shortly after the Fellowship had escaped from Moria, they rested in a dell and "about it stood fir-trees [*Abies alba*], short and bent, and its sides were steep and clothed with harts-tongue [*Asplenium scolopendrium*] and shrubs of whortle-berry" (LotR 2: VI). It was here that Aragorn gently removed Frodo's jacket and tunic, discovered his mithril corslet, and treated the wound he had received from the orc chieftain. Sam and Frodo also encountered "a thick growth of gorse [*Ulex europaeus*] and whortleberry" (LotR 4: VII) shortly before they rested on their journey to the Cross-roads. Clearly, whortleberry was common in open, acidic habitats of Middle-earth, as it and related shrubby species

FIGURE 7.10 *Bilberry or whortleberry.*

of *Vaccinium* are in temperate regions of the Northern Hemisphere today. Not surprisingly, given its latitude, these shrubs also occurred in Beleriand during the First Age, and they are mentioned in the unfinished *Of Tuor and His Coming to Gondolin.* As Voronwë led Tuor to the Hidden Kingdom, they encountered a company of orcs encamped on the highway leading north to Tol Sirion from the Crossings of Teiglin. Trusting in the mantle of Ulmo, they crossed the road and at once heard a wild cry—the orc watchers had heard them, although they were not seen. Desperately, Tuor "stumbled and crept forward with Voronwë at his side, up a long slope deep in whin and whortleberry among knots of rowan and low birch" (UT 1: I). They hid beneath a large boulder, panting "like tired foxes" and the sounds of the searching orcs slowly grew faint. They had escaped capture and soon afterward would cross the Ford of Brithiach and approach the Encircling Mountains that protected the realm of Turgon. This close encounter is presented in beautiful detail, and the plants—whortleberry (*V. myrtillus*), whin (*Ulex europaeus*), forming low thickets with scattered birch (*Betula pendula*) and rowan (*Sorbus aucuparia*)—allow us to picture it clearly.

Etymology: The English name "bilberry" is of Scandinavian origin, as seen in its similarity to the Danish *bøllebær* (literally meaning ball-berry) from the Old Norse *bollr.* "Whortleberry" is a dialectal variant of "hurtleberry": from Middle English *hurtilberye,* the first element coming from Old English *horte. Vaccinium* apparently was the classical name of bilberry, and the word is derived from *bacca,* meaning "berry," a reference to the characteristic fruits of these shrubs.

Distribution and Ecology: Species of the genus *Vaccinium* (i.e., the blueberries, cranberries, bilberries, and their relatives) are widely distributed from arctic to tropical regions; the genus is large, containing from 150 to as many as 500 species, the variation in number due to the fact that the group is an artificial assemblage, delimited arbitrarily from many similar (and related) plants such as *Gaylussacia* (huckleberries), *Agapetes* (diverse in the Indomalesian region, many with showy red flowers), and many other genera (e.g., *Cavendishia, Macleania, Satyria*) of the Andean region. The genus is, therefore, of uncertain delimitation. Blueberries are also botanically difficult at the species level as a result of hybridization and polyploidy (i.e., genome duplication) or merely because they grow in difficult-to-access tropical montane regions and are thus poorly collected and understood. The bilberry or whortleberry (*V. myrtillus,* but the common names are inconsistently applied and the same names are used for the related *V. uliginosum*) is widely distributed, occurring across Europe and northern Asia and also growing in the mountains of western North America. *Vaccinium uliginosum* has a boreal distribution, growing across northern regions of North America, Europe, and Asia. Blueberries occur in numerous plant communities (e.g., bogs, swamps, wet shorelines, taiga, deciduous or coniferous forests, tundra, muskegs, alpine thickets and meadows, heaths, and moors) and are usually restricted to acidic soils. The north temperate species of *Vaccinium* are distinguished by characters such as stature, surface texture, and angularity of the twigs; whether the leaves are deciduous or evergreen; their shape, size, coloration, pubescence (hairs simple, unicellular, nonglandular hairs or multicellular, gland-headed), and margin (entire vs. variously toothed); whether or not the calyx is articulated from the floral pedicel; shape of the corolla; number of petals and the extent of their fusion; whether the stamens stick out from or are enclosed within the corolla; presence or absence of paired appendages on the anthers; and the color of the berries and number of seed-containing cavities (locules) within them. The usually white flowers function in an inverted position, like downward oriented bells or urns, and attract various bees, with nectar and pollen being the rewards. The pedicels curve upward after pollination so the blue, blue-black, or red berries develop in an erect position in which they are more visible to the dispersal agents: many different birds and mammals.

Economic Uses: Blueberries are widely cultivated for their edible fruits, mainly cultivars of the highbush blueberry (*V. corymbosum*) but to a lesser extent also the sweet lowbush blueberry (*V. angustifolium*); the fruits of many wild populations of these species and many others are also flavorful and gathered. In addition, the cranberries (especially *V. macrocarpon*, but also *V. oxycoccos*) are economically important crops. The fruits of bilberry or whortleberry (*V. myrtillus*) are much gathered in Europe for use as fresh fruit, in pies and jams, or for wine-making. We note that *V. myrtillus* is a member of *Vaccinium* sect. *Myrtillus*, a group that has flowers borne singly in the axils of the lowermost leaves; these flowers have pedicels that are not articulated with the calyx tube, stamens each with paired appendages, and fruits (in cross-section) with five seed-containing locules. The economically important highbush blueberry (and its lowbush relatives) are in sect. *Cyanococcus*, with flowers in clusters, having pedicels articulated with the calyx tube and stamens lacking paired appendages, and berries seemingly with 10 seed-containing locules. Finally, a few blueberry species are used as ornamentals.

Description: The arctic to temperate species are evergreen or deciduous, often rhizomatous shrubs, lianas, or small trees, with simple, nonglandular and/or gland-headed hairs. Leaves alternate, spirally arranged, simple, ovate to obovate, pinnately veined, the secondary veins forming loops, the apex acute to acuminate to rounded, the base acute or cuneate to rounded, the margin entire to serrate; stipules absent. Inflorescences axillary or terminal, short to elongate racemes or reduced to a solitary flower. Flowers radially symmetrical, bisexual, the sepals 4–5, slightly fused; the petals 4–5, usually strongly fused and forming a cylindrical, globose, urn-shaped, or less commonly bell-shaped corolla, white or greenish to pink; the stamens 8–10, their anthers with or without a pair of projections (awns or spurs), opening by pores, and each pore at the apex of a short to elongate tube; and 4–5 fused carpels, the style elongate, with truncate to capitate stigma, the ovary inferior. Fruit a blue, purple, black, or red berry with an apical "crown" formed by the persistent calyx, with few to numerous small seeds (Figure 7.10).

BIRCHES (*BETULA* SPP.)

(In the Birch family [Betulaceae].)

> *The next day at the hour of sunset Aragorn walked alone in the woods, and his heart was high within him; and he sang, for he was full of hope and the world was fair. And suddenly even as he sang he saw a maiden walking on a greensward among the white stems of the birches; and he halted amazed, thinking that he had strayed into a dream.* (*LotR Appendix A: I (v)*)

Birches are widely distributed in Middle-earth and are therefore frequently referenced in J. R. R. Tolkien's writings. Their first mention is in the Green Hill Country of the Shire, on the night that Frodo, Sam, and Pippin began their journey to Rivendell, and their branches, above the heads of the hobbits, made "a black net against the pale sky" (LotR 1: III), reminding us that we are in "tree-tangled Middle-earth" (see the song to Elbereth, LotR 2: I, and discussion in T. Shippey's, 2002, *J. R. R. Tolkien: Author of the Century*). They are last mentioned in the quote cited, which describes the first meeting of Aragorn and Arwen, occurring at sunset in a white birch forest in Rivendell. Aragorn had been told of his lineage only the day before and was singing the part of the *Lay of Lúthien* that tells of the meeting of Lúthien and Beren in the beech forest of Neldoreth, and, for a moment, thinking that she was Lúthien Tinúviel, he called out to her "even as Beren had done in the Elder Days." The emphasis here, perhaps, is on the hope for the renewal and purification of Middle-earth, expressed in the association of Arwen and the white trunks of the birches, even though many years of darkness and struggle remained. The smooth bark characteristic of birches is reflected in the name, Skinbark, a birchlike ent, who lived on the mountain-slopes west of Isengard and had been wounded by Saruman's orcs. Many of his tree-herds also had been destroyed, and Treebeard told Merry and Pippin that in response "he has gone up into the high places, among the birches that he loves best" (LotR 3: IV). A white birch forest also surrounded the open, grassy summit of the Hill of Anwar (= Halifirien), the holy mountain in the Ered Nimrais (= White Mountains) where Elendil was once buried. Their beautiful trunks, surrounded by silence, added to the awe and feeling of protection of any who visited the place (see UT 3: II; Watts, 2007). Birches are also a striking element in the landscape of Beleriand and play an important role in the narrative of *The Silmarillion*. For example, with the aid of Círdan, Eärendil built Vingilot, the ship that would carry him to Valinor (Figure 7.11), and its timbers were "hewn in the birchwoods of Nimbrethil" (SILM 24; also alluded to in Bilbo's poem in LotR 2: I). We are told that on the high ridges of Amon Rûdh grew "knots of birch, rowan [*Sorbus aucuparia*] and ancient thorn trees" (*Crataegus* spp.; CoH VII), and birches grew about a clear pool near the entrance to Mîm's cave. Birches also grew around the beautiful lake, Tarn Aeluin, surrounded by wild heaths in the highlands, where Barahir and his 11 companions (one of whom was Barahir's son Beren) camped, hiding from Morgoth after the ruin of Dorthonion (as recorded in Lays IV: lines 190–198). White birches are beautifully illustrated in Ted Nasmith's painting of Nienor being carried to Ephel Brandir by the men of Brethil (opposite p. 224 in the illustrated *Silmarillion*). Finally, although birches are not mentioned in *The Hobbit*, Tolkien's beautiful color illustration "Rivendell" clearly shows a silver birch (*Betula pendula*) in the right foreground.

Etymology: The English name "birch" comes from the Old English, *bierce* or *birce*, and is of Germanic origin, derived from the Proto-Indo-European **bhergo*, from the root **bhereg-*, meaning to shine or gleam, in reference to the smooth, shiny, and often white bark of many species of birch. The Latin name, *Betula*, is derived from Gaulish *betu*, referring to bitumen, because a tarlike substance can be extracted from the bark of these trees. The Sindarin word *brethil* means silver birch and was the most commonly used name for these trees in the First through the Third Ages; *nim* is Sindarin for white, and thus *Nimbrethil* means "white birch" and is also the name of the birchwoods

FIGURE 7.11 *Birches.*

in Arvernien. Giraudeau (2011) lists *chwind* and *whinn* as other Sindarin names for birches. The Quenya word for birch may be *silwin* and/or *hwindë*.

Distribution and Ecology: Birches belong to the genus *Betula*, a group of ca. 40 species distributed broadly across temperate and boreal regions of the Northern Hemisphere. The silver birch (*B. pendula*) and downy birch (*B. pubescens*) are widespread in Europe, while paper birch (*B. papyrifera*) occurs across cool temperate to boreal North America. Gray birch (*B. populifolia*), river birch (*B. nigra*), sweet birch (*B. lenta*), and yellow birch (*B. alleghaniensis*) are common trees in eastern North America, and water birch (*B. occidentalis*) is widespread across western North America. Numerous species (e.g., *B. chinensis*, *B. dahurica*, and *B. platyphylla*) occur in eastern Asia. Leaf form, bark, and characteristics of the fruiting catkins are often useful in identification, but species are often difficult to distinguish as a result of hybridization. Birches grow in cool, moist woods, moorlands, bogs, fens, and along streams and lakeshores. Their flowers are wind-pollinated, and the small, winged fruits are wind-dispersed.

Economic Uses: The wood is used in veneers and to make furniture, tool handles, spools, toothpicks, baskets, or woodpulp for paper, and the twigs are used to make brooms; the waterproof bark of some species was once used in making canoes and tepee covers, provided writing material, and is also useful tinder. The bark can be processed to release birch-tar, which was once widely used as an adhesive and disinfectant and is still used medicinally and in leather processing. The sap of some species can be processed to make birch syrup or beer. Many species are popular ornamental trees because of their striking bark (smooth, peeling, and often white, gray, or yellowish). Birches are rapidly growing plants and are typically pioneer species, colonizing disturbed and/or open habitats. They are also ecologically important in providing browse for wildlife.

Description: Deciduous trees or shrubs, the bark more or less smooth, often peeling, dark brown, yellow to chalky white, with dark, often horizontal lenticels, and branches sometimes with the taste or odor of wintergreen. Leaves alternate, more or less two-ranked, simple, ovate to elliptic, pinnately veined, with secondary veins running into the teeth; the apex acuminate or acute to rounded, the base acute or cuneate to rounded or cordate, the margin usually double serrate (i.e., with small and large teeth) with variably developed simple and gland-headed hairs, petiolate; stipules present. Flowers inconspicuous, radially symmetrical, unisexual, lacking a perianth, in groups of three associated with each three-lobed, bract complex, densely clustered in dangling, elongate, staminate catkins (formed in the previous growing season, exposed in winter) and erect, ovoid to cylindric, carpellate catkins (with staminate and carpellate catkins on the same plant); staminate flowers each with two or three stamens, and the filament of each stamen divided nearly to the base; carpellate flowers with two fused carpels, with two styles, the ovary nude (as the perianth is absent). Fruit a samara with two lateral, membranaceous wings (Figure 7.11).

BLACKBERRIES, RASPBERRIES, BRAMBLES (*RUBUS* SPP.)

(The Rose family [Rosaceae].)

They will be harvesting and blackberrying, before we even begin to go down the other side at this rate. (Hobbit IV)

Blackberries and raspberries are infrequently mentioned in Tolkien's writings, although they were as widespread and common in Middle-earth (including Beleriand, in the First Age) as they are in Europe and throughout the Northern Hemisphere today. The evidence for this abundance is the much more frequent use of the alternative common name, "bramble" for these difficult plants, alluding to the fact that they often form tangled, almost impenetrable thickets in disturbed or open habitats, which are especially difficult to walk through because of their sharp, straight to recurved prickles. When hiking, it is thus unpleasant to pass through areas where they are abundant, even though the fruits provide a nice treat (if in season!). Shortly after leaving Rivendell, Bilbo was somewhat dismayed at the slow progress they were making toward the Lonely Mountain (see quote) and wistfully thought that it would be blackberry season back in the Shire (Figure 7.12) before he and the dwarves even got to the eastern slopes of the Misty Mountains. It would be unlike hobbits to miss out on the pleasure of gathering wild blackberries, and, in *The Lord of the Rings*, the hobbits enjoyed a blackberry tart, among other delicious items, at the Prancing Pony in Bree. From *The Hobbit* we know that Bilbo kept raspberry jam in his larder! However, travelers in Middle-earth frequently encountered blackberries, as brambles, when traveling, and such encounters were usually unpleasant. Brambles are first mentioned when Frodo, Sam, Merry, and Pippin are lost in the Old Forest, when they were led in the wrong direction by deep folds in the ground, which were "choked with brambles" (LotR 1: VI). These must have ripped at their clothing and scratched their arms and legs, adding to the misery of that hike. Brambles are often associated with disturbed habitats, and we see this aspect of their ecology when they are encountered in the vicinity of Isengard, where the trees had been felled and the landscape ravaged by Saruman, and most of the wizard's valley "had become a wilderness of weeds and thorns. Brambles trailed upon the ground, or clambering over bush and bank, made shaggy caves where small beasts housed" (LotR 3: VIII). They are sun-loving plants and are early successional; that is, they are among the early colonizers of disturbed habitats, assisting in the recovery of biodiversity and providing food and cover for wildlife. Likewise, as Gollum led Frodo and Sam up into the hills of Ithilien, as they approached the Cross-Roads they passed through habitats long impacted by environmentally destructive activities of Sauron—and they again encountered "thickets and wastes of brambles" (LotR 4: VII). However, we see this evidence of environmental damage most keenly within Mordor itself. When Frodo and Sam scrambled on the eastern slopes of the Ephel Duath they found themselves in a polluted and dying land. But it was not yet dead. In the valleys there were areas in which "low scrubby trees lurked and clung, coarse grey grass-tussocks fought with the stones, and withered mosses crawled on them; and everywhere great writhing, tangled brambles sprawled," and the two remaining members of the quest certainly experienced their "hooked barbs that rent like knives" (LotR 6: II). Here, the plants were in a desperate struggle for life, living in a poisoned environment, and all plant life had become harsh and twisted as a result of this struggle, as clearly seen in the writhing brambles with their hooked barbs—almost the only plant still thriving there. Tolkien's highlighting of brambles in such environments is quite traditional—weedy species of *Rubus* have long been associated with ecological degradation (see, for example, Isaiah 5:6, 7:23, 9:18, and 10:17). Blackberries/brambles thus present a remarkable contrast: the same plant that produces delicious berries (eliciting memories of summertime blackberry

FIGURE 7.12 *Blackberries, raspberries, brambles.*

picking and also providing food for wildlife) also forms tangled, prickly thickets that especially flourish in habitats damaged by human activities (or, in Middle-earth, the activities of Maiar, such as Saruman or Sauron, who have left the path of wisdom).

Etymology: The name "blackberry" (a reference to the color of the fruits) is derived from Old English *blacberian*, from *blæc* (black) + *berie* (berry); the word *berie* is from Proto-Germanic **basjom*, which is of unknown origin. The common name "bramble" goes back to the Old English *bræmbel*, from earlier *bræmel*, from Proto-Germanic **braemaz* (thorny bush), from the Proto-Indo-European root **bh(e)rem-* (to project, a point), presumably in reference to the sharp prickles characteristic of these plants. The name is now applied to various species of *Rubus*, but originally the name meant any thorny or prickly plant (Watts, 2007). Raspberry is derived from "raspis berry," possibly from *raspise*, a sweet rose-colored wine (mid-fifteenth century), from Anglo-Latin *vinum raspeys*, and likely related to Medieval Latin *raspecia, raspeium* (raspberry). The scientific name *Rubus* is derived from the Latin, *ruber* (meaning red), in reference to the color of the mature fruits of some species (but others have fruits that are only red when immature and turn black at maturity). *Pûg* is the Sindarin word for blackberry, while in Quenya the name may be *piucca*.

Distribution and Ecology: Blackberries (also brambles and raspberries, all species of the genus *Rubus*) are widely distributed but are especially diverse in the Northern Hemisphere; some species, however, also grow in the West Indies, Central and South America, Africa, the Pacific region, and Australia. The total number of species in the group is unknown and, in fact, difficult to determine because of extensive hybridization, polyploidy (i.e., chromosome doubling, etc.), and asexual reproduction, but it is likely somewhere between 250 and 700. Thirty-seven species (broadly delimited) are recognized in North America, north of Mexico, while around 75 species and species complexes have been recognized in Europe. The group is most diverse in Eastern Asia, and more than 200 species are known from China and nearly 40 from Japan. Species are distinguished by characters of their habit (stems erect or creeping, biennial or perennial, evergreen or deciduous), leaves (simple or compound, number and arrangement of leaflets, extent of lobing, leaflet shape and size, development of marginal teeth, form of the stipules), pubescence (development of both unicellular, nonglandular hairs, and multicellular, gland-headed hairs), armature (presence or absence of prickles, and variation in their form), flowers (color and size of the petals, number of carpels), and fruits (whether or not the cluster of drupelets separates from the receptacle, size and color of the cluster of drupelets). Blackberries grow in an exceptionally broad array of plant communities, including open woodlands, savannas, swamps, bogs or fens, prairies, meadows, in rocky or sandy habitats, lakeshores and stream banks, and various wet to dry disturbed habitats such as clearings and roadsides. Their showy flowers attract a variety of insects (especially flies, bees, wasps, butterflies, beetles), and the fleshy, black or red fruits are dispersed by birds and a few mammals (such as bears).

Economic Uses: Blackberries and raspberries are economically important because of their flavorful, edible berries (actually, botanically speaking, clusters of drupelets), and the fruits are either cultivated or gathered from the wild, and eaten fresh or made into jams, pies, or beverages. The blackberries are those species in which the drupelets adhere to the receptacle, so when a fruit is picked the receptacle (tissue bearing the cluster of drupelets) is picked along with the drupelets. In contrast, in the raspberries, the mature cluster of drupelets falls free from the receptacle. This distinction, although used in distinguishing these two fruit conditions, does not reflect evolutionary relationships. Important raspberry species include *R. idaeus* (red raspberry) and *R. occidentalis* (black raspberry); many cultivated plants represent complex hybrids of these two and also other species. Popular prickleless blackberry cultivars have been derived from hybrids of *R. ulmifolius* (elm-leaf blackberry) and *R. hastiformis*, although many blackberries are gathered from wild species. A few species, such as *R. odoratus* (purple-flowering raspberry), are grown as ornamentals because of their large flowers. Finally, some species have been used medicinally (especially to alleviate labor pains).

Description: Perennial, herbs to subshrubs (or shrubs); stems erect, arching or creeping, and of two types: primocanes (each a biennial or perennial vegetative stem, prior to being sufficiently developed for flowering) and floricanes (each a biennial or perennial stem after it has begun flowering); hairs variably developed, some not glandular and others gland-headed; both stems and leaves often with sharp-pointed modified hairs (i.e., prickles), and these sparse to dense, erect to curved, weak to stout. Leaves evergreen or deciduous, alternate, spirally arranged, occasionally simple or more commonly variously compound (with 3–9 leaflets), lobed or unlobed; the leaves or leaflets variously shaped, with pinnate or palmate venation, the apex acute or acuminate to rounded; the base cordate to acute or cuneate (and asymmetrical in lateral leaflets); the margin crenate, dentate, or serrate, or doubly so; the lower surface without prickles or with prickles on midvein similar to those of the stem, without hairs or with unicellular, nonglandular hairs and/or multicellular, gland-headed hairs, the leaves petiolate. Stipules present, often adnate to the petiole, with the leaves of the primocanes often differing from those of the floricanes. Inflorescences terminal or axillary, the flowers numerous to few, in cymes or reduced to a solitary flower. Flowers usually bisexual, radially symmetrical, more or less erect, with a flat floral cup (= hypanthium) bearing five sepals, five distinct white to pink or magenta petals, and numerous stamens; the carpels usually numerous, distinct, and with or without hairs; the ovaries superior, each with a slender to distally thickened style. Fruit a cluster of drupelets, these golden yellow to red or black, globose to cylindric, and each with a single small, hard pit (Figure 7.12).

BRACKEN (*PTERIDIUM AQUILINUM*)

(In the Hay-scented Fern family [Dennstaediaceae].)

> *She [Niënor] stood still a moment as in wonder, and then, in a swoon of utter weariness, she fell as one stricken down into a deep brake of fern. And there amid the old bracken and the swift fronds of spring she lay and slept, heedless of all. (CoH XV)*

Bracken is larger than most temperate ferns and often forms dense stands (called *brakes*). As in most ferns, the young leaves are coiled, forming fiddleheads (or crosiers). Tolkien, as narrator, describes them as the "tight-curled fronds of spring" (LotR 3: XI) in the valley where Gandalf and his traveling companions camped after their confrontation with Saruman. Their stature is emphasized several times, as when, after escaping from the goblins, Bilbo and the dwarves hiked through the dense understory of a pine forest and found that "they were pushing through a sea of bracken with tall fronds rising right above the hobbit's head" (Hobbit VI) (Figure 7.13). In fact, the tendency of this tall fern to form dense stands means that it provides good cover, hiding Beleg "in bracken cowering . . . through the leaves peering" as he watched Túrin "as he tottered forward neath the whips of the Orcs" (Lays I: lines 1013–1016) when he passed in captivity through Taur-nu-Fuin, being taken by an orc company to Morgoth's stronghold. The value of bracken as fragrant bedding material will be recognized by many hikers and backpackers, and we see this usage in Treebeard's bed, which is described as "covered deep in dried grass and bracken" (LotR 3: IV). In a more tragic tale, Niënor, suffering from the effects of her encounter with the dragon Glaurung, collapsed and slept in a deep brake of fern (i.e., bracken thicket) near the forest of Brethil. She soon after was found by Túrin, on the very mound where the elf maiden Finduilas was buried.

Etymology: The English name "bracken" was derived from the Middle English, *braken*, a name of Scandinavian origin, and related to the Swedish, *bräken*, referring to a large fern, especially the plant now known under the scientific name *Pteridium aquilinum*. "Brake" (a shortened form of bracken) is also of Middle English origin, referring both to this large and often weedy fern and to an open area overgrown with ferns and shrubs. *Pteridium* is derived from the Greek, *pteridion*, a diminutive of *pteris*, meaning fern. The specific epithet *aquilinum* is derived from the Latin *aquila*, eagle, possibly an imaginative allusion to its leaves, appearing like eagle's wings, or to the pattern of vascular bundles of the petiole that, when cut in cross-section, appear as a two-headed spread eagle (but see also Vickery, 1995).

Distribution and Ecology: Bracken (*Pteridium aquilinum*) occurs almost worldwide and is morphologically variable (divided into two subspecies and 12 varieties, a few of which are sometimes recognized as distinct species). It grows in the understory of dry to moist woods or in open areas and often invades pastures, fields, and roadsides, forming large colonies due to its aggressive rhizomatous growth, suppression of germination of other species, and resistance to fires (due to its deeply buried system of rhizomes).

Economic Uses: In many regions bracken is a common weed of pastures, and its presence is especially problematic because it is toxic to both animals and humans if eaten. Ironically, the young emerging leaves—fiddleheads—were once widely consumed, especially in Eastern Asian cultures, but they contain thiaminase and other mutagenic and carcinogenic compounds. The plant was once used as roofing thatch, as a quick fuel source, as a source of potash (used in soap-making and the glass industry), and in tanning leather. It is still used as a source of a yellow dye, and the dried leaves provide bedding for livestock. Ecologically, bracken is important in providing cover for wildlife.

Description: Herbaceous, moderate-sized to large fern, emitting a pleasant, somewhat sweet fragrance, with underground, slender, and long-creeping rhizomes that lack scales. Leaves (or fronds), alternate, widely spaced, emerging individually from the soil, the blade triangular in shape, compound, divided pinnately two to three times, the ultimate segments pinnately lobed, some divisions (pinnae) with nectar glands at the base; the margins entire, revolute, the lower surface with usually sparse to dense hairs; the petiole elongate, grooved on the upper surface, glabrous or with short hairs. Spore-bearing regions (i.e., the sori) on the underside of the leaf, elongated, positioned along the leaf margin, covered when young by an outer false indusium (the inrolled leaf margin) and an inner true indusium (i.e., an elongate flap). Spores borne in stalked, globose containers (sporangia) that open by means of a vertical annulus

FIGURE 7.13 *Bracken.*

(line of thick-walled cells) that is interrupted by the stalk and that curves back when dry, ripping open the sporangium and dispersing the tiny spores (Figure 7.13).

BUTTERCUPS (*RANUNCULUS* SPP.)

(The Buttercup family [Ranunculaceae].)

> *Old Tom in summertime walked about the meadows*
> *gathering the buttercups, running after shadows.* (*TATB 1*)

All but one of the references to buttercups in J. R. R. Tolkien's legendarium are connected with Tom Bombadil, perhaps because they are common plants of wetland habitats such as moist river margins, marshes, and wet meadows, and Tom is closely associated with the wetland habitats of the Withywindle River (as is his wife, Goldberry). In the introductory quote (from the poem *The Adventures of Tom Bombadil*), we see Tom gathering buttercups, probably in a meadow near his home. Later in the poem, at his wedding with Goldberry, Tom goes without his hat and feather and instead is "crowned all with buttercups," while Goldberry wore a garland of forget-me-nots (*Myosotis* spp.) and flag-lilies (*Iris pseudacorus*). In the poem *Once upon a Time* (included as an appendix in the expanded edition of *The Adventures of Tom Bombadil*, 2014, edited by Christina Scull and Wayne Hammond), the buttercups are described as sending "up their light in a stream of gold," an allusion to their bright yellow flowers. It is interesting that Tom's boots are yellow, matching the color of the cheerful buttercups. Another reference to *Ranunculus* may be the use of the common name "celandine" for one of the wildflowers of Ithilien because this name often is applied to *R. ficaria* (see LotR 4: VII).

Etymology: The name buttercup arose through the merger of two older common names: gold-cups and butter-flowers. The word "butter" is from Old English *butere*, which was early borrowed from Latin *butyrum* (butter) and Greek *boutyron* (perhaps with an original literal meaning of cow-cheese) and perhaps derived from a Scythian word. The word "cup" comes from Old English *cuppe*, which was derived from Late Latin *cuppa*, through Germanic languages, and originally is from Proto-Indo-European **keup-* (a hollow). The scientific name *Ranunculus* is derived from the Latin *rana* (frog) and *unculus* (little) and is an allusion to the wet habitats, loved by frogs, in which some species grow.

Distribution and Ecology: Buttercups (belonging to the genus *Ranunculus*, a group of about 400 species) are distributed nearly worldwide (except for the lowland tropics), with 131 species documented in Europe, 77 in North America (north of Mexico), 125 in China, and 25 in Japan. The group is in need of study and may not be monophyletic; species recognition is often problematic as a result of asexual reproduction and hybridization. Buttercups grow in various aquatic habitats, as well in terrestrial communities such as meadows, alpine slopes, tundra, pastures, marshes, bogs, fens, stream banks and lakeshores, swamps, moist rocky ledges, moist woods, and in various open, disturbed habitats. Their flowers are pollinated by various insects, especially bees, flies, and beetles, while the small fruits are dispersed by wind, water, rain-wash, or externally on the fur or feathers of animals.

Economic Uses: A few species (e.g., *R aconitifolius*, *R. acris*, *R. asiaticus*) are cultivated as ornamentals because of their showy yellow or white flowers. The plants contain various alkaloids and ranunculin (a lactone glycoside) and are poisonous (if eaten, as well as causing contact dermatitis); they have an acrid taste and irritate the mouth.

Description: Annual or perennial herbs, sometimes producing rhizomes; hairs, when present, simple; stems usually erect to prostrate. Leaves usually alternate, in a basal rosette, along stems, or both; simple or compound, often lobed or dissected, with palmate venation; the apex acute to rounded; the base narrowly acute to cordate; the margin, entire, serrate or crenate, with or without a petiole; stipules absent. Inflorescences terminal or axillary, few- to many-flowered cymes, or reduced to a solitary flower, sometimes with large bracts. Flowers radially symmetrical, bisexual, more or less erect, with 3–5 sepals, these persistent or quickly deciduous, with usually five to numerous petals, these distinct, usually yellow, but occasionally white, red, or green, with a nectar gland, with usually 10 to numerous stamens, these distinct; carpels numerous, distinct; the ovaries superior. Fruits a cluster of achenes, these sometimes with various projections.

CAMPION (*SILENE* SPP.)

(The Carnation or Pink family [Caryophyllaceae].)

> *Then Tuor went up the wide stairs, now half-hidden in thrift and campion, and he passed under the mighty lintel and entered the shadows of the house of Turgon; and he came at last to a high-pillared hall. (UT 1: I)*

The name "campion" occurs only once in J. R. R. Tolkien's legendarium—see the introductory quote from the beautiful but unfortunately unfinished tale *Of Tuor and His Coming to Gondolin.* Tuor had been forced from Dor-lómin, his homeland, by Easterlings in the service of Morgoth, and, following the unconscious promptings of Ulmo, the Lord of Waters, he went in search of the Gate of the Noldor and eventually reached the Great Sea—and "none, save the Eldar, have ever felt more deeply the longing that it brings" (UT 1: I). Tuor was the first human to see the ocean, and, upon reaching the coast, he stood atop a sea cliff in Nevrast, alone, with arms outstretched, looking toward the setting sun. Journeying southward along the coast, he eventually found the ruins of Vinyamar, once the dwellings of Turgon, but now long-abandoned. Following the beckoning of a bevy of swans that had landed on the highest terrace, he climbed the wide stairs of Turgon's high-pillared hall, now overgrown with sea campion and thrift (Figure 7.14). It was there that he found on a wall behind the throne a shield, great mail shirt, helmet, and a long sword in a sheath—these had been left for him by Turgon, many years before, by the command of Ulmo. Wearing them, he was ready to meet the Lord of Waters (Figure 7.65) and receive his message for Turgon, now dwelling in the Hidden Kingdom. It is appropriate from a mythological standpoint that the preparation for this meeting with Ulmo, King of the Sea, is associated with the sea campion (*Silene uniflora*; often called *S. vulgaris* subsp. *maritima*), a branched herb with succulent leaves and white flowers characteristic of coastal rocky habitats and cliffs (often growing with thrift, *Armeria maritima*, another maritime species). Of course, it is ecologically appropriate as well. However, another connection is with Tuor, himself; he is a champion for

FIGURE 7.14 *Campion.*

the cause of men and elves in the First Age of Middle-earth, reflecting the meaning of campion, the name of this plant.

Etymology: The English name "campion," applied to all species of the genus *Silene*, is derived from Middle English *campiun*, from Old French *champiun* (champion), from the Latin *campio, -onem*, a combatant in the arena or *campus* (field of athletic contest, place of fighting in single combat). In addition, species of campion, especially the rose campion (*S. coronaria*), were used to make garlands used in crowning athletic champions. The word "campion" thus is related to champion. The scientific name *Silene* is the feminine form of Silenus, the intoxicated foster father of the Greek god Bacchus, who was described as covered with foam, and this is perhaps an allusion to the sticky secretions that cover many species.

Distribution and Ecology: Campions (i.e., members of the genus *Silene*) are a large group of annual to perennial herbs, possibly including as many as 700 species. They are widespread in the Northern Hemisphere, with ca. 166 known from Europe, 70 in North America (north of Mexico), ca. 110 in China, and 11 in Japan, occurring in a variety of open, wet to dry, often disturbed habitats. The species are distinguished by features such as habit, pubescence, leaf shape and arrangement, inflorescence structure, floral characters (e.g., form of the fused sepals, variation in color and form of the clawed petals, number of stigmas), and seed characters. Their showy flowers are pollinated by a variety of insects, especially various bees, butterflies, and moths, and the small seeds are dispersed by wind or rain wash.

Economic Uses: Many species of campion have showy flowers and are cultivated as beautiful horticultural herbs.

Description: Annual to perennial herbs; stems unbranched or branched, erect to prostrate, the hairs various, but often glandular (and producing a sticky secretion). Leaves opposite or occasionally whorled, simple, arising from a slightly swollen node, ovate to obovate, often narrowly so, sometimes linear, with pinnate venation, but often only the midvein visible; the apex usually acute to acuminate or obtuse; the base narrowed and leaves of a pair slightly connate basally; the margin entire; and most leaves along the stem without a petiole, although basal leaves may be petiolate; stipules absent. Inflorescences usually terminal, the flowers few and in cymes or reduced to a single flower. Flowers usually bisexual (but occasionally unisexual), radially symmetrical, held erectly to pendulous, with the five sepals fused, forming a cylindric to bell- or urn-shaped, frequently inflated calyx; the five petals distinct, white to red or purple, clawed (i.e., jointed, with a narrow base, broad) and usually two-lobed (but sometimes apically dissected) apical portion, and with two small appendages at the joint and 10 stamens; the carpels, 3–5, fused, the ovary superior, and with 3–5 styles. Fruit a capsule, opening longitudinally, forming 3–5 segments, and often also apically splitting into small teeth; seeds small, sometimes winged (Figure 7.14).

CEDAR (*CEDRUS LIBANI*)

(The Pine family [Pinaceae].)

> *Sam gathered a pile of the driest fern, and then scrambled up the bank collecting a bundle of twigs and broken wood; the fallen branch of a cedar at the top gave him a good supply. (LotR 4: IV)*

In *The Lord of the Rings*, cedars (*Cedrus libani*) are only mentioned as a component of the aromatic forests of Ithilien, and they may have been restricted to this region in the Third Age. Along with species such as bay (*Laurus nobilis*), cypress (*Cupressus sempervirens*), olive (*Olea europaea*), and terebinth (*Pistacia terebinthus*), their presence in Ithilien indicates clearly that the forests in this region were similar to those growing in the eastern Mediterranean today. This link of Ithilien with the eastern Mediterranean region also fits geographically, and we could picture Ithilien as approximately in the region of the southern Balkan Peninsula and the Greek islands—not that far from the nearest populations of the Mediterranean cedar today, which occur along the southern Mediterranean coast of Turkey. Sam and Frodo encountered cedars soon after entering Ithilien (see introductory quote), and a grove of old cedars grew atop the steep slope above the small lake were they camped. Sam used dry wood from a fallen cedar branch (from a tree in this grove; Figure 7.15) in building the fire on which he cooked a couple of rabbits. Cedars certainly contribute to the beauty of this locality, which almost made Sam and Frodo forget about the danger of orcs. We are called by Tolkien to imagine these massive old trees, with their green to greenish-blue tufted needles; their broad, flat crowns; and beautiful, somewhat drooping, layered, horizontal branches.

Etymology: The English common name cedar is derived from Old English *ceder*, blended in Middle English with Old French *cedre*, both from Latin *cedrus*, which is also the source of the scientific generic name *Cedrus*. *Cedrus* is taken from Greek *kedros* (cedar), of uncertain origin, and probably first used for juniper (*Juniperus* spp.) but now applied to the Mediterranean cedar—the two associated because of the similarity of their aromatic woods. In England, the name always has been applied to cedar, but in North America, more recently, the application of the name again has changed, at least in part, and it is now also used for junipers (*Juniperus*, the red cedars) and related trees such the white cedars (*Chamaecyparis* and *Thuja*) and incense cedar (*Calocedrus*). The genus *Cedrus* is a member of the Pinaceae, whereas *Calocedrus, Chamaecyparis, Juniperus,* and *Thuja* are in the Cupressaceae.

Distribution and Ecology: The distribution of true cedars (species of the genus *Cedrus*) is quite limited. The Himalayan cedar or deodar (*C. deodara*) grows in the western Himalaya from eastern Afghanistan to western Nepal. The only other species in the genus is the Mediterranean cedar (*C. libani*), which has a disjunct distribution (of partially differentiated populations) around the Mediterranean Sea: the Atlas cedar (*C. libani* subsp. *atlantica*) occurs in the Atlas and Rif Mountains of Morocco and Algeria, the Cyprus cedar or kedros (*C. libani* subsp. *brevifolia*) is restricted to Cyprus, and the cedar of Lebanon (*C. libani* subsp. *libani*) occurs in the eastern Mediterranean, in Turkey, Syria, and Lebanon. Sadly, due to unsustainable harvest and poor environmental practices only a few groves remain in Syria and Lebanon. The fossil record indicates that Mediterranean cypress once occurred in southern Europe but was eliminated from that region during the Pleistocene, and the present fragmented distribution of the genus likely relates to climatic changes associated with periods of cold, glaciated conditions alternating with warmer periods. Cedars usually grow in montane forests and are frequently a dominant and conspicuous tree in such groves. They can be confused with larches (*Larix* spp.), but the latter have deciduous (instead of evergreen) needles and much smaller cones that open to release the seeds (instead of fragmenting).

Economic Uses: Cedar trees produce valuable timber, and the aromatic heartwood is also used as a source of incense (and in the production of resins and essential

FIGURE 7.15 *Cedar.*

oils that are antifungal and repel insects). Famously, the wood of the cedar of Lebanon (*C. libani* subsp. *libani*) was used in the temple of Solomon in Jerusalem and also in ships and temples of the Egyptian pharaohs. Cedar populations in the eastern Mediterranean are now much reduced due to millennia of overharvesting and destruction. These trees are also popular ornamentals, and cultivars with unusual foliage colors (blue, golden) and well as dwarf and weeping forms have been developed.

Description (of *Cedrus* spp.): Erect, evergreen trees with sparse crown and horizontal, spreading to drooping branches; the bark flaking and forming a checkered pattern, to vertically furrowed in age. Twigs of two distinct forms, some branches elongated, and these bearing numerous, short or spur branches in their second and subsequent years of growth; wood lacking resin canals (but developing such canals in the rays, in response to wounding). Leaves in tufts of 15–45 on short shoots or borne singly and spiral on first-year long shoots; each leaf simple, linear, and needlelike, ± terete (three- or four-sided in cross-section), with two resin canals, with a single midvein, the apex acute, the base slightly narrowed, attached to a raised peg on the stem, the margin entire; stipules absent. Plants with pollen and seed cones on the same plant. Pollen cones ovoid-cylindric, yellowish, with each pollen-bearing structure flat, bearing two sporangia. Seed-cones ovoid to ellipsoid or cylindrical, solitary at the tips of the short shoots, maturing over two or three seasons, held erectly and breaking up at maturity, shedding both seeds and cone scales and leaving behind a narrowly conical cone stalk; scales deciduous, flat, strongly obovate and wedge-shaped, each bearing two seeds, and free from the associated tiny bracts (which are hidden when cone is intact). Seeds strongly winged (Figure 7.15).

CLOVERS (*TRIFOLIUM* SPP.)

(In the Legume family [Fabaceae or Leguminosae].)

> *It was the middle of the afternoon before they noticed that great patches of flowers had begun to spring up, all the same kinds growing together as if they had been planted. Especially there was clover, waving patches of cockscomb clover, and purple clover, and wide stretches of short white sweet honey-smelling clover. There was a buzzing and a whirring and a droning in the air. Bees were busy everywhere. (Hobbit VII)*

As Gandalf, Bilbo, and the dwarves approached the homestead of Beorn, a mysterious "very great person," they noticed pastures dominated by flowers, especially various clovers, and heard the buzzing of many bees. These were "bigger than hornets" and had bands of yellow on their black abdomens that "shone like fiery gold" (Hobbit VII); they were actively visiting the clover flowers, gathering nectar. Bilbo noticed at least three kinds—purple clover (*Trifolium pratense*), white clover (*T. repens*), and cockscomb clover (probably *T. incarnatum,* usually called scarlet clover). Soon afterward, they came to a belt of ancient oaks (*Quercus*) and then to a high thorn hedge (*Crataegus monogyna*), where Gandalf instructed the dwarves to await his call. Coming to a wooden gate in the hedge, Bilbo and Gandalf could see beyond gardens, a few outbuildings, a long wooden house, and rows of bee hives. Beorn was a skin-changer (and could transform into a huge black bear), and he lived mainly on foods such as bread, butter, cream, twice-baked cakes, honey, nuts, and fruits—and, of course, mead (a fermented beverage made from honey and water). It is not surprising that Beorn grew these particular species of clover; although the flowers of all clovers are pollinated by bees, white clover and cockscomb clover are especially attractive to honeybees and are excellent honey plants (resulting in the production of especially flavorful honey). Purple clover is less frequently visited by honeybees because their fairly short mouthparts cannot easily reach the nectar in its larger flowers; thus, purple clover is more frequently visited by long-tonged bees, such as bumblebees. However, the bees kept by Beorn were especially large, so likely all three clover species were effectively used. Beorn probably also used his clover meadows for fodder, as he kept ponies, horses, cattle, and sheep. Finally, we note that the name Beorn is etymologically ambivalent, taken from Old English (with the meaning of a warrior or man of valor), and such a derivation certainly fits his description in *The Hobbit*. Yet, the Old Norse *björn*—clearly a cognate of the Old English *beorn*—means bear, and thus the Beorn we meet in *The Hobbit* also is ambivalent—taking the form of a brave and strong black-haired man or a fearsome black bear (see also Shippey, 2000). Both the bear and the man, however, liked honey, which was provided by his bees and clover pastures.

Etymology: The word clover is derived from Middle English *clovere* and Old English *clafre, clæfre* (clover), from Proto-Germanic **klaibron* (clover), of uncertain origin. The scientific name *Trifolium* comes from the Latin *tres* (three) + *folium* (leaf), and the word *folium* is related to the Greek *phyllon* and may be derived from Proto-Indo-European **bhol-yo-* (leaf). The Quenya word for clover may be *camilot*.

Distribution and ecology: Clovers are members of the large genus *Trifolium*, a group of ca. 300 species that is nearly cosmopolitan in distribution but most diverse in temperate regions of the Northern Hemisphere. About 100 species are native or naturalized in North America, and 99 are reported from Europe; the group is less diverse in eastern Asia, with only 13 species in China and 4 in Japan. Species are distinguished on characteristics such as habit (annual to perennial, producing rhizomes or not, plant height), variation in pubescence of various structures, leaf arrangement, leaflet size and shape, stipule form, inflorescence form (including number of flowers), and flower form (especially size relationships of the calyx and corolla, form of the calyx, corolla color and size). Clovers are sun-loving species, often of disturbed habitats, occurring in dry to wet meadows, prairies, grassy places, rocky slopes, thickets, open woodlands, margins of woods and clearings, pastures, fields, lawns, waste ground, and roadsides. The flowers are typical of those of faboid legumes (i.e., members of Fabaceae subf. Faboideae; see description) and are pollinated by various bees. The small fruits are few-seeded (and may remain closed or open to release the seeds); in some species, dispersal may occur by external transport (on the fur of animals or the clothing, shoes, etc. of humans) as a result of the adherence the fruits and associated calyces, and these have the potential of being carried long distances. Additionally, other species show modifications of the calyx for wind dispersal. The individual seeds show no obvious adaptations for dispersal and likely many are dispersed quite near the parent plant or moved only short distances by wind or rain-wash. Of course, many species are widely grown as forage plants and are purposefully transported by humans, although unintentional transport certainly also is common (e.g., fruits or seeds carried on farm equipment, bales of hay, etc.).

Economic uses: Several species of clover are important as forage crops for livestock (for livestock grazing, fodder, hay, etc.), and they have the additional advantage of being able to fix atmospheric nitrogen, thus enriching the soil. The economically most important species, at least in Europe and North America, are white clover (*T. repens*), red or purple clover (*T. pratense*), scarlet clover (*T. incarnatum*), and Alsike clover (*T. hybridum*). All these species are native to Europe (and in some also associated regions such as western Asia or northern Africa), but are now widely grown and naturalized. Clovers are important honey plants; their flowers produce nectar and are pollinated by various short- to long-tongued bees: white clover and scarlet clover are especially valuable in this regard. Some species are used as survival food, but they are hard to digest raw and, depending on the species, may contain isoflavones (affecting mammal reproductive systems), coumarins (reducing coagulation of blood), or cyanogenic glucosides (causing cyanide poisoning). They are, therefore, occasionally used medicinally and should only be eaten in very small quantities.

Description: Annual to perennial herbs, with or without simple hairs; stems erect to prostrate, sometimes rhizomatous. Leaves alternate, spirally arranged, compound, with usually three leaflets, these ovate to obovate, narrow to quite broad, with pinnate venation, with veins entering the teeth, the apex acute to rounded or emarginate, the base cuneate to acute, the margin serrate to nearly entire; stipules present, entire, serrate or dissected, fused to the petiole. Inflorescences terminal or axillary, forming heads, short racemes or spikes of small flowers. Flowers bilaterally symmetrical, bisexual, held horizontally, with five fused, narrow-lobed sepals, forming a symmetrical or asymmetrical calyx, and with uppermost petal differentiated in size and shape and forming a banner (or standard) larger than the others and positioned on the outside in bud; two lateral, clawed, wing petals, and the two lowermost petals fused together, forming a straight, clawed keel, these usually all the same color (although flower color many change with age) and white to red, purple, or yellow, with 10 stamens, 9 of which are fused together by their filaments, and with a single carpel, the ovary superior, and the single style straight to slightly curved. Fruit a small legume or indehiscent pod, opening by both sutures, just the ventral suture, or indehiscent, with 1–8 small seeds.

COFFEE (*COFFEA ARABICA* AND *C. CANEPHORA*)

(The Coffee or Madder family [Rubiaceae].)

> *Some called for ale, and some for porter, and one for coffee, and all of them for cakes; so the hobbit [Bilbo] was kept very busy for a while. (Hobbit I)*

Hobbits and the other races of Middle-earth seem to much prefer beer (as seen in the quote) or other alcoholic beverages, such as wine or cider, to either coffee or tea. This is historically appropriate, since neither coffee nor tea became popular in Europe until the seventeenth century. Either of these caffeinated beverages, therefore, would be very anachronistic in the Middle-earth of the Third Age, which according to J. R. R. Tolkien should be considered as the European world at some point in the distant past (see Letters: No. 165). In contrast, beer and wine have been consumed for thousands of years in Europe. Coffee and tea originated well outside Europe—coffee in the Middle East and tea in China. Coffee is not mentioned at all in *The Lord of the Rings* or *Silmarillion*, and the word occurs only twice in *The Hobbit* (both in chapter 1, where coffee was one of the beverages called for by the dwarves; Figure 7.16). It is rather surprising that Tolkien did not remove the references to coffee when he removed the reference to tomatoes from *The Hobbit* (i.e., third edition, in 1966, Ballantine; see also Anderson, *The Annotated Hobbit*, p. 41). In fact, we now know (thanks to the detailed work of John Rateliff, 2007, revealed in *The History of the Hobbit, Part Two: Return to Bag-End*, see pp. 774–775) that in the abandoned "fifth phase" revisions of 1960, which were essentially a complete rewriting of *The Hobbit* in order to make that book more like *The Lord of the Rings* in tone, Tolkien did replace the references to coffee with cider (and, of course, also replaced tomatoes with pickles). However, he eventually decided that such an extensive rewrite of *The Hobbit* was inappropriate, and, only a few years later, when making the revisions connected with the third edition, he retained the mention of coffee in chapter 1 but still replaced the word tomatoes with pickles. Tolkien was working from memory, as the 1960 rewrites were unavailable when he worked on the revisions for the 1966 edition, so perhaps he merely forgot that he had decided to remove the mention of coffee. Yet this seems unlikely because he did delete tomatoes. More likely, he considered the presence of coffee in Middle-earth as representing an independent, and earlier, introduction from the mountains of northeastern Africa—a plant brought into lands controlled by Gondor as a result of its trade with Haradwaith and Khand. Transport from the Americas, as would have been necessary in the case of tomatoes (which are South American plants) would have been much more difficult within the imaginative landscape of the Third Age. Additionally, he may have thought that coffee (in contrast to the tomato) was more in keeping with the essentially English culture of the Shire.

Etymology: The English word "coffee" is derived from the Italian *caffe* (or perhaps the Dutch *koffie*) from the Turkish *kahveh* and the Arabic *quhwah* (i.e., coffee, the drink, not the plant, which is called *bunn*). The scientific Latin name *Coffea* is likewise ultimately derived from Arabic.

Distribution and Ecology: Arabian coffee (*Coffea arabica*) is native to the mountains of Ethiopia but was early taken to the Arabian Peninsula and is now widely cultivated in ecologically suitable tropical regions. The less important Congo coffee (*C. canephora*) grows in west-tropical Africa. The genus *Coffea* includes ca. 100 species that occur in tropical Africa, Madagascar, and southeastern Asia. Coffee is a shrub or small tree of the understories of moist, often montane forests. *Coffea arabica* (a tetraploid) has smaller leaves and usually larger fruits than *C. canephora* (a diploid). The fragrant white flowers would appear to be insect-pollinated, but selfing is frequent in *C. arabica* and wind pollination in *C. canephora*.

Economic Uses: Coffee and tea are now the most important caffeine-containing beverages in Europe and the Americas. Coffee, a drink made from the roasted, ground, and brewed seeds, originated in the Arabian Peninsula and from there spread throughout the Middle East. Coffee drinking did not become common in England until the late seventeenth century. The drink is stimulating mainly due to the presence of the alkaloids caffeine and theobromine, which in the plant serve to deter herbivory.

Description: Evergreen shrubs or small trees, glabrous, with branches of two kinds, some erect and others held more or less horizontally. Leaves opposite and decussate (clearly so on the erect branches, but often appearing two-ranked on the horizontal branches), simple, ovate to elliptic, pinnately veined, with the secondary veins

FIGURE 7.16 *Coffee.*

forming loops, with tiny cavities (i.e., *domatia*, minute homes for mites) on the lower surface (at junction of the secondary veins and midvein), the apex acuminate, the base acute or cuneate to rounded, the margin entire, sometimes undulate, with short petioles; stipules present and interpetiolar (i.e., with the stipules of the two opposite leaves fused), forming a triangular structure on the stem, between the two opposite leaves. Inflorescences axillary, in small clusters on the horizontal branches. Flowers radially symmetrical, bisexual, the sepals five, fused, quite small, the petals five, fused into an elongated tube, but with well-developed and spreading lobes; the stamens five, fused to the corolla tube and with short filaments and elongated anthers, and two fused carpels; the style single, elongated, with two slender stigmas; and the ovary inferior. Fruit a drupe, ovoid to ellipsoid, bright red, with two pits, and seeds (Figure 7.16).

CRESS, WATERCRESS (*NASTURTIUM OFFICINALE*, AND RELATED SPECIES)

(In the Mustard family [Brassicaceae or Cruciferae].)

> *The falling stream vanished into a deep growth of cresses and water-plants, and they could hear it tinkling away in green tunnels, down long gentle slopes towards the fens of Entwash Vale far away. (LotR 3: II)*

Aragorn, Gimli, and Legolas encountered watercress, along with other unnamed wetland species, in a stream at the edge of the grassy plains of Rohan as they followed the track of the orcs that had captured Merry and Pippin (Figure 7.17). This is the only occurrence of this widespread European species in J. R. R. Tolkien's legendarium. Although the name "cress" can be applied to species of several genera of the mustard family (Brassicaceae), such as *Arabis, Cardamine, Lepidium, Rorippa,* there is little doubt as to the identity of the plant mentioned here because the habitat described is so characteristic of the watercress (*Nasturtium officinale*). This is an important edible leafy herb that is native to Europe, western Asia, and northern Africa, and it is one of the oldest used of our leafy vegetables (in salads and as a garnish for meats). Its presence in Middle-earth is thus not surprising, and it was probably as widespread in the Third Age as it is now.

Etymology: The English common name "cress" is applied to several species of pungent herbs of the mustard family (Brassicaceae), and watercress, an aquatic and/or wetland species, is one of these. The name "cress" is derived from Old English *cresse*, from Proto-Germanic **krasjon-*, possibly from the Proto-Indo-European root **gras-* (to devour), a reference to the edibility of this long-used group of European leafy vegetables. The scientific Latin name is *Nasturtium* and is derived from *nasus* (nose) + *tortus* (twisted), a reference to the pungent smell of their leaves and shoots, which contain mustard oils.

Distribution and Ecology: The watercresses (i.e., the five species of *Nasturtium*) grow natively in North America, Europe, northern Africa (Morocco), and western Asia. They are wetland and/or aquatic herbs, typically of springs, streams, ditches, lake and pond margins, swamps, wet meadows, and marshes—especially in calcareous situations. *Nasturtium officinale* and *N. microphyllum*, which are native to Europe and western Asia, have been introduced and are widely naturalized in North America, but

FIGURE 7.17 *Cress, watercress.*

two other North American species are native; *N. officinale* is also widely naturalized elsewhere, for example, in tropical and southern Africa, South America, eastern Asia, Australia, and New Zealand. Their small flowers are insect-pollinated, and the seeds likely are water-dispersed. The tissues contain glucosinolates, which are converted to mustard oils by the enzyme myrosinase (when the foliage is damaged or chewed), and the presence of these pungent compounds deters herbivores.

Economic Uses: Watercress is widely grown as a leafy vegetable; both leaves and young shoots are used and are popular because of their peppery flavor. The plants also have long been used medicinally.

Description: Perennial, rhizomatous herbs of aquatic and wetland habitats, with or without simple hairs; stems erect to prostrate and rooting at the nodes. Leaves alternate and spirally arranged, borne along the stem, simple (especially in deeply submerged plants) to pinnately compound (when emergent, with 3–17 leaflets), with pinnate venation, the leaflets orbicular to narrowly to broadly ovate, oblong, or linear, with apex rounded, obtuse to acute, the base decurrent to nearly cordate, the margin entire or undulate to rarely dentate, the leaves petiolate; stipules absent. Inflorescences terminal or axillary racemes. Flowers bisexual, radially symmetrical, held erectly and not associated with bracts, with four green sepals; four usually white, obovate petals, six stamens (and two of these shorter than the other four); the carpels two, fused, the ovary superior, with a single short style. Fruit elongated, with two valves separating from the central portion (i.e., a thickened rim [the replum]) surrounding a thin layer of tissue (the septum) and releasing the several to numerous seeds (Figure 7.17).

CYANOBACTERIA, BLUE-GREEN BACTERIA

(Various genera of Cyanobacteria, a subgroup of the Bacteria.)

> *The only green was the scum of livid weed on the dark greasy surfaces of the sullen waters. (LotR 4: II)*

The quote, from the chapter "The Passage of the Marshes," describes the characteristic green scum seen in waters polluted by excess nutrients, an all too common condition in degraded freshwater habitats near agricultural or urban areas. As Frodo and Sam are being led through the Dead Marshes by Gollum, they are surrounded by "dead grasses and rotting reeds" (LotR 4: II), and nutrients, especially phosphorous, have been released in this decomposition. With so little uptake by the emergent aquatic plants (whose vigor has been damaged by the harmful environmental actions of Sauron), nutrient levels in the water have increased, leading to a population explosion of various blue-green bacteria and producing a bright green to blue-green frothy scum on the water's surface. Such mats have a bad odor, as noticed by all three travelers, and they can cause human health problems (e.g., skin irritation, gastrointestinal discomfort, and perhaps even neurological disorders) due to the toxins produced by such bacteria. Frodo fell into such polluted water and nearly drowned (Figure 7.18). Sauron's activities certainly follow those of Morgoth—and in the Spring of Arda the Valar knew that he was again at work in the world when "Green things fell sick and rotted, and rivers were choked with weeds and slime" (SILM 1). It should make us stop and think. Why have we—like Morgoth and Sauron—so damaged the once beautiful wetlands near our towns and farms?

Etymology: Under certain conditions, cyanobacteria form a floating film on the surface of water, and this is referred to as "scum" in the introductory quote. The word "scum" is derived from Middle English *scum*, which was derived from either Old English **scum* (foam) or Middle Dutch *schume* (foam, froth), from Proto-Germanic **skuma-*, perhaps taken from a Proto-Indo-European root **(s)keu-* (to cover, conceal). The scientific name cyanobacterium is derived from Greek *kyanos* (blue) and *bacterion* (a small rod) in reference to their frequent bluish coloration (due to the presence of the pigment phycocyanin) and the rod-shape characteristic of many bacteria.

Distribution and Ecology: The Cyanobacteria or blue-green bacteria (formerly called blue-green algae) are cosmopolitan in distribution, and their many species live in an exceptionally wide array of terrestrial and aquatic habitats, from deserts to freshwater or marine environments and from polar to tropical regions. Some are closely associated with other organisms, such as those that live in sponges, together with fungi (forming lichens), with the floating aquatic fern *Azolla*, in the coral-like roots of cycads, or even on the hair of sloths. Most amazingly, the chloroplasts (i.e., photosynthetic organelles surrounded by two membranes and containing chlorophyll a or chlorophyll a and b) within the cells of plants are descended from a free-living cyanobacterial ancestor that long ago became an obligate endosymbiont. Without chloroplasts, plants could not take up carbon dioxide from the atmosphere and fix that carbon using the energy of sunlight, forming glucose and other organic molecules, with oxygen gas given off as a waste product of these reactions. Thus, we can thank free-living cyanobacteria and the chloroplasts (= endosymbiotic cyanobacteria) within plant cells for the oxygen that we breathe (and the organic molecules that we eat). In fact, the biochemical activities of this ancient group have dramatically altered the composition of the Earth's atmosphere, which originally did not contain oxygen. They are also important because of their ability to take nitrogen gas from the atmosphere, converting it into a chemical form that can be used by plants. However, on the negative side, many cyanobacteria produce various toxins, causing illness and death if ingested, and they frequently become ecologically dominant in waters polluted by excess nutrients. Species of *Anabaena*, *Aphanizomenon*, and *Microcystis* are abundant in such waters.

Economic Uses: Blue-green bacteria are occasionally used as food sources (e.g., *Aphanizomenon flos-aquae* and *Arthrospira platensis*), but others are important as sources of environmental toxins that may kill cells, cause respiratory or gastrointestinal problems, damage to the liver, or affect the function of the nervous system. β-methylamino-L-alanine (or BMAA) is a neurotoxin produced by many cyanobacteria, and its ingestion is suspected as a contributing factor in the development of neurodegenerative diseases such as amyotrophic lateral sclerosis (ALS), Parkinson's disease, and Alzheimer's disease. Finally, populations of various cyanobacteria can dramatically increase in waters polluted by excess nutrients, and such bacterial blooms, resulting in the buildup of a thick layer of scum on the water's surface, harm fish, other aquatic organisms, and plants.

Description: Cyanobacteria exist as single cells, but may also form colonies consisting of branched or unbranched filaments, flat sheets of cells, or globose masses. Their cells are prokaryotic in their level of organization; that is, they have no membrane-bound nucleus and have their DNA organized as a single circular strand (neither forming a complex with proteins nor packaged into chromosomes). They are covered with a thick, gelatinous cell wall and lack flagella, but they can glide over surfaces, and filamentous species can generate a wavy motion. They contain a well-developed internal

FIGURE 7.18 *Cyanobacteria, blue-green bacteria.*

membrane system (i.e., photosynthetic lamellae) and, unlike other bacteria, possess chlorophyll a and are therefore photosynthetic, fixing atmospheric carbon dioxide and releasing oxygen in the production of the sugar glucose; many species also fix atmospheric nitrogen. Some species can float due to the presence of gas vesicles (Figure 7.18).

DAFFODILS (*NARCISSUS* SPP.)

(The Amaryllis or Daffodil family [Amaryllidaceae].)

> *Beautiful she is, sir! Lovely! Sometimes like a great tree in flower, sometimes like a white daffadowndilly, small and slender like. (LotR 4: V)*

Sam, in his attempt to describe Galadriel to Faramir, used a series of contrasts, and, in this quote, we see him contrast a great tree in flower (perhaps a mallorn, since he had visited Lothlórien, and those were the largest trees he had seen) with a lovely spring-blooming herb (the daffodil). It is unclear precisely which species of *Narcissus* he had in mind, but the fact that it was white-flowered rules out the numerous yellow-flowered species, and his statement well matches the paperwhite (*N. papyraceus*). Although "daffodil" is the most widespread common name (in English) for this plant, there are many more obscure names of restrictive occurrence in the British Isles. Here, Sam uses one of these—"daffadowndilly"—which is quite similar to "daffydowndilly," recorded as used in Somerset, and "daffydilly," used in Northamptonshire. Sam stresses the small stature of this flower and its slender, flower-bearing scape. The flower is frequently associated with the liturgical season of Lent, as suggested in the common names *lent lily, lents, lenty cups,* and *lent pitchers.* This association with Lent resonates etymologically since daffodils are emblematic of spring, and the word "Lent" is derived from Old English *lencten* (meaning spring, when the days lengthen). Lent is a season of preparation and self-denial—reminding us of Galadriel's denial of the ring. In that moment of decision, as she raised up her arm, she alone was illuminated and seemed "tall beyond measurement, and beautiful beyond enduring, terrible and worshipful" (LotR 2: VII)—as if she were a great tree—but then she lowered her hand and laughed. She had relinquished her pride and desire for power and was now diminished, merely an elf-woman in a simple white gown—a slender white-flowered daffodil.

Etymology: The English common name "daffodil" is derived from Middle English *affodill,* from Medieval Latin *affodillus,* from Latin *asphodelus,* taken from Greek *asphodelos,* which is of unknown origin. Note that the meaning of the name has shifted, since the name daffodil is now used for members of the genus *Narcissus* (of the Amaryllidaceae), while *Asphodelus* is a genus of Asphodelaceae. However, both genera are petaloid monocots and are members of the order Asparagales. The scientific name *Narcissus* is derived from Greek *narkissos* (an ancient name for a kind of petaloid monocot), which is perhaps derived from a word taken from another language of the Aegean region. In Greek mythology *Narkissos* is a youth who fell in love with his own reflection and was changed into a flower. The name of this flower in Sindarin is *maloglin,* while in Quenya it may be *kankale-malina.*

Distribution and Ecology: *Narcissus* (a genus of ca. 30 species) is distributed throughout the Mediterranean region, but several species are naturalized in Eastern Asia and North America. Species are differentiated by characters such as the shape of the leaves in cross-section, number of flowers per inflorescence, shape and size of the corona (i.e., a petaloid outgrowth of the perianth), and the color of the tepals and corona. Daffodils are plants of moist woods and various open, herb-dominated habitats (e.g., pastures, meadows, rocky hillsides, marshes). Their flowers are pollinated by various insects (bees, butterflies, flies, hawk-moths). The seeds are dispersed by wind.

Economic Uses: Daffodils are very popular, important garden ornamentals because of their beautiful, spring-blooming flowers, and a vast array of hybrids and cultivars exists in the horticultural trade. They are poisonous because of the presence of various "amaryllis" alkaloids.

Description: Perennial herbs, arising from bulbs; roots contractile, pulling the bulb into the soil. Leaves alternate, in two ranks, arising from the bulb; simple, flat to partially terete, fleshy, linear and elongate, with parallel veins, the apex rounded to acute, the base broad, sheathing, the margin entire, glabrous (and often glaucous); stipules absent. Inflorescences terminal, with 2–20 flowers borne in an umbel atop an elongated scape (stem with a single elongate internode) or sometimes reduced to a solitary flower. Flowers radially symmetrical, erect to held horizontally or somewhat nodding, large, with a perianth of six yellow or white tepals, these fused proximally, but forming separate, well-developed, erect to reflexed lobes distally; the apex of the perianth tube with a cup-like to trumpetlike corona, with its margins often frilled, with six stamens, their filaments arising from the perianth tube, and with three fused carpels; the style single, with a three-lobed stigma; ovary inferior. Fruit a capsule; seeds black, globose, often with a fleshy aril.

DAISY (*BELLIS PERENNIS*)

(The Aster or Compositae family [Asteraceae or Compositae].)

"Ssss, sss, my preciouss," he said. "Sun on the daisies it means, it does." (Hobbit V)

The best known occurrence of the daisy (i.e., *Bellis perennis*) in J. R. R. Tolkien's legendarium is its role in the chapter "Riddles in the Dark" (see quote), where "sun on the daisies" is Gollum's correct guessing of Bilbo's riddle in which "An eye in a blue face" (i.e., the sun) saw "an eye in a green face" (i.e., daisies growing in the grass). In this riddle, Tolkien is cleverly expressing the etymology of the word daisy, which is derived from the Old English for "day's eye" (see "Etymology" and also discussion in D. A. Anderson's 2002 *The Annotated Hobbit*). The connection of this species with the sun arises because its flower heads open at dawn and close at dusk, and, in addition, the sun itself sometimes has been compared to an eye. In this riddle, we see Bilbo's love of the sunlit world and of the beauty found even in the commonplace—in a plant such as the daisy, a common weed of pastures, lawns, and roadsides, responding to the sun above. This sunlit world, of course, stands in stark contrast to the dark tunnels in which he was lost, trapped with Gollum, a strange creature who surprisingly also had memories of a sunlit life. A similar expression is used for this plant in the tale of Túrin Turambar (in *The Lay of the Children of Húrin*). In these verses, we see Túrin's friend Beleg, buried under the corpses of Túrin's men, after their betrayal. While Túrin had been carried away by the orc band, Beleg had lain unconscious as the sun rose "and the eye of day was opened wide" (Lays I: line 718). Finally, as related in *The Lord of the Rings*, Sam Gamgee, in an attempt to describe Galadriel to Faramir, used a series of contrasts, saying that she was as "Proud and far-off as a snow-mountain, and as merry as any lass I ever saw with daisies in her hair" (LotR 4: V). Here, the daisy, a common weed, provides a connection with the humble and innocent, yet authentic, joys of life, telling us that Galadriel, despite her deep knowledge and great power (focused by her ring, Nenya), was well grounded, like the daisy, having an accurate (and humble) assessment of her place in the world. Daisies are clearly growing in the grass in the right foreground of Tolkien's painting *Rivendell* (see fig. 108 in W. G. Hammond and C. Scull's 1995 *J. R. R. Tolkien Artist and Illustrator*).

Etymology: The English common name "daisy" is derived from the Old English *dæges eage* (meaning day's eye) because the floral heads open at dawn and close at dusk; Chaucer called it the "eye of the day," while in Medieval Latin it was possibly known as *solis oculus* (sun's eye), although this term could also have been applied to other similar species. The word "day" is derived from Old English *dæg*, from Proto-Germanic **dagaz* (day), from Proto-Indo-European **agh-* (a day), while the word eye is derived from Old English *ege* (Mercian) and *eage* (West Saxon), from Proto-Germanic **augon* (eye), possibly derived from Proto-Indo-European **okw-* (to see). The scientific, generic name *Bellis* is from the Latin *bellus* (meaning pretty, beautiful). The Sindarin name of this flower may be *eirien*.

Distribution and Ecology: The English daisy (*Bellis perennis*) is a widespread species of Europe and adjacent western Asia, but it has now been introduced very widely in temperate regions around the world (including North America). It is rather weedy, occurring in open grassy habitats, especially when disturbed, such as lawns and roadsides. It is one of ca. 15 species of the genus *Bellis*. The English daisy has showy floral heads that have a typical daisy form, with yellow disk flowers and white ray flowers; these are pollinated by various insects, while the small fruits are dispersed by wind or carried externally by animals or humans.

Economic Uses: The English daisy (*B. perennis*) is a lawn weed, but horticultural forms exist and are used as ornamental herbs, especially the doubled cultivars, which have numerous white to red ray flowers. The plant is also occasionally used medicinally, as a tea taken for coughs or oral ulcers or applied externally to wounds or bruises.

Description (of *B. perennis*): Perennial rhizomatous herbs, with simple, nonglandular hairs; stems erect, unbranched. Leaves alternate, spirally arranged in a basal rosette, simple, obovate, pinnately veined, but only midvein prominent; the apex obtuse to rounded; the base decurrent; the margin crenate-serrate, with stiff hairs, with a winged petiole; stipules absent. Inflorescences terminal, borne singly atop the scapelike stem, the flowers densely clustered in each head with conic receptacle bearing 95 to more than 170 flowers, the peripheral ones numbering ca. 35–90; fertile (i.e., carpellate), white (on the upper surface) but often pink- to purple-tinged on the lower surface, bilateral, with an elongated corolla, and these called ray flowers (and closing at night). The central ones, numbering 60 to more than 80, pale yellow, radially symmetrical, with a narrowly funnel-shaped corolla formed by five fused petals and with five, erect to incurved triangular lobes, and these called disk flowers; both the disk and ray flowers surrounded by numerous narrowly ovate bracts, each with a more or less obtuse apex. Each disk flower with the sepals lacking, with five fused petals, five stamens, these arising from the corolla tube and, with their anthers, forming a cylinder around the style, and with two fused carpels, the style elongate, with two style branches, and on each of these the stigmatic region restricted to two lines, and the ovary inferior. Fruit obovoid and slightly compressed, marginally two-ribbed, brown achene.

DANDELIONS (*TARAXACUM OFFICINALE, T. ERYTHROSPERMUM*, AND RELATIVES)

(The Aster or Compositae family [Asteraceae or Compositae].)

Goldberry was there in a lady-smock
blowing away a dandelion clock. *(TATB, Appendix II)*

Dandelions occur only once in J. R. R. Tolkien's legendarium, in the poem *Once upon a Time*, which was first published in *Winter's Tales for Children 1* (edited by C. Hillier) and more recently included as Appendix II in *The Adventures of Tom Bombadil* (2014, as edited by C. Scull and W. G. Hammond). This poem nicely expresses Goldberry's simple enjoyment of the natural world, especially as represented by the world of flora. In the quote we see her blowing away the small plumed fruits of a dandelion head (i.e., a cluster of achenes, each attached to a parachutelike set of radiating bristles by a long stalk). Tolkien's use of the phrase "dandelion clock" refers to another of the common names of this plant, "tell-time," and relates to the childlike custom of telling time by blowing away the wind-dispersed fruits, with the number of puffs representing the number of hours. Other plants mentioned in this short poem include (in order of appearance) hawthorn trees (*Crataegus*, with their snowlike blossoms dropping in May), buttercups (*Ranunculus*), daisies (*Bellis*, i.e., "earth-stars with their steady eyes watching the Sun"), wild roses (*Rosa*), waterlilies (*Nymphaea*), and grasses (Poaceae). In addition, the phrase "lady-smock" may be an allusion to yet another wildflower, the lady's smock (*Cardamine pratensis*, a member of the mustard family, Brassicaceae).

Etymology: The word "dandelion" is derived from the Middle French *dent de lion* (lion's tooth) and earlier from Latin *dens leonis*, based on the resemblance of the irregularly and sharply toothed leaves of this plant to the teeth in a lion's jaw. The scientific name *Taraxacum* is derived from Medieval Latin *taraxacon*, which was based on the Arabic *tarakhshaqun*, which was earlier taken from Persian *talkh chakok* (a bitter herb).

Distribution and Ecology: Dandelions are of nearly worldwide distribution and comprise the genus *Taraxacum*, a group of perhaps 60 species, although many more (ca. 2000) have been recognized; their classification—especially the determination of species limits—is extremely confused because of polyploidy (i.e., genome duplication), hybridization, and asexual reproduction. Dandelions grow in a variety of open habitats, and several species are extremely weedy, occurring in lawns, roadsides, pastures, waste grounds, and disturbed moist habitats. *Taraxacum officinale* (common dandelion) can be distinguished from *T. erythrospermum* (red-seeded dandelion) by its usually shallowly to deeply lobed or toothed leaves (vs. deeply lobed or dissected leaves) and olive-brown to straw-colored or gray achenes (vs. brick red to reddish brown or reddish purple achenes). The two species are native to Eurasia but are widely naturalized. The flower heads of dandelions attract various insects, and the fruits are associated with a highly modified calyx, forming a radiating tuft of bristles and leading to dispersal by wind.

Economic Uses: The leaves of species of *Taraxacum* (especially the weedy *T. officinale* and *T. erythrospermum*), although bitter, are occasionally eaten (as greens), and the flower heads can be used to make wine. These two species are widespread weeds and much money is spent in ridding them from lawns. *Taraxacum bicorne* (Russian dandelion) is cultivated as a rubber source. Several species are used medicinally (as diuretics).

Description: Perennial herbs, with milky sap, with simple, nonglandular hairs, and from a taproot; stems erect, unbranched. Leaves alternate, spirally arranged, in a basal rosette, simple, pinnately veined, oblong to obovate, the apex obtuse or acute to acuminate, the base attenuate to narrowly cuneate, the margin shallowly to deeply lobed or cut, or merely toothed, the lobes (when present) broadly to very narrowly triangular, with acute to long-acuminate apices, with a more or less winged petiole; stipules absent. Inflorescences terminal, borne singly atop the scapelike stem; the flowers densely clustered in each head, with a flattened-globose receptacle bearing 40 to more than 100 flowers that are surrounded by several linear to narrowly triangular bracts. The flowers all alike, bisexual, bilaterally symmetrical (i.e., ligulate), with a highly modified calyx of numerous, radiating, minutely barbed bristles (and parachutelike, in mature fruit, forming a pappus), with a yellow corolla with a basal tubelike portion and a distal tonguelike structure extending outward from one side of the flower and ending in five minute teeth, with five stamens, these arising from the corolla tube, and with their anthers forming a cylinder around the style, and with two fused carpels, the style elongate, with two style branches, the inner surface on each of these stigmatic, and the ovary inferior. Fruit narrowly obovoid, minutely ribbed achenes, the fruit body separated from the parachutelike pappus bristles by an elongate beak.

EBONY (*DIOSPYROS* SPP.)

(The Ebony family [Ebenaceae].)

> *[H]is bow was made of dragon-horn,*
> *his arrows shorn of ebony.* *(LotR 2: I)*

Ebony only occurs once in the writings of J. R. R. Tolkien's legendarium, in which this hard black wood is merely a part of a long and quite fantastical description of the gear of the great mariner Eärendil, who was the father of Elrond and had sailed to Valinor with the help of Elwing, his wife, who had brought him the Silmaril. There, he presented to the Valar the prayers of both elves and humans who in Middle-earth still suffered under the overwhelming power of Morgoth. Bilbo described Eärendil's arrows as cut from ebony in the long poem (of his own composition) that he recited in the Hall of Fire, and certainly this unusual and expensive wood fits with the imagery of the poem. Readers of *The Lord of the Rings* may easily miss the force of Aragorn's comment that he thought it rather cheeky of Bilbo to compose verses about Eärendil in the house of Elrond, unless they clearly remember Aragorn's brief telling of this story when he and the hobbits camped at Weathertop! The full tale, of course, is related in *The Silmarillion*. What is the source of this dark wood? The wood of *Diospyros lotus* (Caucasian persimmon) is light colored, so although this species occurs in Europe, it is not likely the source of the "arrows shorn of ebony." The source must have been from another species, perhaps *Diospyros ebenum*, the ebony of commerce, which is today native to India and Sri Lanka. The wood, therefore, would not have been common in the Middle-earth of the Third Age, which likely is precisely why this word was chosen by Bilbo as he composed the poem.

Etymology: The word "ebony" is perhaps an extended form of Middle English *ebon*, or derived from *hebenyf*, which perhaps is a Middle English misreading of Latin *hebeninus* (of ebony), from Latin *ebenus*, and from Greek *ebenos* (ebony), which probably was derived from an Egyptian or other Semitic source. The scientific Latin name *Diospyros* is based on the Greek *dios* + *pyros* and means something like "divine fruit" although a literal translation would be "Zeus-grain," an ancient fruit name of uncertain application that was taken up for these plants by Linnaeus.

Distribution and Ecology: Species of *Diospyros* (ebony, the persimmons, ca. 480 species) are distributed pantropically, with a few species extending into temperate regions in eastern North America, southern South America, southern Europe, and eastern Asia. Two species occur natively in North America (*D. texana, D. virginiana*), one in Europe (*D. lotus*), three in Japan, and ca. 60 in China. The ebony of commerce, *D. ebenum*, is native to India and Sri Lanka. Most are trees of wet to seasonally dry forests at low to moderate elevations. The species are distinguished by numerous characters relating to the shape, size, and pubescence of the leaves, flowers, and fruits. Their urn-shaped flowers are pollinated by various insects, while the berry fruits are eaten (and the seeds dispersed) by birds and mammals.

Economic Uses: Ebony trees (*D. ebenum*, and some related species) are the source of a hard, black wood used for furniture and musical instruments. Some species of *Diospyros*, such as *D. blancoi* (velvet-apple), *D. digyna* (black sapote), *D. kaki* (Japanese persimmon), and *D. virginiana* (persimmon), produce edible fruits (which are sweet at maturity, but very astringent when unripe).

Description: Evergreen or deciduous trees or shrubs, the bark blocklike and furrowed; hairs simple to variously branched, sometimes glandular; all tissues with black or dark-colored naphthaquinones (or related compounds). Leaves alternate, two-ranked, simple, ovate to obovate, pinnately veined, the secondary veins forming loops, the apex and base variable, the margin entire, with or without hairs on the lower surface, but usually with a few scattered nectar glands, with short to long petiole; stipules absent. Inflorescences axillary, in small cymes or reduced to a solitary flower. Flowers radially symmetrical, unisexual, with the staminate flowers on one plant and the carpellate on another, with the 3–7 sepals persistent, fused, and often expanding in size as fruit develops; the 3–7 petals also fused, forming an urn-shaped to bell-shaped corolla, with six to numerous stamens in the staminate flowers (represented by sterile stamens in the carpellate flowers) and usually 3–8 fused carpels, with fused styles, the ovary superior. Fruit a yellow, orange, red, purple, brown, or black berry.

ELANOR

(A species of Anagallis unique to J. R. R. Tolkien's legendarium; the Primrose family [Primulaceae].)

> *There at last when the mallorn-leaves were falling, but spring had not yet come, she laid herself to rest upon Cerin Amroth; and there is her green grave, until the world is changed, and all the days of her life are utterly forgotten by men that come after, and elanor and niphredil bloom no more east of the Sea. (LotR Appendix A I(v))*

Elanor is one of the best-loved flowers of J. R. R. Tolkien's legendarium, and this beautiful starlike flower is associated with the most sacred and beautiful of the elven-dominated localities in Middle-earth—Doriath and Gondolin (in the First Age) and Lothlórien (in the Second and Third Ages)—places where elven horticultural magic had transformed the landscape. For example, we see elanor blooming in the grass between the Fifth and Sixth Gates of Gondolin, where Tuor and Voronwë encountered 200 elven archers (Figure 7.19). How should we envision this special flower, and is it based upon anything growing in our own world? Frodo (and his companions) first saw these flowers on the grassy slopes of Cerin Amroth, noting the "small golden flowers shaped like stars" that grew intermixed with nodding flowers of "white and palest green." Haldor, who was guiding them to the city of the Galadhrim, then said "Here ever bloom the winter flowers in the unfading grass: the yellow *elanor*, and the pale *niphredil*" (LotR 2: VI). Much later, after the quest was ended, Frodo suggested to Sam that he name his eldest daughter Elanor, after "the sun-star," saying—"remember the little golden flower in the grass of Lothlórien" (LotR 6: IX). Thus, a careful reading of *The Lord of the Rings* (and also *The Silmarillion* and *Unfinished Tales*) would lead one to the conclusion that these radially symmetrical (and thus starlike) flowers were golden yellow. This conclusion is even reflected in the translation of the name itself: sun-star (see also "Etymology"). Interestingly, in a letter to Amy Ronald, Tolkien compared this plant to a pimpernel (i.e., a member of the genus *Anagallis*) but "perhaps a little enlarged"—and this connection fits with its description as herbaceous, the starlike form of its flowers, and its occurrence in open habitats. (We thus reject other suggestions as to the identity of this plant; e.g., *Hypoxis* of the Hypoxidaceae or *Crocus* of the Iridaceae, both petaloid monocots.) However, he goes on to say that it bore "sun-golden flowers and star-silver ones on the same plant, and sometimes the two combined." So, although it may often have been golden yellow, we see that, like the scarlet pimpernel (*Anagallis arvensis*, a widespread Eurasian species, also naturalized in North America), elanor had flowers of more than one color—in the scarlet pimpernel they can be salmon, red, or blue, while in elanor they were golden yellow, silver, or a blending of these two colors. It is critical to add that Tolkien, in this letter, stated that this imagined plant (like several others in his legendarium, e.g., niphredil) was "lit by a light that would not be seen ever in a growing plant" in our everyday world (Letters: No. 312). Lothlórien possessed a certain magical quality, and, as Sam explained, the magical power was deep down, resisting an easy understanding, and he felt as if he were inside a song. In Lothlórien, the aesthetic qualities of the natural world are enhanced, and this is evident even in its name, which means "Lórien of the Flower"—evoking the beautiful gardenlike dwelling of the Vala Irmo (Lórien), the master of visions and dreams, who with his spouse Estë brings refreshment to those in Valinor. The grass of Lothlórien is "unfading"—which cannot be said of our lawns—and elanor, seen in the light of Valinor, has significance much beyond that of a pimpernel, however beautiful. Note that the blossoms of elanor are golden yellow and silver—just as the mellyrn of Lothlórien combine gold (as seen in their flowers and winter-held leaves) and silver (as seen in their bark and fruits). Even the torches set up by Gildor and his high-elven companion at Woodhall burned "with lights of gold and silver" (LotR 1: III), and Frodo in the Hall of Fire (in the House of Elrond), under the influence of elven singing, felt "an endless river of swelling gold and silver" flowing over him (LotR 2: I). It seems clear that the golden

FIGURE 7.19 *Elanor.*

and silver flowers of elanor (and the many other references to gold and silver in connection with things elvish) are meant to take us back to the special light of Valinor, once produced by its Two Trees—back to Telperion, with his dark green leaves, silver beneath, and white flowers that produced a "dew" of silver light, and Laurelin, with her lighter green leaves, golden along the margins, and flame yellow flowers that produced a "rain" of golden light. Perhaps the high elves, who longed for the sea and their home in Valinor, found comfort in the gold and silver of elanor, a plant that exemplified their appreciation of the value and beauty of the natural world—and its preservation—which they could achieve only with difficulty in Middle-earth but are fully realized in Valinor. Finally, we note that Cerin Amroth, the center of elven-power in Middle-earth, is where Aragorn and Arwen first pledged to marry while walking barefoot "on the undying grass with elanor and niphredil" (LotR Appendix A1 (v)) (see also niphredil, Figure 7.48). It was on that hill that she rejected the Twilight (i.e., the life of the Eldar), choosing instead a mortal existence in order to be with Aragorn. It is fitting, then, that at the end, after the death of her husband, Arwen returned to Lórien and dwelt alone until winter came. There, upon Cerin Amroth, lying among the elanor and niphredil, she gave up her life.

Etymology: *Elanor* is Sindarin for sun-star (from *el*, star + *anor*, sun). The scientific Latin name *Anagallis* is derived from Greek *ana* (again) and *agallein* (to delight in), which is an allusion to the flowers closing and opening in response to changing environmental conditions (closing when cloudy, opening when sunny). The species of *Anagallis* are usually called pimpernels, a word derived from Anglo-Norman *pimpernele*, which is in turn taken from Old French *pimprenelle*, from Medieval Latin *pipinella*, from Latin *piperinus* (pepperlike), referring to the fruits of these plants, which are shaped like the fruits of the spice plant pepper (*Piper*).

Distribution and Ecology: *Anagallis* is a genus of ca. 20 species that are widely distributed (Eurasia, Africa, and the Americas) and grow in various open habitats. *Elanor*, in the context of J. R. R. Tolkien's legendarium, is of broad, yet quite restricted distribution; it grows on Tol Eressëa and was brought to Númenor at the time of the wedding of Erendis and Aldarion (the sixth king of Númenor). It is best known, however, from the population that grew on the hill Cerin Amroth in Lothlórien (see introductory quote), although in the First Age it also grew in Beleriand—both in Doriath (in open, grassy areas) and Gondolin (in the lawn near the Sixth Gate, as seen by Tuor and Voronwë, as well as on the grass-covered grave of Glorfindel, beside the pass over the mountains that encircled that city). Elanor is always restricted to open, grass-dominated habitats; in both Doriath and Lothlórien it grew intermixed with the white-flowered niphredil, while in Gondolin it grew with evermind (simbelmynë).

Economic Uses: Species of *Anagallis* are occasional ornamental herbs and once were prized as medicinal plants. Elanor is a beautiful ornamental herb and certainly added to the beauty of the places where it grew—but it is much more than that. The lovely elanor is a sacred plant, growing only in those localities that are representative of elven culture in its highest development—places in Middle-earth such as Doriath, Gondolin, and Lothlórien.

Description (of elanor): Perennial herbs; stems erect or ascending, branched; hairs absent (i.e., plant glabrous) and leaves and stems with inconspicuous resin canals. Leaves opposite or occasionally whorled, simple, ovate to elliptic, with obscure pinnate venation, the apex acute to obtuse, the base ± rounded, the margin entire, and sessile (without a petiole); stipules absent. Inflorescences of axillary, solitary flowers. Flowers bisexual, radially symmetrical, held more or less erectly and showy, with five slightly fused, ovate sepals; the petals five, usually yellow, but occasionally silver (and rarely having both colors in a single flower), fused, forming a corolla with a short tube that expands distally forming a rim comprised of five well-developed, ovate lobes with acute apices (and thus starlike), and five stamens, these included and fused to the corolla-tube; the carpels five, fused, the ovary superior, with a single style and tiny stigma. Fruit a globose capsule, with many tiny seeds (Figure 7.19).

ELMS (*ULMUS* SPP.)

(The Elm family [Ulmaceae].)

Alas! For the gulls. No peace shall I have again under beech or under elm. (LotR 5: IX)

Elms are especially common in bottomland and floodplain forests, and we see this habitat preference reflected in their occurrence in Tolkien's Middle-earth (Figure 7.20). Bilbo Baggins observed scattered "oaks and elms, and wide grass lands, and a river running through it all" (Hobbit VII) as he was carried by a large eagle toward the Carrock in the midst of the Anduin River. And in *The Lord of the Rings*, Frodo, Sam, and Pippin saw elm trees in the low ground near the Stock-brook, growing along with oaks (*Quercus*) and ashes (*Fraxinus*); it was autumn and their leaves were "turning yellow" (LotR 1: IV) as is characteristic of elms. Elms also occur in rich deciduous forests, and we see this aspect of their ecology stressed in their association with elves and their love for the forests of Middle-earth. At Pelargir, which is near the delta of the Anduin River, Legolas heard the crying of the gulls, and their calls awakened the sea-longing that lay deep in his heart. As stated in the preceding quote, he realized deep down that he would never again be at peace in tree-tangled Middle-earth, under beech (*Fagus*) and elm (*Ulmus*). These two trees, so characteristic of forests in North Temperate regions, here represent all the forest trees of Middle-earth. The connection of beech and elm with elves is also evident in *The Lay of Leithian* where Thingol, the greatest King of the Eldar during the First Age, is described as

lord of the forest and the fell;
and sharp his sword and high his helm,
the king of beech and oak and elm. (Lays III: lines 70–72)

FIGURE 7.20 *Elms.*

Etymology: Our word "elm" is not different from the Old English *elm*, and these are derived from Proto-Germanic **elmaz*; the scientific Latin name *ulmus* is a cognate. Various Proto-Indo-European words for elm have been hypothesized. The probable early Quenya names for elm-tree are *alalmë* and *lalmë*. The Sindarin form of the word is *lalf, lalorn,* or *lalven* (pl. *lelvin*).

Distribution and Ecology: About 30 species of elms (i.e., the genus *Ulmus*) are distributed widely across the Northern Hemisphere; 10 species (native as well as naturalized) occur in North America, 5 in Europe, 21 in China, and 3 in Japan. The most common and widespread species of North America is the American elm (*Ulmus americana*); the most common European species are the wych elm (*U. glabra*) and field elm (*U. minor*). The Siberian elm (*U. pumila*) is also widespread, growing natively from Central Asia to eastern Siberia, China, India, and Korea, and widely naturalized in North America. Elms grow in deciduous woodlands, floodplain forests, and swamps, as well as in disturbed habitats such as pastures, old fields, and fencerows, and they are frequently planted as street trees. The species sometimes hybridize and are distinguished by characteristics such as presence or absence of corky wings on the twigs; form of the overwintering buds; leaf persistence, size, shape, pubescence, and form of the marginal teeth; blooming time; inflorescence form; number of flowers per cluster; floral characters (such as the pubescence and number of calyx lobes), and the fruits (e.g., size, shape, development of the wing, presence or absence of a marginal hair-fringe). Elms have small, rather inconspicuous flowers that are pollinated by wind, and their flattened, winged fruits are wind dispersed. Elms have been decimated in both Europe and North America by the introduced fungi *Ceratocystis ulmi* and *C. novo-ulmi* (which cause Dutch elm disease and were

introduced from Eastern Asia) carried from tree to tree by bark beetles (*Scolytus* spp.). Sadly, in large parts of their former range, elms are now represented only by root sprouts or young individuals. Attempts to develop disease-resistant cultivars are currently under way.

Economic Uses: Elms produce valuable timber that is rot-resistant under water. Before the advent of the Dutch elm disease, they were widely planted as street trees in urban areas, and the American elm produces a beautiful vase-shaped canopy. The Chinese elm (*U. parvifolia*) is often cultivated for its ornamental, multicolored, flaky bark.

Description: Deciduous or less commonly evergreen trees or shrubs, the bark smooth to more commonly roughened, with anastomosing vertical ridges and furrows, sometimes with flat plates that slough off, the twigs with or without simple hairs, sometimes with corky ridges. Leaves alternate, two-ranked, simple, ovate to obovate, pinnately veined, the secondary veins running into the teeth, the apex acuminate or acute to obtuse, the base asymmetrical, the margin serrate or doubly serrate, the upper surface roughened or smooth, and the lower surface glabrous or variously hairy; stipules present. Inflorescences axillary, short to elongate clusters, or reduced to a solitary flower, arising from branches of the previous season, appearing in spring before the leaves emerge or in the fall with the leaves. Flowers radially symmetrical, bisexual, stalked or not, the sepals (or tepals) 3–9, fused, with or without hairs, the petals absent, the stamens 3–9 and with two fused carpels, the style strongly two-branched, and the stigmas two, elongate along inner margin of each style branch, the ovary superior. Fruit winged or with a fringe of hairs, and flattened; a samara, with a single seed (Figure 7.20).

EVERMIND, SIMBELMYNË, UILOS

(A species of Anemone unique to J. R. R. Tolkien's legendarium; the Buttercup family [Ranunculaceae].)

> *Upon their western sides the grass was white as with a drifted snow: small flowers sprang there like countless stars amid the turf. 'Look!' said Gandalf. "How fair are the bright eyes in the grass! Evermind they are called, simbelmynë in this land of Men, for they blossom in all the seasons of the year, and grow where dead men rest." (LotR 3: VI)*

In the *Nomenclature of the Lord of the Rings* (see *The Lord of the Rings: A Reader's Companion*, by W. G. Hammond and C. Scull, 2005), evermind is said to grow in turf like the pasque-flower (*Anemone pulsatilla*) but to have white flowers like the wood anemone (*A. nemorosa*). Tolkien considered it to be an imagined kind of anemone, and we thus interpret this herbaceous plant as a species of the genus *Anemone* (see also anemones). In addition to its conspicuous white flowers, each with 6–9 radiating tepals (= perianth parts), it likely had much dissected leaves in which each segment was quite narrow; so, except for its flowers, it blended well into the grasses among which it grew. It was restricted to open, grass-dominated habitats and was unusual in blooming throughout the year. At two of the three localities at which this plant is recorded (i.e., just outside Edoras in Rohan [LotR 3: VI: simbelmynë; Figure 7.21] and at Amon Anwar at the Rohan-Gondor boundary [UT 3: II: alfirin]), it occurred on burial mounds, and its beautiful white flowers served as a remembrance of the lives of those on whose tombs it grew—the kings of Rohan, including the mound raised for King Théoden, and Elendil, the founding king of Arnor and Gondor, who with Gil-galad overthrew Sauron on the slopes of Mount Doom. Their white flowers, in such situations, in addition to remembrance, may also have honored their virtuous actions and given hope to those pondering their lives. These two localities are the only ones recorded for the Third Age; however, in the First Age, evermind also grew in Gondolin—the most elven of cities in Beleriand. Tuor, upon entering the Hidden Kingdom with Voronwë, was led to the Fifth Gate, and he saw "beside the way a sward of grass, where like stars bloomed the white flowers of uilos, the Evermind that knows one season and withers not" (UT 1: I). The Sindarin name, *uilos,* for this plant is only used once in Tolkien's legendarium—in this passage from the *Unfinished Tales.* The flowers brought to Tuor the feeling of wonder and a lightening of his heart as he approached the Gate of Silver, which was made of white marble topped with a silver trellis. This gate was adorned with an image of Telperion, the White Tree of Valinor, and thus evermind is associated with the White Tree, as the yellow elanor, which grew beside the Sixth Gate, the Gate of Gold, is connected with Laurelin the Golden, the younger of the two trees. The white starlike flowers of evermind were sacred to the elves, in part because in them they recalled the beauty and power of the White Tree (from which comes the light of the moon). Again, this plant with its snow-white flowers is connected with remembrance—but this time the memories (being elven ones) go back very deeply, to the time before the creation of sun and moon, before the destruction of the two trees by Morgoth, to that mythic time when the gold and silver beams of the Two Trees were mingled.

Etymology: The name *simbelmynë* in the Language of Rohan is translated as evermind (in the initial quote). The language of the Rohirrim is based on Old English, and the name *simbelmynë* comes from the Old English *simbel* (ever) and *myne* (mind, referring to memory), and the name thus means much the same as forget-me-not, although the flower (of course) is quite different. In Tolkien's legendarium, the same plant is also given the Sindarin name *uilos* (meaning snow-white) and *alfirin* (immortal, probably a reference to its blooming in all seasons), although the name *alfirin* was also applied by Legolas (in LotR 5: IX) to a quite different plant (which had golden, bell-shaped flowersl; see treatment of **alfirin**).

Distribution and Ecology: In Tolkien's legendarium, evermind is recorded as growing only at three localities: (1) the burial mounds of the kings of Rohan near Edoras (as referenced in the initial quote), (2) the burial mound of Elendil at Amon Anwar (at the boundary of Gondor and Rohan), and (3) in the grass beside the Fifth Gate of Gondolin (in Beleriand) and also on either side of the road connecting the Fifth and Sixth Gates, where it grew intermixed with elanor.

Economic Uses: The plant is a beautiful ornamental herb with white flowers.

Description: See description under **Anemones**, from which it differs principally in blooming during all seasons of the year and having white flowers (Figure 7.21).

FIGURE 7.21 *Evermind, simbelmynë, ulios.*

FERNS (*ADIANTUM, ASPLENIUM, ATHRYIUM, BLECHNUM, DRYOPTERIS, POLYPODIUM, POLYSTICHUM, THELYPTERIS, WOODSIA*, AND OTHERS)

(Various genera of Polypodiales.)

> *[M]usic in Doriath awoke,*
> *and there beneath the branching oak,*
> *or seated on the beech-leaves brown,*
> *Daeron the dark with ferny crown*
> *played on his pipes with elvish art*
> *unbearable by mortal heart.* (*Lays IV: lines 39–44*)

Most references to ferns in Tolkien's writings can be applied to bracken fern (*Pteridium aquilinum*), which is a large fern (by temperate zone standards) that typically forms dense stands and often is a conspicuous element of forested or open environments, but many other ferns grow in the open to forested environments of the Northern Hemisphere, and these are occasionally mentioned in Tolkien's works, as when Frodo awakes on a bed "of fern and grass" the morning after his meeting with elves in the Shire (LotR 1: IV). And ferns are even represented among the botanical names of the Bree folk, as we see for example in Bill Ferny. But perhaps the closest connection of such woodland ferns (e.g., the genera listed in the heading) to a central element of Tolkien's legendarium is provided in the introductory quote, in which Daeron, a Sindarin elf who was the greatest minstrel east of the Sea (because his music was inspired by his love for Lúthien), is described as wearing a ferny crown. He observed her secret meetings with Beren in the forest of Neldoreth and reported them to Thingol, and later, when Lúthien asked for his help in fleeing Doriath to aid Beren in his quest for a Silmaril, he again betrayed her, leading to her imprisonment in the great beech Hírilorn. The first element of his name, *dae*, means shadow in Sindarin, and this is alluded to in the above-mentioned poem, in which he is described as "the dark" and his name possibly means "shadow-one"—perhaps because he often sat hidden in the shaded understory or shadowy canopy of the trees: "haunting the gloom of tangled trees" (Lays III: line 842), piping while Lúthien danced. Certainly, he watched in hiding with "fiery eyes" while Lúthien and Beren as "two lovers linked in dancing sweet . . . where lonely-dancing maid had been" (Lays III: lines 841, 845, 847). It is fitting that this elf who had spent so much time in the shaded forest understory would adorn his head with nothing colorful, but merely the green fronds of shade-tolerant ferns that grew in such habitats.

Although not frequently mentioned in Tolkien's writings, a diversity of ferns certainly must have occurred in the swamps, meadows, and forests of Middle-earth, just as they are of frequent occurrence in the Northern Hemisphere today. Tolkien recognized that ferns are part of the large monophyletic group that we call the embryophytes: plants that have an embryo in their life cycle, whether they are dispersed as spores (such as the mosses or ferns) or by seeds (as in herbs such as grasses, or coniferous and flowering trees). This is expressed beautifully and mythologically in the first chapter of *The Silmarillion* where it is stated that because of the work of Yavanna and the light of the Lamps of the Valar "there arose a multitude of growing things great and small, mosses and grasses and great ferns (Figure 7.22), and trees whose tops were crowned with cloud as they were living mountains, but whose feet were wrapped in a green twilight" (SILM 1). Here, plant life is pictured as a scaled order, from tiny mosses to gigantic trees—and ferns are part of this amazing diversity.

Etymology: The word "fern" is derived from Old English *fearn* (fern), from Proto-Germanic **farno-*, and possibly is taken from Proto-Indo-European **por-no-* (feather, wing), alluding to the feathery leaves or fronds of many ferns. The Quenya name may be *filqë* or *filinqë*.

Distribution and Ecology: Ferns, belonging to the large order Polypodiales (with about 260 genera and more than 7,000 species), are of nearly cosmopolitan distribution, including many temperate as well as tropical species. They are ecologically diverse. Many occur in shaded forest habitats, but others love full sun; some are terrestrial whereas others are epiphytic; and they grow in swamps, mesic forests, or even open, dry habitats. Some grow in acidic soils, while others prefer limestone outcrops. All produce minute spores on the lower leaf surface (and thus the obvious, visible fern plant is called a sporophyte plant; i.e., a spore-producing plant). The spores are dispersed by wind, and landing on at least temporarily moist soils, germinate to produce tiny gamete-producing plants (i.e., gametophytic plants, producing sperm and eggs). The sperm swim, or are splashed by raindrops, reaching the eggs (held in special chambers called archegonia) and the gametes fuse; the resulting zygote develops into a visually obvious sporophytic plant, continuing the life cycle. The Polypodiales are the most diverse group within the leptosporangiate ferns, which are those ferns that have stalked and thin-walled sporangia with a cluster of thick-walled cells in the sporangial wall that change shape upon drying and forcefully rip open the globose sporangium, in the process scattering the spores.

Economic Uses: Several genera are used as ornamentals because of their interesting, feathery, and beautiful leaves. The young leaves of *Matteuccia struthiopteris*, a fern of alluvial woods that is broadly distributed in the Northern Hemisphere, are called "fiddleheads" and are eaten (fresh-cooked or canned) as a vegetable.

Description: Usually terrestrial or epiphytic herbs, often with rhizomes, and these sometimes variously hairy. Stems erect to horizontal, bearing closely clustered to well-separated leaves. Leaves all alike or the reproductive leaves differing from nonreproductive ones, often compound, once or twice pinnately so, occasionally palmately compound, or even simple, and often the leaf or the leaflets pinnately lobed, the veins forming an open branching system or a network, with or without hairs or scales on blade and/or petiole; the young leaves coiled, forming fiddleheads. The lower leaf surface with sporangia (i.e., spore-producing chambers; these are stalked and opening by a vertical row of thick-walled cells, releasing minute spores that are all alike) clustered together, forming structures called *sori*, and these variously shaped (e.g., round, slightly to markedly elongated) and positioned (scattered, near major veins, near or along margin, etc.) and each sometimes protected, especially when young, by a variously shaped flap of leaf tissue (the indusium) or merely by the inrolled leaf margin (Figure 7.22).

FIREWEED (*CHAMERION ANGUSTIFOLIUM*)

(In the Evening Primrose family [Onagraceae].)

> *No tree grew there, only rough grass and many tall plants: stalky and faded hemlocks and wood-parsley, fire-weed seeding into fluffy ashes, and rampant nettles and thistles. (LotR 1: VI)*

FIGURE 7.22 *Ferns.*

Fireweed is only mentioned only once in *The Lord of the Rings*, and it is one of the weedy species growing in the Bonfire Glade (Figure 7.23), just within the Old Forest, where it occurred with wood parsley (*Anthricus sylvestris*), poison hemlocks (*Conium maculatum*), nettles (*Urtica dioica*), and thistles (*Cirsium vulgare* and *C. arvense*). To Frodo and his three companions, it seemed a charming and cheerful place after their hike through the dense forest. The fireweed was in fruit, shedding its plumose seeds. The plant here occurred in its most characteristic habitat, a disturbed and previously burned woodland clearing. Of the species growing in the clearing, it was certainly the most beautiful—even in fruiting condition—but in flower it would have been spectacular, with showy pink to rose-purple flowers held horizontally in erect racemes. It is of interest that fireweed rapidly colonized burned ground after the bombings in London during World War II, and it was one of the first plants to appear after the eruption of Mt. St. Helens in 1980. Fireweed thus contributes beauty even to much damaged and disturbed environments.

Etymology: The common name, "fireweed," is of American origin, was first used in the late eighteenth century, and refers to the fact that this weedy species rapidly colonizes burned-over clearings. Tolkien used this name (see quote), although the plant is usually called "rosebay willow-herb" in the British Isles, in reference to its flowers that vaguely resemble those of the rose and its bay- or willowlike leaves, and this name goes back to Gerard's Herbal of 1597. The scientific name *Chamerion* is derived from the Greek *chamai* (low-growing) and *nerion* (oleander); the specific epithet *angustifolium* means narrow-leaved.

Distribution and Ecology: Fireweed (*Chamerion angustifolium*) is widely distributed across cool temperate and boreal regions of the Northern Hemisphere. It is an early successional species, occurring in wet calcareous to slightly acidic soils and vigorously colonizing recently burned sites; the plant is common in open fields, pastures, disturbed forests, and clearings. The seeds remain viable in the soil for many years, allowing them to rapidly germinate after fire. Plants may be either diploid (subsp. *angustifolium*, with 36 chromosomes) or tetraploid (subsp. *circumvagum*, with 72 chromosomes), with the diploids occurring at higher latitudes than the tetraploids. The genus *Chamerion* contains eight species and often has been included within *Epilobium*. However, it is differentiated from this closely related genus in having alternate (vs. opposite) leaves and bilaterally symmetrical (vs. radial) flowers that lack an obvious hypanthium (vs. hypanthium extending beyond the ovary apex). The showy flowers are insect-pollinated, while the fluffy seeds are dispersed by wind.

Economic Uses: The young shoots of fireweed may be eaten. These beautiful plants are sometimes used as ornamentals, but they quickly can get out of control in garden settings.

Description: Tall, perennial, rhizomatous herbs; stems with or without simple hairs. Leaves alternate, spirally arranged, simple, narrowly ovate to elliptic or oblong, pinnately veined, the secondary veins forming loops, the apex acute to acuminate, the base more or less acute, the margin entire to minutely serrate, glabrous to minutely hairy, very shortly petiolate to sessile; stipules absent. Inflorescences terminal racemes. Flowers conspicuous, slightly bilaterally symmetrical, bisexual, held horizontally, with the floral tube (hypanthium) disklike, not extending beyond ovary, bearing four slender, green to reddish green sepals, four broadly elliptic, clawed, pink to rose-purple

FIGURE 7.23 *Fireweed.*

petals, and eight deflexed stamens, with four fused carpels; the style elongate, initially deflexed, and at maturity four-branched apically; the elongate stigmas extending along the inner surface of the branches; the ovary inferior. Fruit an elongate, narrowly cylindrical, four-valved capsule, splitting to the base with intact central column to release the numerous, plumose seeds (i.e., each seed with a prominent tuft of elongate white hairs) (Figure 7.23).

FIRS (*ABIES* SPP.)

(In the Pine family [Pinaceae].)

> *Dwalin and Balin had swarmed up a tall slender fir with few branches and were trying to find a place to sit in the greenery of the topmost boughs. (Hobbit VI)*

Firs are frequent trees in Tolkien's legendarium, and these tall trees are especially characteristic of montane habitats, forming dense and dark forests if growing thickly. A firwood was encountered by Frodo, Sam, and Pippin in the Green Hills, less than 20 miles from Hobbiton, and later Aragorn and the four hobbits hiked through firs in the Trollshaws, just before they found the three trolls, turned to stone. Firs also grew in the Misty Mountains, intermixed with pines (*Pinus*) and larches (*Larix*), and, in *The Hobbit*, one provided an escape for Dwalin and Balin when Gandalf, Bilbo, and the dwarves were attacked by wargs and goblins (Hobbit VI, see quote; firs are clearly shown in Tolkien's illustration "The Misty Mountains Looking West from the Eyrie Towards Goblin Gate"). Southern Mirkwood, which long had been the hidden dwelling of Sauron, was dominated by fir trees. As Haldir pointed out to the company (from their view atop Cerin Amroth), "It is clad in a forest of dark fir, where the trees strive one against another and their branches rot and wither" (LotR 2: VI). Under Sauron's evil influence the trees had become selfish, the forest perverted, in striking contrast to the forests of Lothlórien, in which "no blemish or sickness or deformity could be seen in anything that grew upon the earth" (LotR 2: IV) as a result of the power wielded by Galadriel. Firs also dominated the mountainous forests in the vicinity of Dunharrow (south of Edoras, in Rohan), and a small fir-wood, the Dimholt, grew near the Dark Door that led to the Paths of the Dead (Figure 7.24). As Aragorn led his followers "under the gloom of black trees," closer to the door, Gimli said "My blood runs chill." The others were silent and "his voice fell dead on the dank fir-needles at his feet" (LotR 5: II). Certainly, firs are not dangerous or evil, and Merry and Pippin encountered tall, firlike Ents in Fangorn, but firwoods are often dark and inhospitable, providing at least the potential for danger. Firs also grew in Beleriand and are beautifully shown in several of Ted Nasmith's paintings (see illustrated *Silmarillion*, opposite p. 137)—one of our favorites shows the meeting of Finrod Felagund and the people of Bëor in the foothills of the Blue Mountains, below the springs of Thalos.

Etymology: The English name "fir" is derived from the Old Norse *fyri* and Old Danish *fyr*, from Proto-Germanic **furkhon* (all names for fir), and ultimately from Proto-Indo-European **perkwu-* (originally meaning oak, or oak-forest). During the past 5,000 years, conifers and birches gradually became more common than oaks in

FIGURE 7.24 *Firs.*

montane European forests, leading (it is presumed) to the shift in meaning in this series of related words. The scientific name, *Abies*, is the classical Latin name for the European silver fir (*A. alba*). The Quenya name for fir may be *alalmë*, but this word is also applied to pines.

Distribution and Ecology: Firs, a group of about 40 species, are broadly distributed across the Northern Hemisphere, growing in cool, humid forests. Many are species of montane habitats. Seven species occur in Europe, northern Africa, and western Asia, with the European silver fir (*Abies alba*) the most widely distributed. Fourteen species occur natively in North America; the most common of these are Pacific silver fir (*A. amabilis*), balsam fir (*A. balsamea*), white fir (*A. concolor*), grand fir (*A. grandis*), and subalpine fir (*A. lasiocarpa*). The center of diversity is in Eastern Asia, where ca. 20 species occur. Firs are rather uniform in appearance, and related species hybridize when co-occurring, so identification is often difficult. Important characters in differentiating species include the positioning of the leaves on the twigs, their color, shape of the leaf apex, number of rows of stomata, distribution of their resin canals, and the color of pollen and seed cones. The pollen grains are dispersed by wind, and the erect cones fragment at maturity, leading to dispersal of the winged seeds by wind.

Firs are easily confused with spruces (*Picea* spp.), a related coniferous genus with a similar conical growth form. Spruces can be distinguished from firs by their ridged twigs that have the leaves borne on short, hard stalks. In addition, their cones are pendulous, not held erectly as those of firs. It is puzzling that spruces are not recorded as occurring anywhere in Middle-earth. Indeed, they do not occur in Tolkien's legendarium.

Economic Uses: Firs are a source of timber used in poles, paper pulp, and plywood. They also provide aromatic resins and fragrant essential oils. Because of their evergreen condition and columnar to conical shape, firs are important ornamentals, and they are frequently grown in plantations for Christmas trees. Canada balsam, a viscid, clear to pale yellow turpentine, is extracted from Balsam fir (*A. balsamea*) and was once an important mounting fixative for microscope slides. They also provide cover and food for wildlife.

Description: Evergreen trees with a conical crown composed of regular tiers of short, horizontal branches, the trunk usually single, straight, with bark smooth and bearing resin pockets when young, but often becoming furrowed or scaly with age. Twigs all elongate, smooth or grooved; wood with resin canals. Leaves alternate, spirally arranged, but sometimes appearing two-ranked, simple, linear and needle-like, more or less flattened, with resin canals, with a single midvein, the apex acute to rounded or notched; the base expanded, leaving a flat, round scar on the twig (when detached), the margin entire; stipules absent. Plants with pollen cones and seed cones on the same tree. Pollen cones ovate to oblong-cylindric, small, yellow to red, green, blue, or purple, with each pollen-bearing structure flat, bearing two sporangia. Seed cones ovoid to oblong-cylindric, erect, borne in axils of leaves on shoots at the top of the crown, maturing in a single season, fragmenting (scale by scale) at maturity with cone-axis persisting; scales flat, fan-shaped, each bearing two seeds, free from the associated, narrower bracts (which may be hidden or exposed). Seeds strongly winged (Figure 7.24).

FLAG-LILIES OR YELLOW IRIS (*IRIS PSEUDACORUS*)

(In the Iris family [Iridaceae].)

> *The lake had become a great marsh, through which the river wandered in a wilderness of islets, and wide beds of reed and rush, and armies of yellow iris that grew taller than a man and gave their name to all the region and to the river from the Mountains about whose lower course they grew most thickly. (Footnote 13: "The Gladden Fields," in UT 3: I)*

Flag-lilies (yellow irises) were characteristic of wetland habitats in Middle-earth of the Third Age, as they are in Europe today; they were especially abundant in the Gladden Fields, a large marshland near the junction of the Gladden and Anduin Rivers and gave their name to this wetland (see the quote from the *Unfinished Tales*). The Gladden Fields are well known as the place where Isildur lost the One Ring and was soon after killed by orcs (Figure 7.25). Yellow irises had also been a conspicuous element in the wetland flora of Beleriand. As Voronwë, the elven mariner, described to Tuor, in Nantathren, the region where the Narog joined the Sirion River, the rivers "haste no more, but flow broad and quiet through living meads; and all about the shining river are flaglilies like a blossoming forest" (UT 1: I). Their beauty is enchanting, and truly, in this region, as stated by Voronwë, "Ulmo is but the servant of Yavanna" because the beauty of these wetland plants, which she brought into existence, surpassed that of the river itself, which was sacred to Ulmo. Flag-lilies grow in shallow water, and their leaves are elongated, sharply pointed and swordlike, flattened into distinctive fans, with the erect stems bearing large, pale to intensely yellow flowers. Since they spread by rhizomes, they can form dense mats, excluding other species, and thus often dominate wetlands. Given their beauty, it is appropriate that Goldberry, daughter of the Withywindle, when she first met Frodo and his companions, is described as wearing a gown "green as young reeds" and "her belt was of gold, shaped like a chain of flag-lilies" (LotR 1: VII). This wetland flower is quite appropriate for a water-sprite or nymph, its flowers echoing the color of her hair and reflecting the sunlight.

Etymology: In his writings, Tolkien uses the names "flag" or "flag-lilies," "yellow iris," and "gladden," "gladdon-swords," or "iris-swords" to refer to these plants, members of the genus *Iris*, especially *I. pseudacorus*. The English word "iris" is derived through Latin from the Greek, *Iris*, the name of the goddess of the rainbow, a messenger of the gods; the word was used for any brightly colored circle, and thus also is applied to the iris of the eye, due to its many colors, as well as to this group of lilylike plants, presumably because of their many-colored flowers. "Flag" is another name commonly applied to these flowers and is derived from the Middle English *flagge*, used of plants growing in moist places. Yellow-flowered irises grew abundantly in the Gladden Fields, where Isildur was slain and the One Ring was lost, and the word "gladden" is a modification of the dialectal English "gladdon," derived through Middle English, *gladen* to Old English *glaedene*, which was probably taken from Latin, *gladiolus*, a small sword, in reference to the swordlike leaves of irises (and their relatives). *Loeg Ningloron* is Sindarin for Gladden Fields, and *ninglor* is Sindarin for iris. The common name "gladdon," today, is often applied to *I. foetidissima*, a species with vivid purple flowers; however, Tolkien

FIGURE 7.25 *Flag-lilies, yellow iris.*

specifically indicated that in his legendarium, in agreement with early usage, the name should be applied to *I. pseudacorus* (see Letters: No. 297).

Distribution and Ecology: Irises, broadly considered, constitute the genus *Iris*, a group of about 280 species, and are widely distributed in the Northern Hemisphere (occurring in Eurasia, northern Africa, and North America). They grow in a wide variety of habitats, from shaded forests or swamps to sunny meadows, but most are species of various wetlands, such as marshes, fens, swamps, the margin of lakes or rivers, or wet ditches. The yellow flag, *I. pseudacorus*, is, not surprisingly, a plant of wetland habitats; it is widely distributed in Europe and also extends into western Asia and northern Africa. In addition, it has become widely naturalized in North America and Australia. The showy flowers of *I. pseudacorus* are pollinated by bees and flies; its corky seeds, once released from its capsules, float and are water dispersed.

Economic Uses: Irises or flag-lilies are popular garden ornamentals because of their large, attractive, and variously colored flowers, which have six tepals, the outer three, the *falls*, are often reflexed (and sometimes bearded), and the inner three, the *standards*, usually more or less erect. The rhizomes of some species were once used medicinally (and are strongly purgative). *Iris pseudacorus*, a beautiful yellow-flowered species, likely is the source of the fleur-de-lis.

Description (of *I. pseudacorus*): Rhizomatous, perennial herb, the flowering stems erect, cylindrical; roots fleshy. Leaves alternate (and the largest leaves basal, others smaller, along upper stem), two-ranked, and thus ± forming a fan, with the leaf blades flattened in the same plane as the ranks (i.e., equitant); simple, elongate, and sword-shaped (about as long as the inflorescences), with parallel veins (each leaf with a prominent median thickened line), the apex acute, the base broad and sheathing, the margin entire, both surfaces lacking hairs; stipules absent. Inflorescences terminal, few-flowered cymes. Flowers radially symmetrical, erect, large, usually bright yellow, with a perianth of six tepals, fused into a tube at base, the three outer tepals (sometimes called sepals) recurved, the base narrowed and forming a claw that expands into the broader limb, each with a darker yellow blotch and often thin, radiating, usually brown lines in the center; the three inner tepals (sometimes called petals), smaller, erect-spreading, narrowly ovate to obovate. Stamens three, appressed to the style branches. Carpels three, fused, the style dividing into three, petaloid branches, these arching outward, hiding the stamens, and divided at the apex into two, fringed lobes (style crests); ovary inferior. Fruit an obscurely three-angled capsule, the seeds brown, corky, D-shaped, and flattened (Figure 7.25).

FLAX (*LINUM USITATISSIMUM*)

(The Flax family [Linaceae].)

> *He chose for himself from the pile a brooch set with blue stones, many-shaded like flax-flowers or the wings of blue butterflies.... "Here is a pretty toy for Tom and his lady! Fair was she who long ago wore this on her shoulder. Goldberry shall wear it now, and we will not forget her!" (LotR 1: VIII)*

After his rescue of the four hobbits in the Barrow Downs, Tom Bombadil went up to the burial mound and gathered together most of the treasures that had been hidden within; he piled them in the grass, making them available to "all kindly creatures" so that the curse of the mound would be broken. From among the treasures he picked out a beautiful brooch with stones of many shades of blue that are compared to the colors of flax flowers. These vary in color from blue or purple-blue to pale blue or lilac, and can even be pink or white. Tom gave the brooch to Goldberry, and it is clear that both remembered and honored its original owner. Who was she? In Appendix A of *The Lord of the Rings* it is stated that this mound may have been the grave of the last prince of Cardolan (one of the three northern kingdoms of the Dúnedain) who had died in the war of 1409 (Third Age), which was between the Dúnedain of the north (of the allied kingdoms of Arthedain and Cardolan) and the Witch-King of Angmar, the chief city of which was Carn Dûm. During this war, the Dúnedain were defeated, the tower of Amon Sûl (at Weathertop) destroyed, and the remnant of the people of Cardolan retreated into the region of the Barrow Downs. The brooch, therefore, may have belonged to the wife of this fallen Cardolanian prince. This makes sense because Merry, while in the barrow, had taken on the memories of one of those buried there and, upon emerging, said: "The men of Carn Dûm came on us at night, and we were worsted. Ah! The spear in my heart!" This episode reminds us that the hatred of Sauron and the chief of the Ringwraiths (i.e., the Witch-King) for the Men of the West and their elven allies was one with deep historical roots—and it is intriguing to think that perhaps Tom had long ago assisted the Dúnedain, just as he was now helping Frodo and the other hobbits. A second question remains: why did Tom select for Goldberry this particular brooch from among the many beautiful objects that once belonged to this Cardolanian prince and his wife? We think it was because of the beautiful array of blue stones that were set within the brooch. Blue is a symbol of purity, fidelity, and faithfulness, and Tom's choice of this particular treasure may represent his faithful commitment to and love for his wife Goldberry. Finally, there is another, more practical reference to flax in *The Lord of the Rings*. It occurs in the description of the items packed by Sam at Rivendell as he prepared to head southward with those accompanying Frodo on his quest. One of these items was linen, which is made from the woven fibers of the flax plant.

Etymology: "Flax" is derived from Old English *fleax*, from Proto-Germanic **flakhsan*, probably from the Proto-Germanic base **fleh-* (to plait) and corresponding to the Proto-Indo-European **plek-* (to weave, to plait), but possibly from **pleik-* (to flay, as in stripping flax fibers, in its preparation). "Linen" (a noun, meaning cloth from woven flax fibers) was earlier used as an adjective (i.e., made of flax) from *lin* (flax, linen thread or cloth), from Proto-Germanic **linam*, which probably is an early borrowing from Latin *linum* (flax, linen), which in turn was likely taken up from a non-Indo-European language. The scientific name *Linum* is the ancient Latin name of this plant.

Distribution and Ecology: Flax (*Linum usitatissimum*) is known only as a domesticated plant (either in cultivation or as an escape) and may have been derived from the wild species *L. angustifolium* (i.e., pale flax; also sometimes recognized as *L. usitatissimum* subsp. *bienne*). Flax is apparently native to the Middle East, where it was domesticated (before 6000 BC) but was very early (before 4000 BC) taken into Europe. The species is now widely naturalized. Flax is a member of the genus *Linum*, a group with more than 230 species, which is broadly distributed in temperate and subtropical regions; 36 species occur in Europe, and about an equal number grow in North

America (north of Mexico), only 6 occur in China, and 2 in Japan. The flowers are pollinated by a variety of insects (especially bees and flies); the seeds have a mucilaginous testa (when wetted) and may be dispersed by water, rain wash, wind, or external transport on animals.

Economic Uses: Flax is an important source of fiber and is used in textiles (linen), twine, canvas, paper, and more. Its use as a fiber plant is now less important than it once was because of the increased use of cotton (although cotton fibers are much shorter than are those of linen). The seed oil (linseed oil) is valuable and is used in food processing, paints, varnishes, linoleum, inks, and soaps. Crushed flax seeds are occasionally also used medicinally (especially to improve digestion and elimination). Several species of *Linum* provide ornamental herbs (with blue, red, yellow, or white flowers); the flowers of flax itself are variable in color.

Description (of *L. usitatissimum*): Annual herb, glabrous, with erect stems. Leaves alternate and spirally arranged, simple, small, narrowly ovoid or oblong to linear, with three visible veins diverging at leaf base, the apex acute to acuminate, the base narrowly cuneate, the margins entire, without a petiole; stipules absent. Inflorescences terminal, reduced cymes. Flowers radially symmetrical, bisexual, the sepals five, distinct, the petals five, distinct; blue or purple-blue to pale blue, lilac, pink, or white, obovate and apically rounded; the stamens five, slightly fused by their filaments; and the carpels five, fused, their styles five, ± distinct; the ovary superior. Fruit a globose capsule, with up to 10 seeds.

FORGET-ME-NOTS (*MYOSOTIS* SPP.)

(In the Borage family [Boraginaceae].)

> *Old Tom Bombadil had a merry wedding,*
> *crowned all with buttercups, hat and feather shedding;*
> *his bride with forgetmenots and flag-lilies for garland*
> *was robed all in silver-green. He sang like a starling,*
> *hummed like a honey-bee, lilted to the fiddle,*
> *clasping his river-maid round her slender middle.* (TATB 1)

All references to forget-me-nots in Tolkien's legendarium (and there are not many) are connected with Goldberry (Figure 7.26). At her wedding with Tom Bombadil (see quote), she wore a garland of blue forget-me-nots (*Myosotis*) and yellow flag-lilies (*Iris pseudacorus*). Most species of *Myosotis* prefer damp to wet habitats, and some, such as the ones called water forget-me-nots (*M. scorpioides, M. caespitosa,* and *M. secunda*), are native to England and grow in wet places along streams or ponds or in marshes and bogs, and these would be especially appropriate for Goldberry, a river-maid. In *The Lord of the Rings*, the four hobbits saw Goldberry almost as soon as they had crossed the stone threshold of Tom's house. She sat at the far side of the room, and "her long yellow hair rippled down her shoulders; her gown was green, green as young reeds, shot with silver like beads of dew; and her belt was of gold, shaped like a chain of flag-lilies set with the pale-blue eyes of forget-me-nots" (LotR 1: VII). She is surrounded by vessels of green and brown earthenware in which white water-lilies (*Nymphaea alba*) floated, and she seemed to the hobbits to be sitting in the middle of a pool. She is the "daughter

FIGURE 7.26 *Forget-me-nots.*

of the River" (i.e., the Withywindle; LotR 1: VII) and has often been interpreted as the spirit of the Withywindle; in one of his letters, Tolkien states that she represents seasonal changes in such river lands (Letters: No. 210). Such an interpretation certainly supports her association with species that grow in and along rivers, such as reeds (*Phragmites australis*), waterlilies, flag-lilies, and forget-me-nots. Yet the connection may be deeper. The green color of the reeds lining the river's banks is evident in her gown, the white of the waterlilies reflects her complexion, the yellow of flag-lilies her hair, but what about the pale blue of the forget-me-nots? Could this also be descriptive, referring to the color of the River Withywindle itself, reflecting on its surface the blue of the sky or perhaps the blue of its aquatic depths? Or does the color blue, as used here, reflect ancient connotations (and Christian symbolism), representing heavenly grace and living one's life according to the highest standards. Blue also is a symbol of purity, fidelity, and faithfulness, and it is an old English custom for a bride to wear blue on her wedding gown—just as Goldberry wore forget-me-nots on her wedding day and perhaps continued to wear them as a symbol of her faithfulness.

Etymology: The English name "forget-me-not" is a translation of the Old French *ne m'oubliez mye* (don't forget me), and similar loan-translations took the name into other languages. The Latin, scientific name *Myosotis* is derived from the Ancient Greek *muosotis*, from *mus* (mouse) and *ous* (ear), in reference to the hairy leaves of these plants, which imaginatively look like mouse ears. The Greek *mus* (like the English mouse) is derived from Proto-Indo-European **mus-* (mouse); the Greek *ous* (like the English ear) is derived from Proto-Indo-European **ous-* (ear).

Distribution and Ecology: Forget-me-nots (members of the genus *Myosotis*) are widely distributed but are most diverse in Europe and adjacent regions of western Asia and in New Zealand. Europe has 41 species, and New Zealand 40, while only 11 occur in North America (north of Mexico; and of these only 4 are native) and only 5 are native to China. Forget-me-nots occur in open, herb-dominated habitats or within open woods, many occur near ponds or streams, in wetlands, or at least in moist habitats, but some grow in dryer soils and are common in weedy areas. Species are distinguished by characteristics such as their habit, life span (annuals to perennials), form and distribution of the hairs on various plant parts, leaf shape and size, inflorescence structure, pedicel length, calyx form and pubescence, color and form of the corolla, and size and shape of the nutlets. The flowers of forget-me-nots are pollinated by various flies and bees, while the nutlets are dispersed by animals (either by being eaten or by sticking onto fur or clothing), or they may be carried by water (since the plants often grow along streams or ponds).

Economic Uses: Various species of forget-me-nots are cultivated as ornamental herbs because of their showy usually blue or pale-blue flowers. In folklore, this wildflower often represents steadfastness and fidelity (which are also associated with the color blue), and it has been used as a sign of faithfulness and enduring love—the wearer of this flower will never be forgotten by her lover.

Description: Annual or perennial herbs, often densely covered with simple hairs (less commonly nearly glabrous); stems erect or basally prostrate. Leaves alternate, spirally arranged, simple, ovate to oblong or obovate, pinnately veined, the apex acute to rounded, the base narrowed and with or without a petiole, the margin entire, both surfaces more or less hairy; stipules absent. Inflorescences terminal, apically curled and one-sided (i.e., in scorpioid cymes). Flowers bisexual, radially symmetrical, erect, with five green, fused sepals, five fused petals, usually blue or pale blue, less commonly white, pink, or purple (and sometimes changing color with age) forming a narrow tube with an abruptly flaring five-lobed apical region or broadly funnel-form, and with the corolla-throat obstructed by five short scales (these often changing color with age and contrasting with the color of the corolla lobes) and five stamens, these also arising from the corolla tube and included within it, and with two fused carpels; the style short, single, the ovary superior and strongly four-lobed. Fruit a schizocarp, comprising four brown nutlets (Figure 7.26).

FRAGRANT TREES FROM ERESSËA

(Various species, perhaps of the Lauraceae.)

> *All about that place ... grew the evergreen and fragrant trees that they [the elves] brought out of the West, and so throve there that the Eldar said that almost it was fair as a haven in Eressëa. They were the greatest delight of Númenor, and they were remembered in many songs long after they had perished for ever, for few ever flowered east of the Land of Gift: oiolairë and lairelossë, nessamelda, vardarianna, taniquelassë, and yavannamirë with its globed and scarlet fruits. (UT 2: I)*

We are told that the bark, flowers, and leaves of these trees exuded a sweet scent, but each was slightly different, so the region around the Bay of Eldanna was "full of blended fragrance" (UT 2: I). At the bay's center was Eldalondë the Green, the most beautiful of all the coastal cities of Númenor. In fact, the visiting elves said that the region was as nearly fair as Tol Eressëa, and thus the elven ships came most often to this Númenórean harbor. These fragrant trees were the "greatest delight of Númenor" (see UT 2: I). As gifts of the elves, they represented the deep connection of the men and women of Númenor with the Eldar of Eressëa. Most had no pragmatic uses—their sole function was the enhancement of environmental beauty, which increased the quality of life of those living in Númenor. However, the first listed, oiolairë, was also the source of the Bough of Return—a branch of oiolairë placed on the prow of the Númenórean ships just before they departed on their voyages to Middle-earth and symbolic of their safe return. These branches usually held their freshness and beauty as long as they were washed by the ocean spray. Yet, on the last voyage before his marriage to Erendis, Aldarion, who was to become the sixth king of Númenor, was forced far northward into regions of extreme cold, and, toward the end of that voyage Aldarion saw that the bough was withered by the frost, and he was dismayed. Perhaps this was a sign that he had voyaged too long—and he certainly was torn between his love for Erendis and his love of the sea. The last listed, yavannamirë, a tree linked with Yavanna, the Giver of Fruits, is singled out as having "globed and scarlet fruits," and these fruits were not only beautiful, but also may have been delicious and eaten by the Númenóreans. Another beautiful plant brought to Númenor by the elves was lavaralda, which is only recorded from the *Lost Road* (see vol. 5 of the *History of Middle-earth*, p. 57) and formed a great hedge at the western end of Elendil's garden. It had long green leaves that were golden on the undersides, and its fragrant flowers were white with a yellow flush and "laid thickly on the branches like a sunlit snow" (p. 58). With the sinking of Númenor, these lovely trees were lost to those living in Middle-earth. For years, they were remembered in song, but now even the songs have faded. Unfortunately, such is the fate of many things in Arda marred. Will we mourn the passing of the many species in our own world—those that are already lost, as well as those that soon will be extinct? Will our future echo that of Númenor?

Etymology: The names of these six trees are translated from the Quenya as follows—*oiolairë* (ever-summer, and *Coron Oiolairë* is also one of the names of the Green Mound of the Trees in Valinor), *lairelossë* (summer-snow-white), *nessamelda* (beloved of Nessa), *vardarianna* (Varda's crown-gift), *taniquelassë* (leaf of Taniquetil), and *yavannamirë* (jewel of Yavanna). Nessa is the spouse of Tulkas and sister of Oromë; she is slender and quick-footed and of all wild creatures most loves deer. Varda lives with Manwë "above the everlasting snow" upon Taniquetil, the tallest peak of the

Pelóri (the mountains surrounding Valinor). She is the Lady of the Stars, and "light is her power and her joy" (as described in the Valaquenta) for in her face still shines the living light of Ilúvatar. Of all the Valar, she is held most in reverence and love by the elves of Middle-earth. Finally, Yavanna, the spouse of Aulë, is the Giver of Fruits, and she loves all things that grow upon the earth. It is apparent that most of these tree names connect either to Valinor or to one of the Valier.

Distribution and Ecology: These six trees were found most abundantly in the western portion of Númenor, in the region of the Bay of Eldanna, which had rich soils, a warm climate, and the highest rainfall of any region of the island. They were brought to Númenor from Tol Eressëa, the Lonely Isle, and very likely also grew in Valinor. Their fragrant flowers probably were pollinated by various bees, and the bright red or blue-black fruits may have been bird-dispersed.

Economic Uses: As stated in the introductory quote from *A Description of the Island of Númenor* (in the *Unfinished Tales*), these six trees (along with mellyrn, with which they grew) "were the greatest delight of Númenor"—their only use was the pleasure that they brought to the people of the island. One possible exception was yavannamirë, the only one of the fragrant trees whose fruits are described, and the only one that is connected to Yavanna—the Giver of Fruits. It is likely that its globose and bright red fruits were edible.

Descriptions: Since these trees are unique to J. R. R. Tolkien's legendarium and are not described in detail, the task of providing technical identifications is problematic. However, the fact that they were evergreen and that their leaves and bark were fragrant suggests that they were members of the Lauraceae because members of this family have aromatic bark and evergreen leaves with scattered pellucid dots (i.e., with very tiny spherical cells containing aromatic terpenoids in the leaf mesophyll). In addition, their small flowers are often white and their fruits are fleshy and indehiscent (and thus match the minimal descriptions provided for lairelossë and yavannamire, respectively). Based on these assumptions, we present the following generalized and somewhat speculative description of yavannamirë, and we assume that the other fragrant trees were quite similar.

Evergreen tree, the bark aromatic and smooth to weakly furrowed. Leaves alternate, spirally arranged, simple, ovate to elliptic, pinnately veined but with a pair of especially prominent secondary veins arising near the base, the apex acuminate, the base acute to obtuse, the margin entire, glabrous, shortly petiolate; stipules absent. Inflorescences axillary, few-flowered. Flowers radially symmetrical, bisexual, the perianth of six petaloid, white tepals, these distinct, with nine distinct stamens, each opening by four flaps and with paired nectar- and odor-producing glands, and with a single carpel. Fruit globose, bright red drupes (in yavannamirë, but probably blue-black in oiolairë and the others), each with a single pit, and usually only one developing from each flower-cluster.

GRAPE VINES, WINE (*VITIS VINIFERA*, AND RELATIVES)

(In the Grape family [Vitaceae].)

> *He followed the two elves, until they entered a small cellar and sat down at a table on which two large flagons were set. Soon they began to drink and laugh merrily. Luck of an unusual kind was with Bilbo then. It must be potent wine to make a wood-elf drowsy; but this wine, it would seem, was the heady vintage of the great gardens of Dorwinion, not meant for his soldiers or his servants, but for the king's feasts only. (Hobbit IX)*

The fact that elves appreciate wine, a beverage made from the fruits of the European grape (*Vitis vinifera* subsp. *vinifera*) that are fermented through the action of yeast (*Saccharomyces cerevisiae*), and even go through the effort and expense to import this flavorful (and, in moderation, also healthful) drink from a distant, grape-growing region (on the northwestern shore of the Sea of Rhûn) is clear from the quote (see also Figure 7.27). This appreciation extends back into the First Age, as we see in *Of Tuor and His Coming to Gondolin* (in the *Unfinished Tales*) when Tuor and Voronwë are offered food and wine (in order to revive their strength) when they reached the Second Gate of the elven city of Gondolin. The same goes for hobbits, wizards, and dwarves! In the first chapter of *The Hobbit*, shortly after the dwarves have unexpectedly arrived at Bilbo Baggins's home, Gandalf appeared and is asked if he would like some tea. He responds, "What's that? Tea! No thank you! A little red wine, I think for me" (Hobbit I). And Thorin quickly added that he wanted wine as well. And, years later, Frodo, after selling his home and just before leaving for Crickhollow, served the last of his wine (Old Winyards, a strong red wine made from grapes grown in the Southfarthing) at a farewell feast for a few of his friends. The races of men, especially those of the Three Houses of the Elf-friends, must have picked up the knowledge of wine-making from the elves, for in the *Akallabêth* it is related that Númenóreans, early in the Second Age, sailed to Middle-earth, and, taking pity on the humans living there in fear of the creatures of Morgoth, taught them many things. They brought them wheat and wine and instructed them in agricultural methods: "in the sowing of seed and the grinding of grain" and certainly also the making of wine. Obviously, wine is part of the cuisine of Gondor, with its Mediterranean flora, even more so than the more northern Shire. Sam and Frodo are given wine (along with bread and butter, salted meats, cheese, and dried fruits) by Faramir at the secret refuge of Henneth Annûn, and Pippin was offered wine (and white cakes) by Lord Denethor soon after his arrival in Gondor, while he was questioned concerning his travels. Wine, along with beer, is an integral part of the fellowship of shared food and drink—and this is illustrated across the many cultures of Middle-earth.

Etymology: The English word "grape" is derived from Old French *grape* (a bunch of grapes), probably by back-formation from *graper* (grasp, pick, as in pick grapes), from Proto-Germanic **krappon* (hook) and Proto-Indo-European **grep-* (hook), and the original idea was perhaps a vine-hook for grape picking. Grapes are not native to England, and the word "grape" replaced the original Old English word *winberige* (wine berry). The word "vine" is derived from the Old French *vigne* (vine, vineyard), from Latin *vinea* (wine, vineyard) and Latin *vinum* (wine), from Proto-Indo-European **win-o-* (wine). The related word "wine" is from derived Old English *win* (wine) from Proto-Germanic **winam*, an early borrowing from Latin *vinum* (wine). The Sindarin word for wine or vine is *gwîn*.

Distribution and Ecology: Grapes are widely distributed across the Northern Hemisphere, although they are most diverse in eastern Asia and eastern North America. Fifteen species occur in North America (north of Mexico), 37 in China, 6 in Japan, and 1 (*Vitis vinifera*) in Europe, although several North American species have naturalized there. They grow as lianas in a wide variety of habitats. Characteristics important in species identification include bark (shredding or not), angularity and coloration of the young branches, leaf shape and extent of lobing, color of the lower leaf surface as well as the form and density of its hairs, size of marginal teeth, and position

FIGURE 7.27 *Grapes.*

and form of the climbing structures (tendrils). The small flowers are pollinated by various insects, and the berries are dispersed by birds.

Economic Uses: The European grape (*V. vinifera*), with some 10,000 cultivars, is the most important species economically and is used both for its fresh fruits (large, oval-berried table grapes) and for wine, sparkling wine, or vinegar (small-berried wine grapes). Grape leaves are also used in cooking and pickling. Cultivated plants are classified as *V. vinifera* subsp. *vinifera* and are grown widely not only in Europe (where the species is native), but also in southern South America, southern Africa, and western North America (where it is introduced), whereas the closely related, wild populations are placed in subsp. *sylvestris* and are now endangered due to disease and habitat destruction. Grapes were domesticated about 9,000 years ago in areas of southwestern Asia and the eastern Mediterranean and are now culturally and also religiously significant. Other, less economically important species include frost grape (*V. riparia*), muscadine grape (*V. rotundifolia*), and fox grape (*V. labrusca*, including cultivars such as Concord grapes), these used for jams, jellies, juice, wine, and grape-flavored candies and soda-pop, and also sugar grape (*V. rupestris*), used in breeding phylloxera-resistant rootstocks. The European wine industry was severely impacted by the introduction of phylloxera root aphids (*Viteus*) from American grapevines in 1867, and now most European grapes are grafted onto phylloxera-resistant rootstocks.

Description: Deciduous lianas, the bark shredding or not; hairs various. Leaves alternate, spirally arranged, usually simple, sometimes often variously lobed, ovate to elliptic, palmately veined, the veins running to the teeth, the apex acuminate or acute to rounded, the base rounded to cordate, the margin variously toothed, often variously hairy on the lower surface, with a petiole; stipules present. Inflorescences terminal, but appearing to be opposite the leaves (due to growth of an axillary branch from the opposing leaf axil), variously cymose, and tendrils also located opposite many of the leaves (and thus in the same position as the inflorescences to which they are developmentally related). Flowers radially symmetrical, usually unisexual, with the staminate flowers and carpellate flowers on the same plant or on different plants, but bisexual in cultivated European grapes; the sepals five, very tiny; the petals five, fused and forming a pale green to yellow cap that falls as the flower opens; the stamens five, and with two fused carpels; the style short, the ovary superior. Fruit a globose to ellipsoid, green, yellow, red, purple-black, or blue-black berry, with usually four seeds, these unusual, one side with a ridge flanked by deep grooves, and the other with a round to linear, depressed to elevated knot (Figure 7.27).

GRASS (VARIOUS GENERA DOMINATING TEMPERATE PRAIRIES OR STEPPES; E.G., *AGROPYRON, ANDROPOGON, BOUTELOUA, BROMUS, FESTUCA, KOELERIA, PANICUM, POA, SCHIZACHYRIUM, SORGHASTRUM, STIPA*)

(The Grass family [Poaceae or Gramineae].)

At the bottom they came with a strange suddenness on the grass of Rohan. It swelled like a green sea up to the very foot of the Emyn Muil. (LotR 3: II)

Here we consider grasses generally, especially those genera (see listed) that dominate grassland communities (prairies, plains, steppes, etc.) or those that are used in lawns (e.g., *Agrostis, Axonopus, Buchloe, Cynodon, Eremochloa, Festuca, Lolium, Paspalum, Poa, Stenotaphrum, Zoysia*). A few grasses, however (i.e., reeds, barley, and wheat) are individually referenced in J. R. R. Tolkien's legendarium, and for these we provide more detailed coverage (see their separate treatments). The words "grass" or "grasses" are used more frequently in Tolkien's writings than any other plant names, and they are most frequently associated with the plains of Rohan (see the preceding quote and Figure 7.28, where we see Aragorn, Gimli, and Legolas following the orc track). The plains of Rohan bring to mind the Eurasian steppes, which extend from Hungary, Romania, Ukraine, and southeastern Russia eastward across Asia, or the North American Great Plains, stretching from southern Alberta and Saskatchewan through eastern Montana and the Dakotas southward across the continent to New Mexico and Texas. However, in the Third Age, expansive grasslands also occurred south and east of the sea of Rhûn. In the First Age, the major grassland region, Ard-galen, occurred east of the Ered Wethrin and north of the Dorthonion highlands. Sadly, it was destroyed by Morgoth in the Battle of Sudden Flame and became a lifeless, ash-covered desert. It was thereafter known as Anfauglith, the Gasping Dust. The great island of Númenor existed during the Second Age, and much of its central region (Mittalmar) was grassland-dominated; grasslands (in part converted into fields of grain) also occupied much of the Orrostar peninsula. Less obviously, from a landscape perspective, grasses occurred throughout Middle-earth, as they do in our own world: in meadows, marshlands, and forest openings, and as well as in the shaded understory. Of course, they also dominated lawns as well as pastures and grain fields in regions of scattered villages and agricultural development, such as the Shire and the Pelennor (surrounding Minas Tirith). The grasslands of Rohan in West Emnet and near the Entwash were tall-grass prairies, and we know this because when Gandalf, Aragorn, Legolas, and Gimli rode through this region on their way to Edoras "the grass was so high that it reached above the knees of the riders, and their steeds seemed to be swimming in a grey-green sea" (LotR 3: V). Typical of such grasslands, the region is described as flat, with the wind passing through it "like grey waves through the endless miles of grass" (LotR 3: V). Scattered wetlands even occurred within this tall prairie as a result of the presence of the Entwash flowing through its midst. These grasslands are central to the legendarium, explaining the horse culture of the Rohirrim and adding to the imagery of the story, as when Aragorn seemed, at his first meeting with Éomer, to have sprung out of the grass (Figure 7.6). In a very different way, the grasses of forest openings are also essential to the story, especially when such openings are aesthetically transformed, as in the realm of Thingol and Melian, in which there was "laughter and green grass" (Lays III: line 486). In the *Lay of Leithian*, such glades are even called "lawns" (see Lays III: lines 512, 598) since their beauty reflects the creative work (or magic) of Melian, just as the lawns around our homes reflect our horticultural expertise. Such woodland lawns are a central element in our image of Lúthien as seen by Beren, who wandered into the woods of Neldoreth after escaping from Dorthonion. He came upon Lúthien "at the time of evening under moonrise, and she danced upon the unfading grass in the glades beside Esgalduin" (SILM 19). Such grasses complement her beauty and are described as green, unfading, and glimmering—and decorated by the red and white "flowering candles" shed by the horse-chestnuts (*Aesculus hippocastanum*) overhead. Beren searched for her for nearly

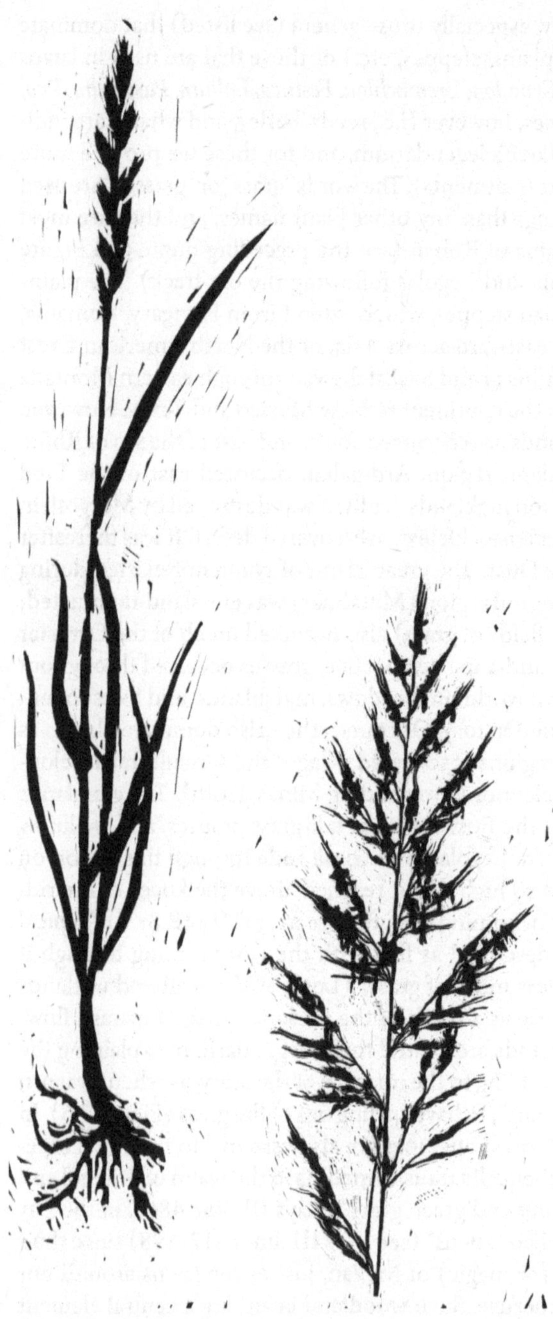

FIGURE 7.28 *Grass.*

a year, repeatedly calling Tinúviel (Nightingale) and eventually Lúthien, "whom no pursuit . . . might hope to win or hold . . . came at the sweet calling of her name; and thus in his her slender hand was linked" (Lays III: lines 800–805; see also 806–809). Without this meeting in a grassy opening, the entire history of Middle-earth would have been much different—and much poorer!

Etymology: The word "grass" comes from Old English *græs, gærs* (herb, plant, grass), from Proto-Germanic *grasan*, which is derived from Proto-Indo-European **ghros-* (young shoot, sprout), from the root **ghre-* (to grow, become green), and this is also the root of the word "green," as emphasized in J. R. R. Tolkien's references to the "green grass" of Doriath. The Sindarin word for grass is *thâr,* while in Quenya the word may be *linquë* or *salquë*. The Sindarin word for grassland is *nan* and for lawn is *parth.*

Distribution and Ecology: Grasses are found nearly everywhere, from extremely cold to very hot and from very dry to aquatic habitats. Grasses are the dominant element in the landscape (resulting in grasslands) in regions where there are periodic droughts, level to slightly hilly topography, frequent fires, and/or particular soil conditions, such as on the North American prairies, South American pampas, African veldt, and Eurasian steppes. Grass flowers are inconspicuous and wind-pollinated, and their small fruits are dispersed by wind, water,

or external transport on animals (i.e., caught in fur or feathers, or in clothing).

Economic Uses: Grasses constitute the plant family of foremost economic importance because of their use as food—about 70 percent of the world's farmland is planted in crop grasses: wheat, barley, and oats in the Near-East; sorghum and pearl millet in Africa; rice in southeastern Asia; and maize/corn in the Americas. These grasses have made possible the rise of civilization in those regions where their cultivation occurs. In terms of current global production, the four most important crops are grasses: sugarcane, wheat, rice, and maize (or corn). Grasses are also used for livestock food and provide the sugar source for the fermentation of several alcoholic beverages (e.g., beer, whiskey, rum).

Description: Usually perennial herbs, often with rhizomes; stems erect to prostrate, jointed, round in cross-section, hollow to solid in internodal regions, short to tall. Leaves alternate, two-ranked, simple and differentiated into a spreading blade, a sheath that closely encircles the stem, with its margins overlapping but not fused, and with a distinct line of hairs or flap of tissue (i.e., the ligule) at the junction of the blade and sheath; the blade linear, flat or folded, with parallel venation. Inflorescences terminal, various. Flowers very tiny, arranged in spikelets, variously flattened, of one to numerous flowers, and each flower surrounded by two bracts (a lemma and palea), with 1–6 usually sagittate anthers and usually two fused carpels, the ovary superior, the two stigmas plumose. Fruit a grain (i.e., small, dry, with a single seed that is fused to the fruit wall) (Figure 7.28).

HART'S-TONGUE (*ASPLENIUM SCOLOPENDRIUM*)

(Spleenwort family [Aspleniaceae].)

> *About it [the stream] stood fir-trees, short and bent, and its sides were steep and clothed with harts-tongue and shrubs of whortle-berry. (LotR 2: VI)*

After escaping from Moria, the remaining members of the Fellowship are led by Aragorn; they followed the road from the Gates southward, along the course of the

Silverlode, toward Lothlórien. Stopping beside a small stream that flowed into the Silverlode, Aragorn tended Sam and Frodo's injuries and discovered Frodo's mithril shirt (Figure 7.29). Firs (likely *Abies alba*) arched over this stream, and growing along its steep banks were whortleberry (probably *Vaccinium myrtillus*) and hart's-tongue fern. What is the significance of their encounter with hart's-tongue at this brief stop? Perhaps it is merely descriptive because this is a common species of moist calcareous slopes and ledges, and it is widely distributed in Europe. However, it is also a beautiful fern and is especially distinctive because of its persistent, elongate, and unlobed leaves—and the species traditionally has been used medicinally. Thus, it may be a signal, telling us that the travelers have reached a wholesome region—after all, soon afterward, the members of the Fellowship heard the rustle of leaves and found themselves at the eaves of the Golden Wood.

Etymology: The common name "hart's-tongue" (a species of spleenwort fern) is derived from the fanciful resemblance of the leaves of this plant, which are simple and unlobed (and thus unusual for a fern), to a deer's tongue. The word "hart" is derived from Old English *heorot* (meaning hart, stag, male of the red deer), from Proto-Germanic **herutaz*, and perhaps from Proto-Indo-European **keru* (horn). The word "tongue" is derived from Old English *tunge*, which is from Proto-Germanic **tungon*, and Proto-Indo-European **dnghwa-*. The scientific Latin name *Asplenium* is based on the Greek *asplenon*, an ancient name for the spleenworts and meaning *a-* (not, without) and *splen* (spleen), a reference to the supposed medicinal properties of these ferns.

Distribution and Ecology: Hart's-tongue (*Asplenium scolopendrium*) is widely distributed across the Northern Hemisphere; it is widespread and common in Europe (where the populations are diploid) and also grows in northwestern Africa, extending eastward to Iran. The fern also occurs in northwestern Asia (China, Japan, Korea, southeastern Russia) and in North America (where it is rare, with a scattered distribution); in both these regions, the plants are tetraploid. This beautiful fern grows in sinkholes, near cave entrances, in limestone ravines, on talus slopes, on limestone cliffs and ledges (often in crevices, with mosses), in rocky areas along streams, and in forests developed on moist, calcareous soils. This fern loves to grow in deep shade and over magnesium-rich limestone. This species is a member of the large genus *Asplenium* (the spleenworts), a group of more than 700 species, which has a nearly worldwide distribution (both tropical and temperate). The group is interesting because both genome doubling (polyploidy) and hybridization are common, and a large number of its species are of hybrid origin.

Economic Uses: Hart's-tongue fern is often used as an ornamental (both outside and indoors) and numerous cultivars have been developed, some of which have crisped leaves. The species is also occasionally used medicinally, especially for digestive disorders.

Description: Herbaceous, evergreen fern, with short, erect, unbranched stem that bears brown scales. Leaves (or fronds), alternate, clustered, simple (and rarely apically divided, cleft), linear to oblong or slightly obovate, with a prominent midvein but the other veins obscure; the apex acuminate to acute, the base cordate, the margins entire and sometimes slightly undulate; the blade with a few scales on the lower surface or nearly glabrous; the petiole elongate, grooved on the upper surface, with scales toward the base. Spore-bearing regions (i.e., the sori) on the underside of the leaf, linear, borne

FIGURE 7.29 *Hart's-tongue.*

along the obscure veins, so more or less perpendicular to the midvein and borne in opposite pairs (one member of each pair on each side of the midvein), with adjacent sori positioned parallel to each other and usually restricted to distal half of the blade, each elongated sorus covered when young by an indusium (i.e., elongate flap), attached along one side. Spores borne in stalked, globose containers (sporangia) that open by means of a vertical annulus (line of thick-walled cells) that is interrupted by the stalk and that curves back when dry, ripping open the sporangium and dispersing the tiny spores (Figure 7.29).

HAWTHORNS, THORNS (*CRATAEGUS* SPP.)

(In the Rose family [Rosaceae].)

> *But Beleg and Gwindor cut the bonds that held him [Túrin], and lifting him they carried him out of the dell, yet they could bear him no further than to a thicket of thorn-trees a little way above. (SILM 21)*

Hawthorns, thorn-bushes, thorn-trees, or simply thorns (all names used in J. R. R. Tolkien's writings) are a characteristic tree both of Beleriand (in the First Age) and from the Shire, Trollshaws, and northern Anduin in the north to Ithilien and Mordor in the south (during the Third Age). The integration of hawthorns into the landscape of the Shire is evident in the hobbits' walking-song (written by Bilbo) in which one line is "Apple, thorn, and nut and sloe, Let them go! Let them go!" (LotR 1: III), and they were likely as common a component of hedges, along with crabapples (*Malus sylvestris*), hazel nuts (*Corylus avellana*), and sloe (*Prunus spinosa*) as they are in rural England today. It is not surprising, therefore, that hawthorns are mentioned at some localities where critical events occurred in the history of both the First and Third Ages. A few of these are mentioned here. As related in *The Hobbit*, Bilbo managed to climb a thorn bush to hide from trolls during his first dangerous encounter on his quest with Thorin and company. The most detailed description of this shrub or tree is in *The Lord of the Rings*, when Aragorn, Gandalf, Théoden, and company, shortly after their confrontation with Saruman, made camp in a hollow "down among the roots of a spreading hawthorn, tall as a tree, writhen with age, but hale in every limb. Buds were swelling at each twig's tip" (LotR3: XI). This camp at Dol Baran was made on March 4, 3019, so of course the thorn tree is not yet in bloom, but spring clearly is not far off, as indicated by the tree's swelling buds. It was at this camp that Pippin looked into the Orthanc-stone. Just a few days later, Frodo and Sam, as they approached the Cross-roads in Ithilien, hid "underneath a tangled knot of thorns" (LotR 4: VII) that had down-slanting branches reaching the ground.

In *The Silmarillion*, three occurrences of hawthorns stand apart from the others and signal turning points in the story. The first is in the chapter on Lúthien and Beren and occurred shortly after their return to Menegroth, after the quest for the Silmaril had been completed and that holy jewel (along with Beren's hand) swallowed by Carcharoth, the ferocious wolf that guarded the entrance to Angband. The joy at Lúthien's return to Doriath was short lived, however, because her father's realm was almost immediately threatened by Carcharoth, who was maddened by the Silmaril. Thingol, King of Doriath, along with Beren, Mablung, Beleg, and Huan (the Hound of

Valinor), set out on a hunt for the wolf. At one point, Huan left their group and entered a hawthorn thicket, but "Carcharoth avoided him, and bursting from the thorns leaped suddenly upon Thingol." Beren, in protecting Thingol, was mortally wounded. At that moment, Huan also leaped from the thicket and attacked Carcharoth: the two fought "and no battle of wolf and hound has been like to it, for in the baying of Huan was heard the voice of the horns of Oromë and the wrath of the Valar, but in the howls of Carcharoth was the hate of Morgoth" (SILM 19). Huan killed the wolf and the Silmaril was removed from its ripped-open belly—and "in fear Mablung took it and set it in Beren's living hand; and Beren . . . held it aloft, and bade Thingol receive it" (SILM 19). The quest had finally been achieved, but at the cost of Beren's life. The second reference to hawthorns is part of the story of Túrin Turambar (see quote at beginning of this entry). Túrin had been captured by orcs who were taking him to Morgoth's stronghold of Angband, but at the edge of the forest Taur-nu-Fuin, a place of dread and dark enchantment, his friend Beleg, an elf of Doriath who had pursued the orcs, and his chance companion Gwindor, found Túrin, unconscious and tied to a tree in an orc camp. They attempted to rescue him as a thunder storm approached. "Then Beleg and Gwindor cut the bonds from the tree, and bore Túrin out of the camp. But he was too heavy to carry far, and they could go no further than to a thicket of thorn trees high on the slopes above the camp" (CoH IX; see also SILM 21) (Figure 7.30). In the hawthorn thicket, Beleg drew his sword to cut the ropes that bound Túrin, but the blade slipped in his hand and he pricked Túrin's foot. Túrin, awakening in rage and fear, killed his best friend, thinking him one of the tormenting orcs. Just then the lightening flashed, and Túrin saw what he had done. He stood "stonestill and silent," (SILM 21) "neath the tangled thorns" (Lays I: line 1324). They buried Beleg in a shallow grave and placed his great bow beside him. The orcs, not finding them, returned to Angband without their prized prisoner, but Túrin—

> a burden bore he   than bonds heavier,
> in despair fettered   with spirit empty
> in mourning hopeless   he remained behind.   (Lays I: 1336–1338)

The third key occurrence of hawthorns is in the tale *Of Tuor and His Coming to Gondolin*, published in the *Unfinished Tales*. Voronwë led Tuor, the cousin of Túrin, to the Hidden Kingdom, which was surrounded by mountainous peaks. Following a dry riverbed, they came to a steep slope "upon which grew a tangled thicket of thorn-trees. Into this thicket the stony channel entered, and there it was still night; and they halted, for the thorns grew far down the sides of the gully, and their lacing branches were a dense roof above it, so low that often Tuor and Voronwë must crawl under like beasts stealing back to their lair" (UT 1: I). With effort they crawled along and eventually came to a great cliff and the mouth of a tunnel—the Guarded Gate. It is fitting that the entrance to Gondolin, the most beautiful of the elven kingdoms in Beleriand, was concealed in a hawthorn thicket because Celtic folklore associates hawthorns with the entrance to the world of fairy. Tuor was admitted, bearing his message to Turgon, King of Gondolin, from Ulmo, Lord of Waters, and eventually would marry Idril, Turgon's daughter. From that union came Eärendil who, with his wife Elwing, would later sail to Valinor with the Silmaril (that had been taken from Morgoth by Lúthien and Beren), bearing the prayer of elves and men: pardon for the Noldor, mercy upon men and

FIGURE 7.30 *Hawthorns, thorns.*

elves, and help in their need. The significance of hawthorns in Tolkien's legendarium is clear, as is seen in their association with these key events in the three great stories of *The Silmarillion*.

Etymology: The English common name "hawthorn" (which is sometimes abbreviated to merely "thorn") is derived from the Old English *hagathorn*, earlier *hæguthorn* (hawthorn), from obsolete "haw" (hedge or encompassing fence) + thorn. "Haw" is derived from Old English *haga* (enclosure, hedge), from Proto-Germanic **hag-* (hedge). "Thorn" is derived from Old English *thorn* (sharp point on a stem or branch, or earlier a thorny tree) from Proto-Germanic **thurnuz*, Proto-Indo-European **trnus* and **(s)ter-n-* (thorny plant), from the root **ster-* (stiff), an allusion to the stiff, sharp-pointed stems (i.e., thorns produced by this plant). Thus, the name hawthorn alludes to the common occurrence of these species in hedgerows and roadsides. The scientific, generic name *Crataegus* is a Latinized form of the ancient Greek *krataigos* (the name of a species of hawthorn growing in the Mediterranean region), from *kratos* (strength) because of its hard wood. The Sindarin word for hawthorn is *toss*, although the word for a sharp-pointed branch (thorn) may be *erék* or *êg*, and the Quenya for hawthorn may be *ektar*.

Distribution and Ecology: Hawthorns, a group of some 230 species (or perhaps even more), are widely distributed in the Northern Hemisphere, with 169 species recognized in North America (north of Mexico), 21 species recognized in Europe, 18 in China, and 2 in Japan. However, the exact number of species, especially in North America, is problematic due to hybridization, polyploidy (i.e., with diploids, triploids, and tetraploids), and asexual reproduction. Hawthorns grow in a wide variety of xeric to wetland habitats; some are sun-loving while others are quite shade-adapted. Characteristics varying among major species groups include the habit; pubescence and orientation of the twigs; length and curvature of the thorns; leaf shape; pubescence (relating to both nonglandular and glandular hairs); surface texture (dull vs. shiny) and extent of leaf-lobing (and whether or not the veins run into the sinuses); inflorescence shape; number of flowers and form of the bracts; and floral characters such as form of the sepal margin, number of stamens and anther color, number of carpels, and color, size and shape of the fruits, including internal characteristics such as the shape and pitting of the segments of the hardened pit. The showy flowers are produced en mass as the shoots emerge in the spring, produce nectar, and attract a great variety of generalist pollinators (e.g., beetles, flies, bees); many are fetid or unpleasant, producing trimethylamine and related compounds, which are also produced by rotting flesh. Thus, although European folklore often associates these beautiful shrubs with spring celebrations and positive attributes such as strength, it is considered unlucky to bring their cut, flowering branches indoors. The small, applelike fruits are dispersed by birds.

Economic Uses: The fruits of some species are used to make jellies, fruit-preserves, or wine, and a few are used as ornamental shrubs because of their showy flowers. Some are used in medicinal teas (for hypertension, heart disease). In Great Britain, hawthorns (usually white thorn, *Crataegus monogyna*) were much used in hedgerows to demarcate property lines. Because the wood is quite hard, it is sometimes used for fence posts or tool handles. Finally, hawthorns are exceptionally valuable in providing food and shelter for birds and mammals; their flowers provide food for many nectar-feeding insects, while their leaves are important for a diverse array of insect larvae.

Description: Usually deciduous trees or shrubs, the bark rough with exfoliating plates or smooth and exfoliating, usually without horizontal lenticels; branches with long and short shoots, and some branches modified, sharp-pointed, forming thorns. Leaves alternate, spirally arranged, simple, but sometimes deeply lobed, ovate to obovate, pinnately veined, with secondary veins running into the lobes or teeth, the apex acute to rounded, the base narrowly cuneate to cordate, the margin serrate to crenate, with variable development of simple, nonglandular hairs, but the marginal teeth often glandular; leaves of nonflowering shoots often different from those of flowering branches; stipules present but quickly deciduous. Inflorescences terminal, but sometimes appearing lateral because of their presence on short shoots, various cymes. Flowers radially symmetrical, bisexual, the sepals five, triangular, the petals five, basally narrowed, white to pale pink; the stamens five to numerous, with anthers globose, white or cream to pink, red, or purple, or pale yellow to yellow or salmon, all borne on a nectar-producing floral cup (= hypanthium), with (1–) 3–5 fused carpels, the styles (1–) 3–5, the ovary inferior. Fruit small to moderate-sized pomes, globose to ellipsoid or pear-shaped, yellow to red, purple, or black, with the core of hard segments (Figure 7.30).

HAZELS AND FILBERTS (*CORYLUS AVELLANA*, *C. MAXIMA*, AND RELATED SPECIES)

(In the Birch family [Betulaceae].)

> *But on a sudden he heard cries, and from a hazel-thicket a young woman ran out; her clothes were rent by thorns, and she was in great fear, and stumbling she fell gasping to the ground. Then Túrin springing towards the thicket with drawn sword hewed down the man that burst from the hazels in pursuit; and he saw only in the very stroke that it was Forweg. (CoH VI)*

Hazels (*Corylus avellana*) are common and widespread in today's Europe, and likely these rather weedy shrubs that commonly grow in hedges or along roadsides were also broadly distributed in the western regions of Middle-earth in the Third Age. Frodo, Sam, and Pippin, while still in the Shire and soon after encountering Gildor and other High Elves, hiked along a lane with "many deep brakes of hazel on the rising slopes at either hand" (LotR 1: III). Later, the four hobbits hid with Strider in "a small patch of thick-growing hazels" (LotR 1: XII) in the Trollshaws, on a slope above the road. Peering through the hazel branches, they could see the road below them and hear the sound of hoofs. They feared the Black Riders, but they heard the faint jingle of bells and Frodo exclaimed: "That does not sound like a Black Rider's horse!" (LotR 1: XII). Soon Strider was leaping down the heather-covered (*Calluna vulgaris* and *Erica* spp.) slope, greeting Glorfindel (see heather, Figure 7.33), who had come from Rivendell in search of them. In contrast to widely distributed hazels, filberts (a closely related species, *C. maxima*) are more restricted and grow in more southeastern regions, occurring today in the Balkan Peninsula. Interestingly, Sam and Frodo encountered filberts in Ithilien, a region far south and east of the Shire (with a much more Mediterranean flora).

Hazels also grew in Beleriand during the First Age and played a role in one of the key events in the life of Túrin while he lived among the outlaws (see opening quote). These multitrunked shrubs often form dense thickets in moist forests or disturbed habitats, and the incident described earlier took place near the homesteads and clearings of the woodmen (close to the River Teiglin), so likely these forests had been impacted by their activities (e.g., cutting of timber, clearing for shifting agriculture, etc.)—an appropriate habitat of hazels and also of hawthorns (*Crataegus* spp.; small thorny trees). Túrin was out walking, just to get away from the squalid camp of the outlaw band that he had recently joined. It was spring, and, as he walked, the sun shone on fresh green leaves and he noticed the woodland wildflowers. "Against his will he remembered the Hidden Kingdom, and he seemed to hear the names of the flowers of Doriath, as echoes of an old tongue almost forgotten" (CoH VI). The dense growth of hazel obscured Túrin's view of the young woman and her pursuer until she burst into the clearing, her clothing torn by hawthorn branches, chased by Forweg. He was the leader of the outlaw band and had attempted to rape her. In an instant, Turin killed him (Figure 7.31). His death changed the dynamics of the band. As Túrin later announced to the group: "I will govern this fellowship now, or leave it. But, if you wish to kill me, set to! I will fight you all until I am dead—or you" (CoH VI). In the end, he was accepted as the head of the band, taking the place of Forweg, and would eventually lead them to Amon Rûdh.

Etymology: The English name "hazel" is derived from the Old English *hæsel*, from Proto-Germanic **hasalaz* (hazel), and Proto-Indo-European **koselo-* (hazel). The scientific name *Corylus* is derived from the classical Latin *corulus* (hazel), which is a cognate of hazel, being derived from the same Proto-Indo-European word. The name "filbert" comes from Middle English *fylberd* and is derived from Anglo-Norman-French *philber*, in reference to St. Philbert, a seventh-century Frankish Abbott, because hazel nuts ripened near his feast day. The name of these trees in Quenya may be *kottulë* or *kotulwë*.

Distribution and Ecology: Hazels or filberts are members of the genus *Corylus*, a genus of ca. 15 species, which is broadly distributed in North Temperate regions (i.e., Europe, Asia, and North America). Three species occur in Europe (European hazel, *C. avellana*, filberts, *C. maxima*, and Turkish hazel, *C. colurna*), while two grow natively in North America (American hazel, *C. americana* and beaked hazel, *C. cornuta*). Eastern Asia is the most diverse, with seven species growing in China and three in Japan (of these, the most widespread are the Siberian hazel, *C. heterophylla*; Manchurian hazel, *C. mandshurica*; and the closely related Japanese hazel, *C. sieboldiana*). Hazels grow in moist to dry, often open woodlands, thickets, rocky slopes, or streamsides, and some species are rather weedy, occurring in waste places, fencerows, hedges, and along roadsides. Characteristics useful in distinguishing the species include stature, extent of development of marginal leaf teeth, patterns of pubescence, and form of the bracts surrounding the nuts, especially whether spreading or forming a tubular sheath. The reduced flowers are in dangling catkins and are wind-pollinated; the nuts are dispersed by rodents (and humans).

Economic Uses: Many species of *Corylus* are commercially important for their edible, oil-rich nuts, especially European hazelnuts and filberts, although all species have edible nuts. European hazels were cultivated as early as Roman times, and domestication (breeding for desirable characteristics in cultivation) seems to have occurred in several regions. Some cultivars likely are of hybrid origin (i.e., *C. avellana* × *C. maxima*), and the

FIGURE 7.31 *Hazels, filberts.*

filbert (*C. maxima*) itself may be only a cultivar of *C. avellana*. The Turkish hazel (*C. colurna*) is sometimes used as a rootstock to prevent suckering in orchards. Hazels were once very important as a source of straight coppice shoots, which were used for hurdles, fences, thatch spars, wattle-and-daub, legume poles, and firewood. Species of *Corylus* also are often used as ornamentals: cultivars with twisted branches (i.e., *C. avellana* "Contorta") or red foliage (e.g., *C. avellana* "Red Majestic" or *C. maxima* "Purpurea") are popular.

Description: Deciduous shrubs or trees, the bark grayish brown, smooth, with horizontal lenticels, but eventually breaking into vertical strips and scales; young branches differentiated into long and short shoots. Leaves alternate, two-ranked, simple, broadly ovate to elliptic, pinnately veined, with secondary veins running into the teeth, the apex acute to acuminate, the base rounded to cordate, the margin double serrate (i.e., with small and large teeth) and occasionally even lobed, with variably developed simple and glandular hairs, petiolate; stipules present. Flowers inconspicuous, radially symmetrical, unisexual, lacking or with a very reduced perianth, in groups of three (when staminate) or two (when carpellate) and associated with a bract complex, densely clustered in dangling, elongate, staminate catkins formed in the previous growing season, ± exposed in winter, and in short, carpellate catkins (and with staminate and carpellate inflorescences on the same plant); staminate flowers (of each bract complex) congested, appearing as a single flower with usually four stamens, and the filament of each stamen divided nearly to the base; carpellate flowers with two fused carpels, with two styles, the ovary nude. Fruit an ovoid or globose nut, these clustered, and each nut associated with and surrounded by two large, fused bracts, these distally toothed or lobed, sometimes spiny, spreading (and nuts partially exposed) to elongated and apically constricted (concealing nuts) (Figure 7.31).

HEATH (*ERICA* SPP.)

(In the Heath or Blueberry family [Ericaceae].)

But still there lived in hiding cold
undaunted, Barahir the bold,
of land bereaved, of lordship shorn,
who once a prince of Men was born
and now an outlaw lurked and lay
in the hard heath and woodland grey,
and with him clung of faithful men
but Beren his son and other ten. *(Lays III: lines 127–134)*

Heaths (various species of *Erica*) typically co-occur with heather (*Calluna vulgaris*) in heathlands, moors, and acidic woodlands of Europe, and likely they also did so in Beleriand and other northern regions of Middle-earth. As described in connection with heather, wild heaths were common in the highlands of Dorthonion surrounding Tarn Aeluin, where Barahir, Lord of the House of Bëor, and his followers hid from the forces of Morgoth (see SILM 18 and 19) after the Battle of Sudden Flame (Figure 7.32). During that battle, Barahir had rescued the elven king Finrod Felagund, allowing him to escape and return to his hidden fortress of Nargothrond. In thanks, he swore an oath of friendship with Barahir and his kin, and, in token of this, he gave Barahir his ring.

In folklore, heath often is associated with good fortune, but the meaning of "the hard heath" here (see initial quote) is almost the opposite. Heathlands develop on poor, acidic soils that are unsuited for the growing of crops, and they reflect Barahir's impoverished condition after that disastrous battle. This once prince of men, allied with the powerful house of Finarfin, found himself almost alone, his allies dead, defeated, or hidden and his followers nothing more than a small band of outlaws. Yet he faced these circumstances with strength and valor, fighting against nearly impossible odds. And although he did not live to see it, his perseverance was rewarded—the ring of Barahir, eventually coming to his son Beren, would play an important role in Beren's quest for a Silmaril and in his gaining the hand of Lúthien Tinúviel in marriage.

Etymology: The word "heath" comes from Middle English *heeth* or *hethe*, from Old English *hæth* (untilled land, wasteland), influenced by the cognate Old Norse word *heithr* (field, heath, moor), derived from Proto-Germanic **haithiz*, from Proto-Indo-European **kait-* (forest, uncultivated land, wasteland). The Latin scientific name *Erica* is derived from the Greek *ereike* (heath, broom).

Distribution and Ecology: Heaths are members of the large genus *Erica*, a group of ca. 860 species that are widely distributed in Europe, Africa, and Asia; a few of these have been introduced and naturalized in North America, Australia, and New Zealand. The group is most diverse in southern Africa, where about 770 species are found; about 20 occur in Europe. Heaths, along with heather (*Calluna vulgaris*), are common low shrubs to trees of wet to dry heathlands, moorlands, bogs, and acidic woodlands across Europe. Their pendulous, cylindric to bell- or urn-shaped flowers usually are pollinated by various insects, especially bees, but some are reduced and wind-pollinated; the fruits can be capsules, releasing tiny seeds dispersed by wind or, in some species, colorful drupes that are eaten and dispersed by birds.

Economic Uses: Several species of *Erica* are used as ornamentals because of their showy pendulous, urn- to bell-shaped flowers.

Description: Evergreen low shrubs to trees, with variously developed simple to branched, nonglandular or glandular hairs. Leaves whorled, simple, tiny, narrowly ovate to oblong or linear, with obscure venation, the apex acute to rounded, the base acute, the margins usually strongly angular-revolute and thus obscuring most to all of the lower surface; stipules absent. Inflorescences axillary, various in form, the floral bracts sometimes mimicking the sepals. Flowers radially symmetrical, bisexual, pendulous, with usually four sepals, usually distinct and shorter than the petals, green; the petals usually four and strongly fused, forming a cylindrical or bell- to urn-shaped corolla, somewhat reduced in wind-pollinated species, white to red or purple; the stamens usually eight, their anthers with or without a pair of appendages (spurs), opening by terminal pores, and usually 3–8 fused carpels; the style elongate, with truncate to capitate or expanded stigma; the ovary superior. Fruit a small capsule or drupe with few to many small seeds (Figure 7.32).

HEATHER, LING (*CALLUNA VULGARIS*)

(The Heath or Blueberry family [Ericaceae].)

Strider sprang from hiding and dashed down towards the Road, leaping with a cry through the heather. (LotR 1: XII)

FIGURE 7.32 *Heath.*

References to heather are frequent and occur in most of the major works of Tolkien's legendarium (i.e., in *The Hobbit, Lord of the Rings, Silmarillion, Unfinished Tales,* and *Lays of Beleriand*). The reason for this is that heather (or ling) is a dominant low shrub of moorlands and heaths, and such open habitats (developing on acidic soils) are widespread in the northern parts of Middle-earth. Heather is specifically referenced by Tom Bombadil, so probably grew on the Barrow Downs near his home—note the line in his song "Wind on the open hill, bells on the heather" (LotR 1: VI), a reference to the pendulous, bell-like flowers of this species. Heather is also mentioned in connection with events occurring in the Trollshaws, near and at Rivendell, at Dol Baran (southern Misty Mountains), and in northern Ithilien during the Third Age, and it also grew in Beleriand (especially in Dorthonion) in the First Age and in the Forostar region of Númenor in the Second Age. In the quote highlighted here, we see Aragorn leaping through a heather thicket in the Trollshaws as he ran to meet Glorfindel (Figure 7.33), shortly before Frodo encountered the Ringwraiths at the Ford of Brunien (near Rivendell). And in *The Hobbit,* we discover that heather grew along the partly hidden path to Rivendell (which was marked with white stones). However, it is in the First Age, as related in *The Silmarillion,* that the most mythologically significant events with a connection to heather (and heathlands) occur. After the Dagor Bragollach (= Battle of Sudden Flame), Dorthonion fell under the control of Morgoth, and the remnants of the people of Bëor, led by Barahir (the father of Beren), were pushed up into the highlands. They lived in the mountainous moors that surrounded a beautiful lake (Tarn Aeluin) where they lived as hunted outlaws: "their bed was the heather and their roof the cloudy sky" (SILM 18). One autumn night, after the death of his father, Beren fled southward, and "the wind that hissed among the heather and the fern found him no more" (Lays III: lines 374–376). Climbing into the Mountains of Terror, he saw afar the forests of Doriath, and, passing through the webs of spiders and the mazes that Melian wove around Thingol's kingdom, he stumbled into the forest of Neldoreth—moving from an open land of heather and heath (*Erica*) to a rich and dense forest of beech (*Fagus*), oak (*Quercus*), and elm (*Ulmus*)—and came at last upon Lúthien Tinúviel.

Etymology: The word "heather" is derived from Middle English words such as *hather, hether, hadder* (and the earlier *hathir*), from Old English **hæddre* (heather), which may have been taken from a Celtic language. "Ling" is derived from Old Norse *lyng,* which also is of uncertain origin. The word "heather" may or may not be related to "heath." The scientific name *Calluna* is based on the Greek *kallyno* (to brush or sweep clean), alluding to its traditional use in brooms. The name in Quenya may be *orikon.*

Distribution and Ecology: Heather or ling (*Calluna vulgaris*) is the only species in its genus and is a characteristic low shrub of moorlands and heaths (but also occurs in bogs and oak-pine woodlands); it is especially common in northern and western Europe, although it is geographically widespread, occurring as far west as the Urals and Asia Minor and extending south into northern Africa. It has been introduced into North America (especially along the Atlantic coast) and also in New Zealand. As is typical of members of the heath or blueberry family (Ericaceae), heather loves acidic soils and open habitats. The flowers are pollinated mainly by various bees (or less commonly by wind), and the tiny seeds are dispersed by the wind.

Economic Uses: In Europe, heather was traditionally used in brooms. It has also been used in basket-making, as a thatch, for mattress stuffing, and as the source of a yellow dye. It is an important food for grouse and is a dominant in heathlands

FIGURE 7.33 *Heather, ling.*

and moorlands managed by grazing, cutting, and fire for grouse hunting. It once was (and occasionally still is) used as a flavoring in beer, in place of or mixed with hops. Heather is used as a honey plant since the flowers are attractive to bees, but the resulting honey is jellylike. Finally, in recent years, heather has become very popular as an ornamental because of its showy, usually pink to rose or purple, more or less pendulous flowers

Description: Evergreen low shrubs, with simple, nonglandular hairs. Leaves opposite and decussate, simple, tiny, narrowly ovate to oblong, with obscure venation; the apex acute to obtuse; the base with a triangular, earlobelike extension on each side of the point of leaf attachment and these basal lobes appressed to the stem; the margins strongly angular-revolute and meeting at a groove along the "lower leaf surface" so that the true lower leaf surface is hidden inside each grooved leaf; stipules absent. Inflorescences each reduced to a solitary, axillary flower, and these clustered near the top of the plant. Flowers radially symmetrical, bisexual, pendulous, with a few sepal-like bracts; the true sepals four, distinct, ovate, longer than the petals, pink to rose or purple; the petals four, partially fused, forming a bell-shaped corolla, pink to rose or purple; the stamens eight, their anthers each with a pair of appendages (spurs), opening by slits, and four fused carpels; the style elongate, with truncate to capitate stigma; the ovary superior. Fruit a small capsule, associated with persistent calyx and corolla and releasing a few small seeds (Figure 7.33).

HEMLOCK, HEMLOCK-UMBELS (*CONIUM MACULATUM*, POISON HEMLOCK, OR PERHAPS *ANTHRISCUS SYLVESTRIS*, COW PARSLEY)

(In the Carrot family [Apiaceae or Umbelliferae].)

The leaves were long, the grass was green,
The hemlock-umbels tall and fair,
And in the glade a light was seen
Of stars in shadow shimmering.
Tinúviel was dancing there
To music of a pipe unseen,
And light of stars was in her hair,
And in her raiment glimmering. *(LotR 1: XI)*

It is remarkable that nearly all references in Tolkien's writings to hemlock, a tall herb with distinctive umbellate inflorescences, are linked to a single dramatic event—the first meeting of Lúthien Tinúviel, the elven-daughter of Thingol of Doriath and Melian of the Maiar, and Beren, the son of Barahir, of the house of Bëor (Figure 7.34). This meeting is described in the *Tale of Tinúviel* (written in 1917); the *Lay of Leithian* (written between 1925 and 1931); in the passage from *The Fellowship of the Ring* (published in 1954) cited above, in which Aragorn chants a portion of the "tale of Tinúviel" while they camp on Weathertop; and in *The Silmarillion* (published in 1977, but written much earlier). In the *Tale of Tinúviel*, these flowers are especially emphasized, and in this early tale Tolkien reports that "the white umbels of the hemlocks were like a cloud about the boles of the trees," which are stated to be elms (*Ulmus*), beeches (*Fagus*), and chestnuts (*Aesculus*) (*The Book of Lost Tales*, part 2, p. 10). It is clear that Tolkien's conception of this meeting changed little if at all over these years. In a letter to his son Christopher (No. 340), written near the end of his life, Tolkien noted that his wife, Edith, was the "source" for Lúthien, and her story "was first conceived in a small woodland glade filled with hemlocks at Roos in Yorkshire," in which Tolkien and his wife spent some time while he was "in command of an outpost of the Humber Garrison in 1917" during the World War I. He added that "In those days her hair was raven, her skin clear, her eyes brighter than you have seen them, and she could sing—and *dance*." Thus, a real Yorkshire meadow, with its hemlocks (*Conium maculatum* or possibly *Anthricus sylvestris*) in full bloom, was taken up to become part of the mythology of Middle-earth.

Because J. R. R. Tolkien used the common name "hemlock," I've taken this plant to be *Conium maculatum* because, within the legendarium, the name "hemlock" is consistently applied to this plant. *Conium* well matches Tolkien's description; however, this species has an unpleasant odor, and it has been suggested that the actual species at Roos in Yorkshire was either hogweed (*Heracleum sphondylium*) or, even more likely, cow-parsley (*Anthriscus sylvestris*): two species that are very similar in appearance to poison hemlock. Both are common in Yorkshire and bloom in late spring. Cow-parsley may be the better choice, since it is taller and its inflorescences more cloud-like. John Garth in his book *Tolkien and the Great War* (p. 238) suggested that the Yorkshire plant was actually *Anthriscus*. He quoted Christopher Tolkien, who reported that his father thought the application of common names among groups of very similar species should be flexible and that, in such cases, technical restrictions to the application of such names were the "pedantry of popularizing botanists—who ought to content themselves with the Linnean names." However, see also the discussion under wood-parsley.

Etymology: The English name "hemlock" was derived from Old English *hymblicae, hymlic,* and *hemlic,* and these are of unknown origin, although the first syllable probably meant something bad (possibly poisonous), and the second is a suffix whose meaning was forgotten before the settlement of Britain by Germanic tribes (see Liberman, 2008, for a lengthy discussion). The word "umbel," a technical term for the inflorescence architecture characteristic of the plant family Umbelliferae (or Apiaceae), in which the flowers have stalks that arise from a common point, is derived from the Latin *umbella* (sunshade or parasol), a diminutive of *umbra* (shadow or shade). Note that the spokes of an umbrella have the same arrangement as the flower stalks in an umbel. The scientific generic name *Conium* is derived from *koneion,* the classical Greek name of the plant, and the specific epithet *maculatum* (spotted) refers to the stem, which is green with scattered purple spots or blotches.

Distribution and Ecology: Poison hemlock (*Conium maculatum*; one of 6 species in the genus), occurs natively nearly throughout Europe and also extends into western Asia and northern Africa; it also has been introduced into North America, where it is now widely distributed. It also has been introduced into temperate South America, Australia, and New Zealand. *Conium maculatum* grows in open woods, open moist habitats, clearings, roadsides or disturbed areas, often near water. The flowers, although individually tiny, are grouped into large, showy umbels and attract a wide variety of nectar-gathering insects (as pollinators). The small fruits are probably dispersed by wind and rain-wash.

FIGURE 7.34 *Hemlock, hemlock-umbels.*

Economic Uses: All parts of the plant are extremely poisonous, containing piperidine alkaloids. Symptoms include degradation of the nervous system, trembling, loss of motor skills, pupil dilation, and eventually coma and death by respiratory failure. The ancient Greeks used hemlock to poison condemned prisoners, as in the well-known case of the philosopher Socrates.

Description (of *Conium maculatum*): Biennial, erect herb to 3 meters tall, with secretory canals and thus a strong, unpleasant, resinous odor; stems hollow, usually with scattered purple spots or blotches. Leaves alternate, spirally arranged, twice- or thrice-pinnately compound with the numerous leaflets ovate, deeply dissected, pinnately veined, glabrous, the apex acute, the base ± attenuate, the margin lobate-serrate, and the leaf base petiolate and sheathing. Inflorescences terminal and axillary, stalked, compound umbels (i.e., umbels of umbels). Flowers radially symmetrical, bisexual, tiny, the sepals five, extremely tiny, the petals five, obovate, the tip inflexed, white, and the stamens five, with globose anthers, all borne at apex of the ovary, which is composed of two fused carpels; the styles two, recurved and basally thickened and nectar-producing; the ovary inferior. Fruit a small, dry, two-parted schizocarp, shortly ovoid, each segment longitudinally five-ridged, and ridges undulate, attached to a central stalk, the seed body with oil canals (Figure 7.34).

HEMP, GALLOW-GRASS (*CANNABIS SATIVA*)

(Hemp or Hackberry family [Cannabaceae].)

[A]nd leaping Beleg
with his sword severed *the searing bonds*
on wrist and arm *like ropes of hemp*
so strong that whetting; *in stupor lying*
entangled still *lay Turin moveless.* *(Lays I: lines 1231–1235)*

Some have connected *Cannabis sativa* (hemp, marijuana) with pipe-weed, but such a linkage is contradicted by a clear reading of prologue *Concerning Pipe-weed,* where the plant is stated to be "a variety probably of *Nicotiana*" (i.e., tobacco). Hemp, in contrast, is actually an important fiber plant, and, in J. R. R. Tolkien's legendarium, it is much more logically linked with rope-making (Figure 7.35). In the initial quote (see also SILM 21) from *The Lay of the Children of Húrin,* we have a clear reference to hemp as a source of a strong and durable fiber used in making ropes. In this dramatic scene, we see the rescue of Túrin who had been captured by orcs and bound with strong cords. Beleg cut the fetters binding Túrin's wrists and arms using his sword Anglachel, but as he cut the ropes around his Túrin's feet, working quickly and in total darkness, he pricked Turin's foot, causing him to awake in fear and accidently kill his best friend. *Cannabis* may also be referenced in the name gallows-weed, which occurs in the eerie poem, *The Mewlips,* in which those searching for the mewlips must travel through "the wood of hanging trees and the gallows-weed" (TATB 9). Gallow-grass is an obsolete name for *Cannabis,* applied because of the importance of hemp in rope-making. The similarity of gallow-grass and gallows-weed is close, and *Cannabis* is often called both "grass" and "weed." The reference to "hanging trees" in this verse is also significant, making it likely that gallow-grass is the source of rope fiber—the two essential components of execution by hanging being a tree and a rope.

Etymology: The English word "hemp" is derived from Old English *hænep* (hemp), from Proto-Germanic **hanapiz*, which was probably a very early borrowing of a Scythian word, which also passed into Greek as *kannabis*. The scientific name, *Cannabis*, is Latin and also comes from the Greek *kannabis*.

Distribution and Ecology: Hemp (fiber plants) and the closely related marijuana (drug plants) both belong to the variable species *Cannabis sativa*, the only species of the genus, which is native to Asia but has been widely naturalized in Europe and eastern North America. Weedy populations exist (in which the achenes have a basal constriction, and thus easily disarticulate and disperse), as well as cultivars, selected either for fiber production (hemp) or drug/medicinal usage (marijuana); both sets of cultivars have nonabscising achenes. The inconspicuous flowers are wind-pollinated, and the achenes are dispersed by wind, water, or can be eaten by birds (and germinate after passing through their digestive track).

Economic Uses: The fiber cultivars (i.e., hemp; often treated as *C. sativa* subsp. *sativa*) have strong and elongate fibers, and their stems are used in making rope, paper, canvas, fish nets, and textiles. Some cultivars have been selected for their seeds, which are the source of nutritious oil; the seeds are also used birdseed mixes. The oil, however, is also used in lacquers, paints, soaps, and as a fuel. Those cultivars that are high in cannabinoids, such as tetrahydrocannabinol (THC), are drug plants (marijuana; often considered as *C. sativa* subsp. *indica*); the THC produced in their glandular hairs affect brain function (binding to CB_1 receptors), and some of these cultivars are also used medicinally (e.g., relief from multiple sclerosis, cerebral palsy, glaucoma, nausea, convulsions).

Description: Annual herbs, with a taproot, with simple, nonglandular and gland-headed hairs; stems erect, simple to well branched. Leaves opposite and decussate below, becoming alternate above, borne along the stem, with cystoliths (microscopic concretions composed of calcium carbonate) at the base of some hairs, palmately compound (and venation also palmate), with 3–9 leaflets, each leaflet ovate to elliptic, often narrowly so, with pinnate venation, the apex acuminate, the base cuneate, the margin serrate, the leaves petiolate; stipules present. Inflorescences axillary, the cymose floral clusters concentrated toward the tips of the shoots; the flowers unisexual, with staminate and carpellate flowers usually on different plants, and radially symmetrical, the staminate flowers with five greenish tepals and five stamens, these opposite the tepals; the carpellate flowers each surrounded by a small bract that is covered with gland-headed hairs, with reduced perianth, appressed to base of ovary, with two fused carpels; the ovary superior, with two elongate stigmas. Fruit an achene, ovoid, more or less enclosed by the perianth (Figure 7.35).

HOLLY (*ILEX AQUIFOLIUM*, AND RELATED SPECIES)

(In the Holly family [Aquifoliaceae].)

The travelers reached a low ridge crowned with ancient holly-trees whose grey-green trunks seemed to have been built out of the very stone of the hills. Their dark leaves shone and their berries glowed red in the light of the rising sun. (LotR 2: III)

FIGURE 7.35 *Hemp, gallow-grass.*

In the Third Age of Middle-earth, European holly (*Ilex aquifolium*) was most common in Eregion, the region in Eriador between the Glanduin and Bruinen Rivers, at the western edge of the Misty Mountains, where the elven rings were made. The name is Sindarin for "Land-of-Holly," and this region (called Hollin in Westron, the language spoken by hobbits and men in Eriador) had been a Noldorin realm in the Second Age, where Celebrimbor, a grandson of Fëanor, had been the greatest of their craftsmen. The elves had used holly to mark the borders of their land, and many of these ancient trees along with their descendants still grew in the Third Age. In the preceding quote, we see the border of Hollin—a low ridge crowned with ancient holly-trees—through the eyes of Frodo as he and the other eight members of the Company journeyed south from Rivendell. The special characteristics of holly are emphasized in this description: their smooth gray trunks, dark green leaves that are held on the tree all winter long, and their bright red fruits, taken by winter-resident birds. Holly also grew along each side of the road leading to the doors of Moria, but nearly all of these trees had died, leaving only two, "larger than any trees of holly that Frodo had ever seen or imagined" (LotR 2: IV), one on each side of the hidden Doors of Durin (Figure 7.36); these two trees were drawn by Tolkien (see *Moria Gate*, colored pencil illustration, in W. G. Hammond and C. Scull, 1995, *J. R. R. Tolkien: Artist and Illustrator*, p. 157; see also the watercolor by Alan Lee in the illustrated version of *The Lord of the Rings*, opposite p. 320). Holly is a tree associated with winter festivities, its lustrous green leaves and bright red berries remind us of the joys of life continuing even in wintertime, and its beauty contributes to the wholesome air of Hollin, a land that had not totally forgotten the elves that once dwelt there. However, it would be a mistake to think that holly, now widespread across Europe, was restricted to Hollin. It probably also occurred in the Ered Luin, in the far west, because holly had grown in Beleriand during the First Age. It is related in the *Children of Húrin* that Túrin and his outlaw band were sheltering in a holly thicket when they saw Mîm and his two sons, each carrying a great sack—an encounter that would eventually lead Túrin to Amon Rûdh and a new (although temporary) home.

Etymology: The English name "holly" is derived from Old English, *holegn*, from the Proto-Germanic **hulin-*, which may be derived from the Proto-Indo-European root **kol-*, prickly, in reference to the spine-margined leaves of the European holly (*Ilex aquifolium*). The generic name, *Ilex*, despite its use by C. Linnaeus for the hollies, is from the classical Latin, *ilex*, the holm oak (*Quercus ilex*), which was possibly a loan word from an extinct non-Indo-European language. The word "ilex," when used by Tolkien in *The Lord of the Rings*, always refers to the holm oak (see oaks). The Sindarin word for holly tree is *ereg* or *eregdos*, as seen in the name Eregion (= Hollin), the former elf kingdom just west of Moria. The common name *ercasse*, perhaps meaning prickly-leaf, was also applied to holly by the Noldorian elves, and in Doriath the name *regorn* (pl. *regin*) was used.

Distribution and Ecology: Hollies, a group of about 400 species, are widely distributed, occurring in open to dense, temperate to tropical forests, and are especially diverse in the Americas and eastern Asia; three species occur in Europe, the most common of which is *Ilex aquifolium*. This species occurs natively in northern Africa, western and southern Europe, and western Asia (extending as far east as Iran); it is naturalized in southeastern Australia, the west coast of the United States, and in Hawaii. In eastern North America, the most common species are American holly (*I. opaca*), dahoon (*I. cassine*), possumhaw (*I. decidua*), winterberry (*I. verticillata*), and yaupon (*I. vomitoria*).

FIGURE 7.36 *Holly.*

More than 200 species can be found in China, and 18 occur in Japan, especially in montane habitats. *Ilex arnhemensis* is native to tropical forests in northern Australia. Their flowers are pollinated by various insects, especially bees. The brightly colored, fleshy fruits usually remain on the plant for a long time and are dispersed by birds. The tiny embryo matures after being dispersed; germination is slow, often requiring 1–3 years.

Economic Uses: Several species of holly, such as European holly (*I. aquifolium*), American holly (*I. opaca*), box-leaved holly (*I. crenata*), yaupon (*I. vomitoria*), and winterberry (*I. verticillata*), are important ornamentals, used because of their often evergreen leaves and bright red fruits, and many cultivars (or hybrids) with variant leaf shapes or habits have been developed. Hollies are traditionally used as Christmas decorations, especially in wreaths, but their use in European winter festivals certainly predates Christianity (holly was considered sacred by the Druids; it was associated with Saturn by the Romans and used as a decoration in the agricultural festival, Saturnalia, held near the Winter Solstice). Some species have spine-margined leaves, making them popular hedge plants. Their wood is white and is used in veneers, inlay work, and in musical instruments. Three species have leaves that are high in caffeine and theobromine and are used in stimulating teas: yaupon (*I. vomitoria*) of the southeastern United States; yerba maté (*I. paraguariensis*) of Argentina, Paraguay, and southern Brazil; and *I. guayusa* of Ecuador, Peru, and Colombia. Some other hollies (e.g., winterberry and gallberry holly [*I. glabra*]) are also occasionally used as teas, although they do not contain caffeine. Finally, hollies are important in providing food for wildlife.

Description: Evergreen or occasionally deciduous trees or shrubs, the bark smooth, gray; hairs, when present, simple. Leaves alternate, spirally arranged, simple, ovate to obovate, sometimes narrowly so, pinnately veined, the secondary veins forming loops, the apex and base quite variable (and those of *I. aquifolium* spinose-acuminate to spinose-acute at apex, and acute to obtuse or cuneate at base), the margin entire to serrate, the teeth sometimes spinose (as in *I. aquifolium* and *I. opaca*), typically dark green and glossy on the upper surface, short to long petiolate; stipules present but tiny, usually brown or black and triangular. Inflorescences axillary, in small cymes, or reduced to a solitary flower. Flowers radially symmetrical, unisexual, with the staminate flowers on one plant and the carpellate flowers on another, and the staminate flowers with conspicuous nonfunctional stamens, the sepals usually 4–6, inconspicuous and fused, the petals usually 4–6, only slightly fused so that each petal forms a well-developed lobe of the corolla, white; the stamens usually 4–6 and slightly fused to the base of the corolla, and usually 4–6 fused carpels; the style very short or lacking, so capitate or discoid stigma sitting atop globose, superior ovary. Fruit a colorful (red, orange-red, purple-black, pink, or white), globose drupe with a broad, more or less circular and persistent stigma and usually 4–6 pits (Figure 7.36).

HORSE-CHESTNUTS, CHESTNUTS (*AESCULUS* SPP.)

(In the Soapberry family [Sapindaceae].)

> *A little way beyond the battle-field they made their camp under a spreading tree: it looked like a chestnut, and yet it still bore many broad brown leaves of a former year, like dry hands with long splayed fingers; they rattled mournfully in the night-breeze. (LotR 3: II)*

Horse-chestnuts only occasionally occur in Tolkien's legendarium, and, when they are mentioned, they are invariably called chestnuts (but do not represent the true chestnut, *Castanea* spp., Fagaceae). Yet they are still significant. Aragorn, Legolas, and Gimli camp at the edge of Fangorn forest under a spreading chestnut (Figure 7.37), after days of following the trail of Merry and Pippin (Figure 7.28). The palmately compound leaves characteristic of horse-chestnuts are clearly indicated in the comparison of its leaves to "dry hands with long splayed fingers." Merry and Pippin see chestnutlike Ents at the Entmoot—and they are similarly described, having "large splayfingered hands" (LotR 3: IV). The striking flower clusters of European horse-chestnut (*Aesculus hippocastanum*) are described in the *Lay of Leithian*: as Lúthien danced to the haunting music of Daeron, "*The chestnuts on the turf had shed their flowering candles, white and red*" (Lays III: lines 515–516). The flowers of this species are initially white with yellow nectar guides, but the guides become strongly red-flushed as the flowers age, a signal to their hymenopteran pollinators. Finally, horse-chestnut trees are beautifully illustrated in the color plate *The Hill: Hobbiton-across-the-Water*, one of Tolkien's illustrations in *The Hobbit* (see also W. G. Hammond and C. Scull, 2012, *The Art of the Hobbit by J. R. R. Tolkien*, fig. 11): their erect inflorescences immediately distinguish them. In *The Scouring of the Shire*, we discover, along with Sam and the other hobbit travelers, that Sharkey's [= Saruman's] men have cut them all down and that the Old Grange had been knocked down, its place taken by rows of tarred sheds. The party tree has also been cut, and its trunk and branches are lying dead in the field. Bag End is a dump, the garden full of huts and sheds, and piles of refuse are everywhere. Sam exclaims, "This is worse than Mordor!" knowing well its beauty before it was "all ruined" and Frodo replies, "Yes, this is Mordor . . . Just one of its works" (LotR 6: VIII). We see here, in the heart of the Shire, the working of evil—environmental destruction, based on the view that the natural world should dominated, exploited for economic gain. This technological approach—seeking to understand through destruction—is very characteristic of Saruman. As Treebeard told Merry and Pippin, Saruman "has a mind of metal and wheels" and "does not care for growing things" (LotR 3: IV). Such an approach eventually leads to the near total environmental devastation (and absence of life) that Frodo and Sam had observed in Mordor. As Frodo observes ironically, Saruman was doing the work of Mordor all along, even when he thought he was working only for himself.

Etymology: The English name "horse-chestnut" may have originated from the erroneous belief that the tree was a kind of chestnut, together with its medicinal use in treating respiratory problems in horses. On the other hand, it has been proposed that "horse" is a corruption of the Welch *gwres*, meaning hot or pungent, and in that case the name would distinguish these trees that bear inedible nuts from the true chestnut, which has sweet, edible nuts of similar appearance. The Latin name *Aesculus* was derived from *esca*, meaning food, and was originally applied to a species of oak with edible acorns. The name, as used for the trees considered here, may have been given ironically because their nuts, although similar in form to acorns or chestnuts, are bitter and unfit for food. The Quenya name perhaps is *mavoisi*.

Distribution and Ecology: Horse-chestnuts, a group of 12 species, have a broad distribution in temperate to subtropical forests of the Northern Hemisphere. The European horse-chestnut (*Aesculus hippocastanum*) is native to a small area region in the Balkan Peninsula but has been widely cultivated (and often has naturalized) in

FIGURE 7.37 *Horse-chestnuts.*

Western Europe, North America, and temperate Asia. Seven species occur natively in North America, with California buckeye (*A. californica*) the most widely distributed in the west, and five species growing in the east: the most widespread being the Ohio buckeye (*A. glabra*) and red buckeye (*A. pavia*). Four species are native to Asia, growing from India to Japan. Species are distinguished by characteristics such as the number, size, and shape of the leaflets; their margins and apices; the number of petals; their shape, color, and pubescence; and the shape and form of the capsules (smooth to very prickly). The showy flowers, borne in erect clusters, are pollinated by insects (bees) or birds, and the large seeds are animal dispersed.

Economic Uses: Horse-chestnuts (also called buckeyes) are widely used as ornamental trees, especially in urban settings; they also provide timber. The seeds are used medicinally, especially for vascular and stomach problems.

Description: Deciduous trees or shrubs, with gray to gray-brown and smooth, scaly, or fissured bark. Leaves opposite, decussate, palmately compound, with 4–11 leaflets, these ovate to obovate, pinnately veined, with secondary veins running to the margin, and often into the teeth, or gradually diminishing toward margin; the apex usually acuminate, the base various, but often narrow, the margin crenate to serrate, with variably developed, simple to branched hairs, sessile to petiolate; stipules absent. Inflorescences terminal, erect, elongate cymes. Flowers bisexual or staminate (and both types together on a single plant), bilaterally symmetrical, large and showy, the sepals five, variably fused, often forming a cup-shaped or tubular structure; the petals four or five, the basal portion of each much narrowed (i.e., clawed) and occasionally with an appendage (or more characteristically, the appendage absent, but the margins of the claw often widened, clasping); the upper petals usually longer and narrower than the lower ones, yellow, red, or white with nectar guides; the stamens 5–8, and a nectar disk positioned between the stamens and perianth, with three fused carpels; the style elongate, with a tiny stigma; the ovary superior. Fruit a globose, thick-walled, smooth to more or less stoutly prickly capsule, opening by three valves to release one or two (rarely three) large, smooth seeds, each of which has a large scar (Figure 7.37).

IVY (*HEDERA HELIX*, AND RELATIVES)

(The Ginseng family [Araliaceae].)

> *The entrance to the path was like a sort of arch leading into a gloomy tunnel made by two great trees leant together, too old and strangled with ivy and hung with lichen to bear more than a few blackened leaves. (Hobbit VIII)*

Mirkwood was once Greenwood the Great, a beautiful forest that was home to numerous "birds of bright song" (see "Of the Rings of Power," in *The Silmarillion*), but as the Third Age progressed, the Dark Lord arose in the southern regions of that forest (at Dol Guldur, where Sauron long dwelt) and a darkness crept northward through the forest, which eventually became a place of fear where fell beasts and evil creatures roamed. The forest was no longer a wholesome place—and this is immediately evident to Bilbo and the dwarves, even at the entrance to the elven path (see initial quote, Figure 7.38). English ivy (*Hedera helix*) has the ability to strangle and kill even full-grown trees by girdling their trunks with its woody branching stems that climb upward using adventitious aerial roots. The ivy's branches spread out in the canopy, cutting off the tree's life-giving access to sunlight. The two great trees—probably oaks—at the entrance have been strangled in this way and are nearly dead, so weakened that they can bear only a few diseased and blackened leaves. Their dead branches have then been colonized by opportunistic epiphytes, such as hanging and beardlike lichens (*Usnea* spp.) that add a sense of foreboding to the scene. The power of Sauron has perverted the once beautiful forest, and the relationships between its plant and animal inhabitants are now out of ecological balance—as was perceived from afar by Haldir, an elf of Lothlórien, when he stated that in southern Mirkwood "the trees strive one against another and their branches rot and wither" (LotR 2: VI). Yet the situation is not hopeless' Mirkwood in places retained some beauty—because the power of Sauron was contested by that of Thranduil and the wood elves—and this beauty was experienced by Frodo when he climbed a tall oak and saw hundreds of velvety black butterflies fluttering above the dark green leaves. Ivy is evident climbing the large tree in the center of Tolkien's ink-wash illustration of Mirkwood, which was included in some editions of *The Hobbit*, and it is even more apparent in the color drawing, *Taur-na-Fúin*, on which this was based, and which he had made for *The Silmarillion* (see figs. 47 and 48 in W. G. Hammond and C. Scull, 2012, *The Art of the Hobbit by J. R. R. Tolkien*). This vine can also be seen in Alan Lee's illustration of the Old Forest in the illustrated edition of *The Lord of the Rings* (opposite p. 128). Each of these two forests, although different (i.e., the first dominated by pines [*Pinus*] and the second by mixed hardwoods such as oaks [*Quercus*] and beeches [*Fagus*]), originally had been beautiful, but both had been turned to horror and evil, Taur-na-Fúin by Morgoth and Mirkwood by Sauron.

Etymology: The name "ivy" is derived from Middle English *ivi,* and Old English *ifig,* from Proto-Germanic **ibahs* (ivy). The scientific name *Hedera* is taken from the classical Latin name of the plant, which is derived from Proto-Indo-European **ghed-* (to seize, grasp, take), perhaps because this liana tenaciously grasps its substrate. The Sindarin word for this plant may be *ethil,* while in Quenya it is perhaps *etl* or *etil.*

Distribution and Ecology: The ivies are a small group (the genus *Hedera*), containing about 10 species, and they are distributed from the Mediterranean region to eastern Asia. The English or European ivy (*H. helix*) is the best known; it grows across Europe and western Asia and has been introduced in North America, Australia, and New Zealand, where it often becomes invasive. It can kill young trees by overgrowing them and can even kill older, established trees by girdling them and outcompeting them for water, nutrients, and sunlight. The plants when young and growing in forest understory conditions are very tolerant of shade. The flowers produce nectar and are pollinated by numerous insects, and the winter-ripening, purple-black or orange-yellow drupes attract a diverse array of birds, which disperse the fruits.

Economic Uses: Many cultivars of English or European ivy (*H. helix*) have been developed, and the species is an important ornamental (used as a liana and often as a building- or wall-covering or as a ground-cover). Ivy has also been used medicinally, and in some people (i.e., those allergic to falcarinone polyacetylenes) causes contact dermatitis. It was once thought to counteract the effects of alcohol and was, therefore, used as a sign for taverns (e.g., the Ivy Bush, an inn of the Bywater region in the Shire).

Description: Perennial, evergreen, creeping or climbing liana, with secretory canals and thus a strong, rather unpleasant, resinous odor, and with stellate hairs

FIGURE 7.38 *Ivy.*

or peltate scales; stems dimorphic, the juvenile with numerous aerial roots (used in climbing); palmate, usually three- to five-lobed leaves; and the reproductive lacking roots and with unlobed leaves. Leaves alternate, usually spirally arranged, simple, sometimes lobed, ovate to elliptic, palmately veined, the apex acuminate or acute to rounded or emarginate, the base obtuse or broadly cuneate to cordate, the margin entire, and the petiole slightly sheathing stem at its base; stipules present. Inflorescences terminal, consisting of a single umbel or an axis bearing several globose umbels. Flowers radially symmetrical, bisexual, the sepals five, extremely tiny; the petals five, ovate-triangular, yellowish green; and the stamens five, with ellipsoid anthers, all borne at apex of the ovary, which is composed of five fused carpels; the style single, short, basally thickened and nectar-producing; the stigmas lobed; the ovary inferior. Fruit a purple-black or orange-yellow, globose drupe with 3–5 seedlike pits (Figure 7.38).

KINGSFOIL, ATHELAS (BASED, AT LEAST IN PART, ON COMFREY, *SYMPHYTUM* SPP.)

(In the Borage family [Boraginaceae].)

> *"These leaves," he said, "I have walked far to find; for this plant does not grow in the bare hills; but in the thickets away south of the Road I found it in the dark by the scent of its leaves." He crushed a leaf in his fingers, and it gave out a sweet and pungent fragrance. "It is fortunate that I could find it, for it is a healing plant that the Men of the West brought to Middle-earth. Athelas they named it." (LotR 1: XII)*

The identity of this healing herb has been much discussed, but it is most likely a species of *Symphytum* (and thus a relative of the medicinal herb comfrey). The plant is described as an "evergreen" herb having "broad and hoary" leaves (Lays III: lines 3118–3121) that have a "sweet and pungent fragrance" (LotR 1: XII), thus allowing Aragorn to find them in thickets south of Weathertop even in the dark. In the healing of Faramir, Aragorn crushed the leaves and placed them in steaming water—and all standing nearby in the Houses of Healing felt their hearts lightened; the fragrance is described as "like a memory of dewy mornings of unshadowed sun in some land of which the fair world in Spring is itself but a fleeting memory." It reminded the healer Ioreth of the roses (*Rosa* spp.) of Imloth Melui when she was a child. In the healing of Merry, the fragrance is "like the scent of orchards, and of heather [*Calluna vulgaris*] in the sunshine full of bees," and for those in Éowyn's room it was an "air wholly fresh and clean and young" (LotR 5: VIII). The preferred habitats, description of the leaves, and medicinal importance all match comfrey well, but that plant dies back in the winter, so cannot be considered evergreen. Additionally, comfrey is only weakly fragrant. Other plants that have been suggested as the inspiration for kingsfoil are briefly considered here.

The wintergreens (*Gaultheria* spp., in the heath family Ericaceae) are evergreen shrubs or subshrubs that are aromatic because they contain wintergreen oil (methyl salicylate). They are used medicinally, especially to reduce inflammation and for pain relief. (Methyl salicylate is also used as a flavoring agent in candy, toothpaste, chewing gum, etc.) Species of *Chimaphila* (also in the Ericaceae) grow in similar habitats, are also called wintergreen (or prince's pine), contain methyl salicylate, and have a variety of medicinal uses. However, neither *Gaultheria* nor *Chimaphila* have broad, grayish white leaves—in fact, their leaves are fairly small, green (or in *C. maculata,* green marked with a white stripe down the midvein), quite leathery, not easily bruised or torn, and they are shrubs or subshrubs, not herbs. Some have compared kingsfoil to sweet basil (*Ocimum basilicum,* of the plant family Lamiaceae). This mint is highly aromatic (containing thymol-essential oils and also linalool, a terpene alcohol) and is used as a culinary and medicinal herb. The name "basil" is derived from Old French, *basile,* from Latin *basilicum* and Greek *basilikon,* meaning "royal plant," from *basileus* (king), perhaps because the plant was used in making royal perfumes. Thus, like kingsfoil, basil is linked to kingship, and both are highly aromatic herbs. However, sweet basil grows only in sunny, open habitats (so would never be found in shaded thickets, as is the case with kingsfoil); also, it is native to southern Asia and, like nearly all of the ca. 65 species of *Ocimum,* is a plant of subtropical to tropical regions. The leaves of *O. basilicum* are intensely green to olive or yellowish green, so cannot be described as hoary. Finally, plantain (*Plantago* spp., especially *P. major,* of the snapdragon family, Plantaginaceae) has been compared to kingsfoil, and both are medicinal herbs. *Plantago major* contains aucubin (and antimicrobial compound), allantoin (see "Economic Uses"), and mucilage, among other compounds, and its leaves can be used as a poultice in treating wounds, stings, and sores to facilitate healing. The leaves of *Plantago major* are broad (although the leaves of some other species in this genus are quite narrow), but they are not gray-green. In addition, none of the species of *Plantago* is aromatic. In summary, we think it most likely that the primary inspiration for Tolkien's athelas was the medicinal herb comfrey—because it best matches the morphological characters of kingsfoil—but he may have added elements of other plants, such as the persistently green leaves of wintergreen and the sweet fragrance of basil (or some other mint).

The Nazgûl are the Ringwraiths, undead and twisted beings unable to find the release of death, always enduring a deadly and wearisome lengthening of days, and totally subservient to the will of Sauron. They destroy the will to live of those around them; coming near these creatures causes the Black Breath or Black Shadow—the result of an evil breath or air that brings only terror, despair, and death. So, a part of Aragorn's cure for those afflicted with this condition is to provide the opposite—good breath or good air—which is the virtue of *athelas,* especially when it is "in the king's hand lying" (LotR 5: VIII). Aragorn used the herb on Weathertop, applying it to Frodo's wound (Figure 7.39), and again after their escape from Moria, using it to soothe the wounds received by Frodo and Sam (Figure 7.29), but it is most prominently treated, as discussed earlier, in the chapter "The Houses of Healing," where he uses athelias in calling Faramir, Éowyn, and Merry back to life, fellowship, and joy.

Etymology: The Westron name, represented by English, "kingsfoil," contains the Old French word, *foil,* as in cinquefoil, and thus means "king's leaf." *Athelas,* the Sindarin name of this plant, may be derived from *athaya* (perhaps meaning helpful, beneficial, or, alternatively, it may be related to the Quenya word *asëa* (leaf) and *las* (leaf). Therefore, the name could mean "beneficial-leaf" or "leaf-leaf," but the latter seems problematic. The name in Quenya, *asëa aranion,* is derived from *aranion* (of the kings), and the name thus translates as "leaf of the kings" or perhaps "beneficial [leaf] of the kings." Tolkien perhaps adapted *athelas* from the Middle English *aethel* (noble by birth or character, a chief or lord), which again connects the plant to nobility and kingship. The precise identity of the plant is much debated, but if linked to the medicinal herb comfrey, the following may be relevant. The English name "comfrey" is derived from Middle English *cumfirie,* and Latin *confervere* (to grow together), an allusion to the traditional use of the herb in healing broken bones. The scientific name of these plants, *Symphytum,* is derived from the Greek *symphusis* (to unite or grow together) and *phyton* (plant), another reference to its folkloric use in healing.

Distribution and Ecology: Kingsfoil or athelias is stated by Aragorn to be introduced from Númenor into Middle-earth by the Men of the West and "grows now sparsely and only near places where thy dwelt or camped of old" (LotR 1: XII). It is more common farther south, in Gondor (and it possibly once grew in Beleriand, see Lays III: 3118–3121, and was used by Lúthien to treat a wound received by Beren). Kingsfoil grew in thickets and woodland glades and also occurred in damp, disturbed sites, such as trail sides. No plant currently exists that matches the healing properties of kingsfoil, but if the connection with comfrey is valid, then related species (i.e., *Symphytum officinale, S. asperum,* their hybrid, *S.* × *uplandicum,* and relatives, totaling some 35 spp.) grow widely in Eurasia and are also cultivated and naturalized in many temperate regions.

Economic Uses: Kingsfoil is a healing herb, effective in treating wounds and, in the hands of the king, a cure for the Black Breath, an illness resulting from close exposure to the Nazgûl. Comfrey (*S. officinale, S. asperum,* etc.) was once highly prized as a medicinal herb and used topically, especially for treatment of bronchial problems, broken bones, sprains, arthritis, burns, skin ailments, wounds, and inflammation. The plants contain allantoin, which promotes healing in connective tissue through proliferation of new cells; allantoin also increases smoothness of the skin and is used in many cosmetics and sun-care creams (as well as in toothpaste, mouthwash, and shampoos). However, comfrey also contains pyrrolizidine alkaloids, which are poisonous, especially if taken internally. Species of *Symphytum* also are grown occasionally as garden ornamentals.

Description (based, in part, upon *Symphytum* spp.): Perennial, evergreen herbs; stems erect, with simple hairs. Leaves alternate, spirally arranged, simple, broadly to narrowly ovate or elliptic leaves, pinnately veined, the secondary veins forming loops, the apex acute, the base narrowed, the margin entire, both surfaces hairy (making them appear grayish white), the upper leaves progressively smaller, and all leaves giving off a sweet and pungent fragrance, especially when crushed; stipules absent. Inflorescences terminal, branched, with each branch apically curled and one-sided (i.e., in scorpioid cymes). Flowers bisexual, radially symmetrical, more or less pendulous, with five green, fused sepals, five fused petals, blue or purple, forming a broadly tubular and bell-shaped corolla, bearing appendages near the summit of the tube, and five stamens, these also arising from the corolla tube, and with two fused carpels, the style short, single, the ovary superior and four-lobed. Fruit schizocarpic, usually comprised of four brown nutlets (Figure 7.39).

FIGURE 7.39 *Kingsfoil, athelas.*

LABURNUMS, FIELD OF CORMALLEN, CULUMALDA (*LABURNUM ANAGYROIDES, L. ALPINUM*)

(The Legume family [Fabaceae or Leguminosae].)

> *For the Field of Cormallen, where the host was now encamped, was near to Henneth Annûn, and the stream that flowed from its falls could be heard in the night as it rushed down through its rocky gate, and passed through the flowery meads into the tides of Anduin by the Isle of Cair Andros. (LotR 6: IV)*

The word "laburnum" is only used once in *The Hobbit*, when Bilbo compares Gandalf's fireworks to "great lilies and snapdragons and laburnums of fire" (Hobbit I), and it is not used at all in *The Lord of the Rings*. However, the latter contains a cryptic reference to these beautiful trees because the field in Ithilien in which the victory over Sauron is first celebrated is called the Field of Cormallen. As explained by Wayne Hammond and Christina Scull (2005, *The Lord of the Rings: A Reader's Companion*, pp. 625–626) in J. R. R. Tolkien's unfinished index to the *Lord of the Rings*, Tolkien defined Cormallen as meaning "golden-circle" in Sindarin and stated that it was called this "after the laburnum that grew there." This unpublished index also included an entry for the word *culumalda* (a word not used in *The Lord of the Rings*) and provided the definition: "a tree with hanging yellow blossoms (prob[ably] a laburnum) growing in Ithilien espec[ially] at Cormallen." If we imagine laburnums, with their long, pendulous racemes of yellow pealike blossoms, growing around the Field of Cormallen, the beauty of the scene is certainly enhanced. The Númenórean tree laurinquë may represent another cryptic reference to laburnums. The laurinquë was grown as an ornamental tree because of its beautiful "long-hanging clusters of yellow flowers" (UT 2: I), and it was erroneously thought by some to be descended from Laurelin, the Golden Tree of Valinor. However, Tolkien's conception of both the golden Laurelin, as discussed in the chapter on the Two Trees, and the beautiful laurinquë of Númenor may be imaginatively derived from his observations of laburnum trees with their beautiful hanging clusters of yellow flowers.

Etymology: Laburnum is both the scientific and common name of these trees (although they are also called golden chain trees), and both are derived from the ancient Latin name *laburnum*, which is of unknown origin, but perhaps was taken from Etruscan. In Quenya these trees are called *lindeloktë* (literally, singing-clusters).

Distribution and Ecology: The two species of *Laburnum* (i.e., *L. anagyroides*, common laburnum and *L. alpinum*, alpine laburnum) are native to mountainous regions of southern Europe (from France to the Balkan Peninsula), but they (along with their hybrid, *L.* × *watereri*) are widely cultivated in temperate regions. The fruits of *L. anagyroides* are hairy whereas those of *L. alpinum* are glabrous. The common laburnum is sparingly naturalized in North America. The showy flowers are pollinated by various bees while the elongate legumes open and toss out the seeds.

Economic Uses: Laburnums are grown as ornamental trees because of their beautiful hanging racemes of yellow pealike flowers; their hard and dark wood is also used in musical instruments and in decorative inlays.

Description: Trees or shrubs, the branches erect, with or without simple hairs, and thorns absent. Leaves alternate, spirally arranged, compound, and each with three leaflets, these ovate to obovate, with midvein and pinnate secondary veins, the apex acute to obtuse, the base similar, the margin entire, the leaf with a long petiole; stipules present. Inflorescences axillary or appearing to be terminal, pendulous racemes. Flowers bilaterally symmetrical, the sepals five, fused, slightly two-lipped, the petals bright yellow, modified, the uppermost forming the standard and positioned on the outside in bud, the two lateral, the wings, and the two lowermost, fused, forming the keel and corolla, thus like a pea-flower; the stamens 10, fused into a tube, with alternating large and small anthers, and the single carpel elongated, with an upcurved style; ovary superior. Fruit an explosively dehiscent, hairy or glabrous legume, with black or brown seeds.

LARCHES (*LARIX* SPP.)

(In the Pine family [Pinaceae].)

> *Fili and Kili were at the top of a tall larch like an enormous Christmas tree. (Hobbit VI)*

Although coniferous forests (i.e., forests dominated by resinous, usually evergreen, cone-bearing trees of the families Pinaceae and Cupressaceae) are a prominent backdrop for the action of *The Hobbit* and *The Lord of the Rings*, larches are seldom mentioned. Larches, unlike most conifers, are deciduous, their leaves turning yellow in the autumn and then dropping, so the trees are bare during the winter. Tolkien alludes to this in his description of Ithilien. As Frodo and Sam hiked through this beautiful region, only recently fallen under the domination of Sauron, it was early spring. They saw bracken fronds (*Pteridium aquilinum*) emerging through the leaf litter and many different wild flowers just coming into bloom. As they walked through a forest of "resinous trees, fir and cedar and cypress" Tolkien states that the "larches were green-fingered" (LotR 4: IV). These trees produce their needlelike leaves on short shoots, and, in the spring, they emerge to form dense clusters, all emerging, fingerlike, at the same time from the winter buds. In the chapter "Of Herbs and Stewed Rabbit," Tolkien provides a detailed description of the hillside forests of Ithilien, so we know that the larches (members of the genus *Larix*) grew along with many other conifers—pines (*Pinus*) and firs (*Abies*) of the Pinaceae, and junipers (*Juniperus*), cedars (*Cedrus libani*), and cypress trees (*Cupressus sempervirens*) of the Cupressaceae. These coniferous trees all produce resins, sticky and aromatic, which contain terpenes such as alpha-pinene, limonene, and terpinolene, contributing to the aroma of Ithilien. Larches also occurred in the Misty Mountains, growing intermixed with pines and firs, and we read in *The Hobbit* (see opening quote) that Fili and his brother Kili climbed one to escape from wolves (Figure 7.40). Larches are mentioned in the poem *Bombadil Goes Boating*, so they also must have grown in the Old Forest, and they also grew along with firs on the western slopes of the mountains of Forostar, the northern peninsula of the great island of Númenor. Despite their scant mention, therefore, larches must have been widely distributed in Middle-earth, especially in its northern and more mountainous regions.

Etymology: Larch is derived from the German, *Lärche*, going back to Latin *larix*, which was possibly a loan word from a Gaulish language. Although not used by J. R. R. Tolkien, another common name of these trees in North America is "tamarack," a name taken from Canadian French, *tamarac*, which is undoubtedly a word of Algonquian origin. The Sindarin name for larch is *fîn*, while the Quenyn names are *finë* and/or *findë*.

Distribution and Ecology: Larches, a group of about a dozen species, are distributed widely across cool forests and bogs of the Northern Hemisphere; they are also frequent in mountainous regions. The European larch (*Larix decidua*) occurs in the Alps and Carpathian mountains; three species grow in North America, the most common of which are tamarack (or eastern larch, *L. laricina*) and western larch (*L. occidentalis*); the common species—Dahurian larch (*L. gmelinii*), Japanese larch (*L. kaempferi*), Siberian larch (*L. sibirica*)—occur in northern and eastern Asia. Species can be difficult to identify and are distinguished by twig pubescence, coloration of new shoots, and especially the characteristics of their seed cones. Their pollen is dispersed by the wind, as are their winged seeds (released from the rather small, erect cones).

Economic Uses: The species of larch provide valuable timber, which is used in construction, plywood, veneer, railroad ties, pilings, posts, and poles; the wood has high resin content and is thus decay- and water-resistant and once was important in boat construction. The sap of larches is used to produce larch gum (mainly arabinogalactan, a biopolymer, with characteristics similar to gum arabic and sparingly used in the food industry). Larches are also important in protecting watersheds and providing wildlife habitat; several species are beautiful ornamental trees.

Description: Erect, deciduous trees with sparse crown and horizontal, spreading branches; the bark gray to brown, scaly or roughened, to vertically furrowed. Twigs of two distinct forms, some branches elongated and these bearing numerous, short or spur branches in their second and subsequent years of growth; wood with resin canals. Leaves in tufts of 10–60 on short shoots or borne singly and spiral on first-year long shoots; each leaf simple, linear and needlelike, flattened, but slightly keeled below, with two resin canals, with a single midvein, the apex acute to rounded, the base slightly narrowed, with margins extending along stem, the margin entire; stipules absent. Plants with pollen and seed cones on the same plant. Pollen cones ovoid-cylindric, yellowish, with each pollen-bearing structure flat, bearing two sporangia. Seed-cones ovoid to globose, erect, terminating the short shoots, maturing in a single season, opening and releasing seeds at maturity; scales persistent, flat, circular to oblong or obovate, each bearing two seeds and free from the associated and narrower bracts (which may be hidden or exposed). Seeds strongly winged (Figure 7.40).

LAVENDER (*LAVANDULA ANGUSTIFOLIA* AND RELATIVES)

(The Mint family [Lamiaceae or Labiatae].)

> *[H]e perfumed her with marjoram*
> *and cardamom and lavender.* (*TATB* 3)

The sole mention of lavender in J. R. R. Tolkien's legendarium is its occurrence in the poem *Errantry* (the third entry in *The Adventures of Tom Bombadil*), which in the preface is said to have been composed by Bilbo in the early days after the return from his journey to the Lonely Mountain. In the poem the mariner perfumed his ship with aromatic herbs: cardamom (*Elettaria cardamomum*, of the Zingiberaceae, the

FIGURE 7.40 *Larches.*

ginger family), marjoram (*Origanum majorana*, of the Lamiaceae, the mint family), and lavender (*Lavandula* spp., also a mint). All three species are edible and used in flavoring foods; however, of these, only lavender produces commercially important and fragrant oil, which is much used in scenting fabrics, bath products, cosmetics, and perfumes. Lavenders are native to southern Europe, so would have been familiar to Bilbo, but perhaps only as an imported spice or oil. Marjoram is a plant of the eastern Mediterranean (and, not surprisingly, was seen by Sam and Frodo in Ithilien), but this fragrant mint also was encountered by Bilbo on the eastern slopes of the Misty Mountains (see Hobbit VI). Finally, cardamom is now native to northern India, and we suspect that in the Middle-earth of the Third Age this was an expensive, imported spice—perhaps included in this poem by Bilbo merely to impart an exotic flavor.

Etymology: The English word "lavender" is derived from the Anglo-French *lavendre*, from Medieval Latin *lavendula* (lavender), which is perhaps from Latin *lavare* (to wash), a reference to its use in bath perfumes and in scenting fabrics. The scientific name *Lavandula* is also derived from *lavendula*.

Distribution and Ecology: *Lavandula*, a genus of 39 species, is widely distributed in the Mediterranean region (Europe, northern Africa, south to Somalia, and western Asia) and is characteristic of dry sunny habitats. Lavenders are widely cultivated, and a few have naturalized in North America and Australia. Species can be distinguished by the degree of leaf-lobing; variation in the leaf margin and hairiness; the size, shape, and coloration of the inflorescence bracts; and number of flowers per cluster. Their showy flowers are visited by various bees, and the fruits (small nutlets) are likely distributed by wind or rain-wash.

Economic Uses: Several species of lavender (and especially *L. angustifolia, L. stoechas*, and *L.* × *intermedia*) are used as a source of an essential oil that is fragrant, antiseptic and anti-inflammatory; the oil is popular in bath products, balms, perfumes, and cosmetics. Lavender is also used as a condiment (in salads, dressings, baked products, and for candied flowers). Lavender flowers are visited by bees, and the resulting honey is of high quality. Finally, several species (and hybrids) are used as garden ornamentals because of their showy blue to purple flowers.

Description: Aromatic, annual or perennial herbs to low shrubs; stems erect to spreading, with various sparse to dense, simple hairs and also gland-headed hairs. Leaves opposite (and decussate), simple, unlobed to pinnately lobed or variously dissected, linear to elliptic, ovate, or obovate, with pinnate venation, the apex acute to obtuse, the base narrowed, with or without a short petiole; stipules absent. Inflorescences terminal, with the flowers appearing to be in whorls along an elongate spicate axis, but actually in paired reduced cymes, each in the axil of a bract, and the inflorescence bracts sometimes expanded and petaloid. Flowers bisexual, bilaterally symmetrical, held horizontally, with the five sepals fused, forming a tubular calyx; the five petals strongly fused, usually blue to purple, two-lipped, the upper lip two-lobed, and the lower lip three-lobed, both spreading, and four stamens, with two of these longer than the others, and the anthers slightly sagittate; the carpels two, the ovary superior, four-lobed, and with a single style, arising from the middle of the four ovary-lobes and apically forked, each style branch ending in a minute stigma. Fruits schizocarps, of four nutlets, glabrous and smooth.

LILIES (*LILIUM* SPP.)

(*The Lily family [Liliaceae].*)

> *Then Aragorn stooped and looked in her face, and it was indeed white as a lily. (LotR 5: VIII)*

Aragorn, seeing the face of Éowyn as she lay in the Houses of Healing, saw that it was as white as a lily (Figure 7.41). The comparison certainly is apt—and it is clear that he was thinking of the plant now called the Madonna lily (*Lilium candidum*), a species now native to Greece and the western Balkans but widely cultivated elsewhere in Europe. (It was early introduced into England and long has been linked with the Virgin Mary.) This lily, with its beautiful white flowers, is even now associated with purity, chastity, and nobility and is appropriately used here in connection with Éowyn. This species would have been familiar to Aragorn; we know that it grew in Ithilien, a region with a diverse and clearly Mediterranean flora. It was seen by Sam and Frodo, who noticed that "many lily-flowers nodded their half-opened heads in the grass" (LotR 4: IV) beside pools and along the streams that flowed down to the Anduin. It also grew in Lebennin, another Gondorian region with a mild, ocean-influenced climate (see Legolas's poem in LotR 5: IX). However, the connection between Éowyn and this beautiful lily arose much earlier; Aragorn commented that when he first saw her and felt her unhappiness and despair, it seemed to him that he "saw a white flower standing straight and proud, shapely as a lily, and yet knew that it was hard, as if wrought by elf-wrights out of steel" but then he adds that perhaps the hardness was due to a frost, which had turned the lily's sap to ice (LotR 5: VIII). This connection of Éowyn with the color white—as seen in lilies, frost, and ice—continued as she was drawn back to life (if not mental health) by the words of Aragorn and the action of the healing herb athelas (Figure 7.41): its influence was that of an air completely fresh and clean, "new-made from snowy mountains high beneath a dome of stars" (LotR 5: VIII). This vision may also have served as an escape from the dishonor and filth she had been forced to endure when Wormtongue had dominated Edoras. Finally, it should be noted that *L. candidum* is not the only lily with white flowers. Others include the Eastern Asian *L. longifolium* (native to the Ryukyu Islands, Japan) or *L. formosanum* (native to Taiwan), and the North American *L. washingtonianum*.

Other lily species also occur in J. R. R. Tolkien's legendarium. Urwen, Túrin's sister and younger by two years, is described as having hair "like the yellow lilies in the grass," and she was a beautiful and happy child: her laughter was "like the sound of the merry stream that came singing out of the hills past the walls of her father's house" (CoH 1), so she was called Lalaith (meaning laughter). Sadly, she died in young childhood from an Evil Breath sent by Morgoth, and Túrin, who had been very close to his sister, suffered the first grief of his young life. His mother, Morwin, met her grief in silence and never comforted her son, who "wept bitterly at night alone" (CoH 1)—and to his mother he never again mentioned his sister's name. This grief was to set a pattern for his tragic life, which was only redeemed through his persistent struggle against an overpowering enemy—Morgoth. Several species of *Lilium* have yellow flowers, such as the European *L. rhodopaeum* and *L. monadelphum*, or the North American *L. parryi*

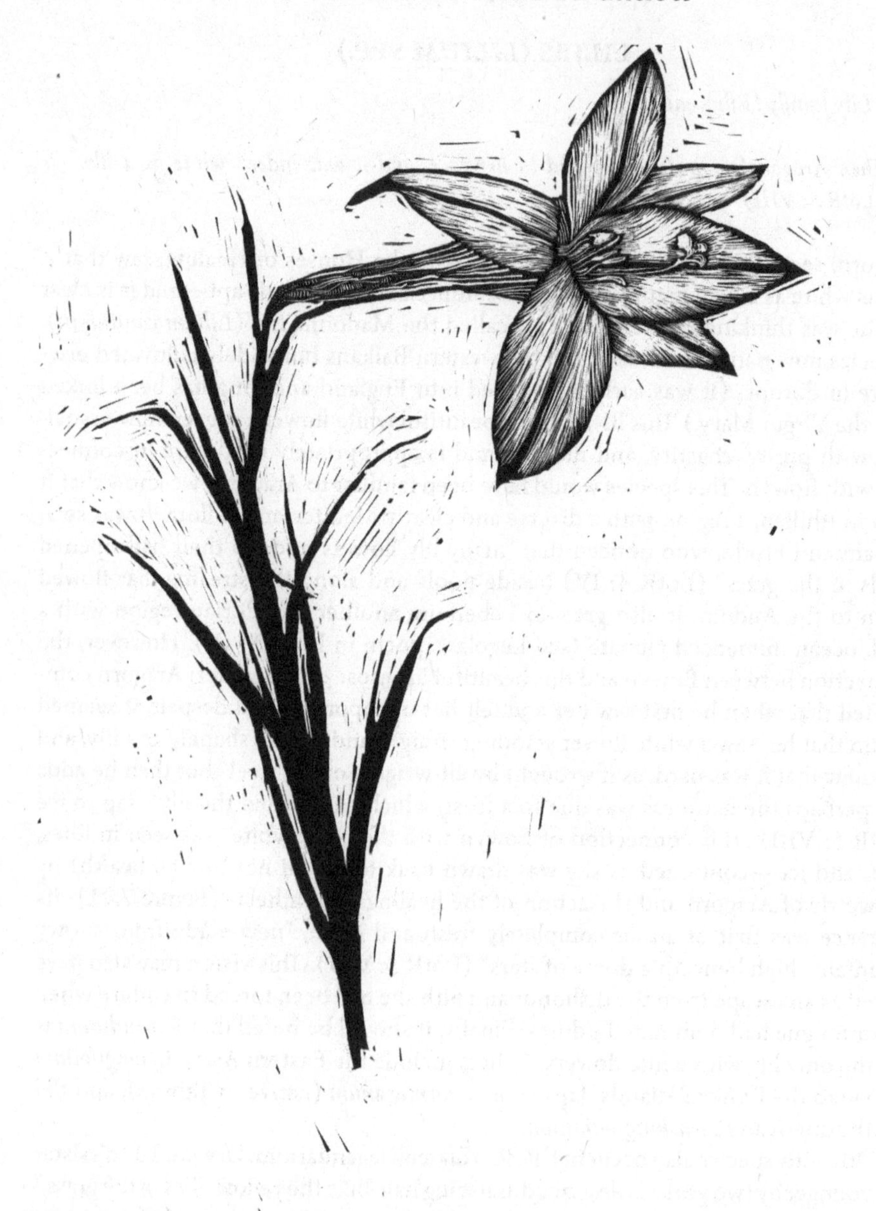

FIGURE 7.41 *Lilies.*

and *L. iridollae*; in addition, several Eastern Asian species are yellow-flowered. We note also that the yellow daylily, *Hemerocallis lilioasphodelus,* has beautiful lily-like flowers and is native to Europe as well as Eastern Asia (and widely grown as an ornamental). Any of these beautiful flowers provides an appropriate visual image of Lalaith, who her father Húrin said was as fair as an elf-child. And this connection is especially appropriate because yellow lilies often are used to represent a lighthearted and cheerful disposition. Finally, in the *Lay of Leithian,* Lúthien's robe is described

as "blue as summer skies" and "sewn with golden lilies fair" (Lays III: lines 27, 29; see also line 2397). These references surely also refer to some species of yellow-flowered lily. Lúthien, even more than Éowyn, exemplifies beauty and purity—and strength. What could Beren have accomplished without her help?

Etymology: The name "lily" is derived from Old English *lilie*, from Latin, *lilia*, plural of *lilium*, which is a cognate of Greek *leirion*, and both may be derived from an Egyptian word. Thus, the English common name and the scientific name, *Lilium*, are closely related. The Quenya name may be *indil* or *insil*.

Distribution and Ecology: Lilies (members of the genus *Lilium*, a group of slightly more than 100 species) are broadly distributed in the Northern Hemisphere but extend southward into the mountains of the Asian tropics. They are most diverse in Eastern Asia (with 55 species occurring in China and 13 in Japan) but are also common in North America (22 species) and Europe (10 species). Lilies grow in a variety of habitats, especially pine or pine-oak forests, various coniferous forests, wet pinelands and savannas, swamps, bogs, thickets, open woodlands and scrub, prairies, montane meadows, rocky slopes, along streams, lakeshores, and disturbed open habitats; many species are characteristic of mountainous regions. Identification of species is often difficult because living material is needed—characteristics of flower orientation; size, shape, coloration, and curvature of the tepals; form of the nectar glands; and pubescence of the style and staminal filaments are quite important, along with form of the scaly bulbs, leaf arrangement, shape, coloration, surface texture, and marginal condition. Their showy flowers are pollinated by various insects, and their seeds are dispersed by wind or water (including rain-wash).

Economic Uses: The various species of *Lilium* are most important as ornamental herbs, widely grown because of their beautiful flowers, which may be pendent to erect, with six large, orange, red, yellow, pink, or white and often spotted tepals. Their bulbs are occasionally eaten or used medicinally.

Description: Perennial herbs, arising from scaly bulbs, these sometimes horizontally elongated, the flowering stems erect, cylindrical; roots thin to thick, some contractile (and pulling bulb downward). Leaves alternate or whorled, distributed along stem, simple, ovate to obovate or more or less linear, short to elongate, with parallel veins, the apex usually acute to obtuse, the base broad, the margin entire, undulate or not, usually glabrous, smooth to papillose or roughened; little "bulbs" occasionally present in leaf axils; stipules absent. Inflorescences terminal, of one to numerous flowers, solitary or racemose. Flowers radially symmetrical, pendent to erect, large, orange, red, yellow, pink, pale green or white, often with magenta or maroon spots, odorless to strongly fragrant, with a perianth of six tepals, these distinct, narrowly ovate, apically acute and narrowed basally, with three outer and three inner ones, arranged in bell- or trumpet-shaped or broadly spreading manner, and with each tepal curved to often ± reflexed, and each with a basal, grooved and sometimes fringed, nectar-producing region, this better developed in the three outer ones. Stamens six, hidden within the perianth or much exserted, the filaments close to the style or spreading, with conspicuous, flexible anthers (attached to the filament near their middle). Carpels three, fused, the style single, usually curved, with a three-lobed stigma; ovary superior. Fruit a slightly six-angled capsule; seeds light brown, flattened (Figure 7.41).

LINDENS (*TILIA* SPP.)

(In the Hibiscus family [Malvaceae].)

> *He [Beren] heard there oft the flying sound*
> *Of feet as light as linden-leaves,*
> *Or music welling underground,*
> *In hidden hollows quavering.*
> *Now withered lay the hemlock-sheaves,*
> *And one by one with sighing sound*
> *Whispering fell the beechen leaves*
> *In the wintery woodland wavering.* *(LotR 1: XI)*

This passage clearly illustrates the importance of plants in Tolkien's legendarium; here we have three important species mentioned—linden-leaves (*Tilia cordata*), hemlock-sheaves (*Conium maculatum*), and beechen leaves (*Fagus sylvatica*)—in his description of the critical meeting of Beren and Lúthien in the beech forest of Neldoreth: the beginning of an amazing love story in *The Silmarillion*. However, our focus here is on the linden tree, which in Germanic folklore often is considered symbolic of truth and love. This poem, as recited by Aragorn to Frodo, Sam, Merry, and Pippin on Weathertop (Figure 7.42) shortly before the attack of the Ringwraiths (in *The Fellowship of the Ring* 1: XI), is similar to (and based upon) the poem told to Túrin by his guardian and guide to lighten his sorrow as they were lost in the bewildering forests of Doriath while attempting to reach Thingol's realm (see *The Lays of Beleriand*, the second version of the poem *The Children of Húrin*, lines 263–270, 398–402). Both poems are based on the poem *Light as Leaf on Lindentree*, which was published in *The Gryphon* (Leeds University) in 1925. Christopher Tolkien (in *The Lays of Beleriand*) reported, based on notes of his father on the earliest of three typescripts of the poem, that it was begun in 1919–20 and retouched in 1924. Thus, the image of Lúthien dancing with "feet as light as linden-leaves" goes back to Tolkien's very early conception of this event, although it is of interest that lindens are not mentioned in the early prose *Tale of Tinúviel*, written in 1917 (see *The Book of Lost Tales*, Part II). The phrase "light as leaf on linden" may have been traditional because it occurs in the postscript to Chaucer's "Clerk's Tale" as advice to wives to "be ay of chiere as light as leef on lynde." In Germanic cultures, linden trees were associated

FIGURE 7.42 *Lindens.*

with judicial courts and festivities, including dancing, and this connection possibly prompted both this traditional usage and Tolkien's link of linden leaves and Lúthien's dancing. Except in various descriptions of the meeting of Lúthien and Beren, lindens are not often mentioned in Tolkien's writings; however, Legolas described the movements of the elf maiden Nimrodel using very similar wording: "And in the wind she went as light as leaf of linden-tree" (LotR 2: VI). And, finally, linden trees also occurred in Fangorn forest, where Merry and Pippin met slender ents that recalled the linden tree.

Etymology: "Linden," the English name of this tree, was originally an adjectival form, meaning "made from lime-wood," and, like the common name lime, is derived from Old English *lind*, through Proto-Germanic **lindjo* to Proto-Indo-European **lent-o-*, meaning flexible, in reference to the tough but flexible fibers (bast) found in the inner bark. The name "basswood," which is more widely used in North America, is derived from bast. The scientific name, *Tilia*, is taken from the classical Latin name of these trees, which is related to the Greek *ptelea* (elm tree), perhaps in allusion to the winged fruits of elms: the fruits of lindens also are associated with a papery wing (a modified bract). The Greek word may be derived from Proto-Indo-European **pteleia*, perhaps meaning broad or broad-leaved.

Distribution and Ecology: Lindens or basswoods (i.e., species of *Tilia*), a group of about 25 species (the exact number uncertain due to hybridization and intergrading geographical entities), occur in moist to dry, cool to warm temperate forests of eastern North America and montane Mexico (one species: American basswood, *T. americana*, incl. varieties *americana* and *caroliniana*), Europe and adjacent western Asia (six species, with large-leaf linden, *T. platyphyllos*; little-leaf linden, *T. cordata*; and silver linden, *T. tomentosa*, the most common), and eastern Asia (a region of exceptionally high diversity, ca. 18 species). Species are often difficult to identify and are distinguished by characters such as the size and shape of the leaf, pubescence of its lower surface, development of the marginal serrations, twig pubescence, number of flowers per inflorescence, the pubescence of the various flower parts, presence or absence of staminodia, and texture and shape of the fruits. The showy, more or less white and fragrant flowers produce copious nectar and are pollinated by various bees. The clusters of small, round, hard fruits are associated with a large, winglike bract and are dispersed short distances by wind.

Economic Uses: Lindens or basswoods are widely used as shade trees. Additionally, their wood is soft, easily worked, and thus used in carvings, musical instruments, and furniture. The fibers of the inner bark are used for rope, baskets, or mats. The trees produce numerous fragrant and nectar-producing flowers, attracting bees, and they are important honey plants. The flowers are used medicinally (especially in herbal teas, for colds, sore throat, coughing, and headache).

Description: Deciduous trees or shrubs, the bark vertically furrowed with inner layer fibrous, tough. Leaves alternate, two-ranked, simple, ovate to elliptic, palmately veined, the secondary veins running to the teeth, the apex acuminate, the base asymmetrical and cordate to rounded, the margin strongly to obscurely serrate or nearly entire, the lower surface with sparse to dense, simple or stellate hairs, and often with dense hair-tufts in vein axils, petiolate; stipules usually present, quickly deciduous. Inflorescences axillary, in cymes, the stalk of which is partly fused to a elongate, papery bract. Flowers radially symmetrical, bisexual, the sepals five, distinct, producing nectar at the base of the upper surface; the petals five, distinct, white to pale green or pale yellow-cream, and elongated; the stamens numerous, distinct or fused into five bundles, with or without petaloid staminodes and five fused carpels; the style single, well-developed, with five-lobed stigma; the ovary superior. Fruit usually a globose nut, with 1–3 seeds (Figure 7.42).

LISSUIN

(An unknown species, never established in Middle-earth.)

> *Their ship was laden with flowers for the adornment of the feast, so that all that sat there, when evening was come, were crowned with elanor and sweet lissuin whose fragrance brings heart's ease. (UT 2: II)*

A ship out of the west arrived on the morning of the wedding of Erendis and Aldarion, who later became the sixth king of Númenor. It was a tall white ship with glimmering silver sails and white birds flying all about it. The elves had come to grace the wedding of Erendis out of their love for the people of the western regions of Númenor, and their ship was filled with flowers for the wedding feast. As we learn from the quote from the unfinished story of *Aldarion and Erendis*, published in the *Unfinished Tales*, among the flowers brought from Tol Eressëa were elanor and lissuin. We know a lot about elanor from other sources (see treatment of that species), but lissuin is only mentioned here. It was perhaps a herb with white fragrant flowers. Lissuin grew on the Lonely Isle and was never established on Númenor. As far as we know, its sole occurrence in Middle-earth was at this momentous wedding. In addition to providing flowers, the elves brought gifts for both bride and groom. To Aldarion, they gave a sapling tree whose bark was snow-white, and the descendants of this tree became the White Tree of Gondor; to Erendis, they gave a "pair of birds, grey, with golden beaks" that sang sweetly to one another, but would keep silent if separated. These gifts contained important messages: for Aldaron, one message, tragically not learned until years later, was that trees should be valued for more than their timber. And for both, another message spoke to the importance of good communication and a life of shared interests—sadly, this was also ignored. The elven-tree was planted in a garden in Armenelos, and the elven-birds sang in its branches. Here we catch a glimpse of a wider world, mysterious and beautiful plants of Eressëa and Valinor, many of which were never seen in Middle-earth.

Etymology: The name *lissuin* is related to the Quenya word *lissë* (sweet).

MALLORN-TREES

(A species unique to J. R. R. Tolkien's legendarium, based in part upon the beech tree, Fagus, but also on cherries, Prunus; in an unknown family, but perhaps in the Ericales.)

> *As Frodo prepared to follow him, he laid his hand upon the tree beside the ladder: never before had he been so suddenly and so keenly aware of the feel and texture of a tree's skin and of the life within it. He felt a delight in wood and the touch of it, neither as forester nor as carpenter; it was the delight of the living tree itself. (LotR 2: VI)*

Mellyrn are not native to Middle-earth. They were brought by elves from Tol Eressëa to Eldalondë on the eastern coast of Númenor during the Second Age, where they formed beautiful groves (along with other trees brought out of the West). Later, "some were given as a gift by Tar-Aldarion, the sixth King of Nûmenor, to King Gil-galad of Lindon. They did not take root in that land; but Gil-galad gave some to his kinswoman Galadriel, and under her power they grew and flourished in the guarded land of Lothlórien beside the River Anduin, until the High Elves at last left Middle-earth" (UT 2: I). However, there is tenuous evidence (see *Unfinished Tales* 1: I, footnote 31) that mellyrn also were introduced into Beleriand in the First Age and grew in the Hidden City, Gondolin. The most detailed description of the mallorn is provided in the *Unfinished Tales,* in the chapter "A Description of the Island of Nûmenor": "Its bark was silver and smooth, and its boughs somewhat upswept after the manner of the beech; but it never grew save with a single trunk. Its leaves, like those of the beech but greater, were pale green above and beneath were silver, glistening in the sun; in the autumn they did not fall, but turned to pale gold. In the spring it bore golden blossom in clusters like a cherry, which bloomed on during the summer; and as soon as the flowers opened the leaves fell, so that through spring and summer a grove of *malinorni* was carpeted and roofed with gold, but its pillars were of grey silver. Its fruit was a nut, with a silver shale" (UT 2: 1). A very similar description is given by Legolas in *The Lord of the Rings,* when he and his fellow travelers, led by Aragorn, reached the eaves of Lothlórien, although there it is added that they heard "an endless rustle of leaves like poplars in the breeze" (LotR 2: VI). Legolas told the travelers that Lothlórien was the most beautiful of all the dwellings of the elves, and most readers would go even further—acknowledging that the mallorn forest of Lothlórien was the fairest in all of Middle-earth. Tolkien produced a colored pencil drawing, *The Forest of Lothlórien in Spring* (see fig. 157 in W. G. Hammond and C. Scull, 1995, *J. R. R. Tolkien: Artist and Illustrator*), which confirms the features already discussed and shows also that the trees had alternate leaves, flowers with five elliptic to obovate petals, developed a fairly superficial root system, and occurred in dense, almost parklike groves with very little undergrowth. This forest is also shown on the cover of this book, illustrating the company shortly after their crossing of the Silverlode. In the interaction of the silver (seen in the trunks, lower surface of the leaves, and the fruits) and the gold (of the autumn/winter leaves and flowers) we perceive the mingling of the lights of the Two Trees of Valinor—Telperion and Laurelin—the ideal of elven beauty as expressed in the created world. Telperion had dark green fluttering leaves that were shining silver beneath, "and from each of his countless flowers a dew of silver light was ever falling," while the leaves of Laurelin were "young green like the new-opened beech" with margins of "glittering gold" and her flowers were like clusters of yellow flame that gave forth warmth and a great light, which spilled on the ground like a "golden rain" (SILM 1). As can be seen, mellyrn beautifully combine the characteristics of Telperion and Laurelin and are thus an essential part of Galadriel's vision of Lothlórien as "a refuge and an island of peace and beauty, a memorial of ancient days" (UT 2: IV), a beautiful and unstained garden brought about through her artistic and horticultural enhancement of the natural world. Thus, as shown in the passage initially quoted (see also Figure 7.43), the value of the mallorn has nothing to do with any practical use, no matter how special; instead, it is an artistic delight—"the delight of the living tree itself"—that brings a

FIGURE 7.43 *Mallorn-trees.*

true appreciation and love of the natural world. As Tolkien stated in his letter to the *Daily Telegraph* in 1972, "Lothlórien is beautiful because there the trees were loved" (Letters: No. 339).

Etymology: *Mallorn* (plural: *mellyrn*) is the Sindarin name of the dominant trees of Lothlórien; the name is based on *mal* (gold) and *orn* (tree), in reference to the color of their leaves in the fall and winter, as well as the color of their flowers (see the Appendix to *The Silmarillion* and the Index to the *Unfinished Tales*). The name in Quenya is *malinornë* (plural: *malinorni*).

Distribution and Ecology: During the Third Age of Middle-earth, mellyrn grew only in Lothlórien, although Galadriel gave a seed to Sam who planted it in the Shire, and in "the Party Field a beautiful young sapling leaped up: it had silver bark and long leaves and burst into golden flowers in April. It was indeed a mallorn, and it was the wonder of the neighborhood" (LotR 6: IX).

Economic Uses: A mallorn is the quintessential ornamental tree—their primary function is the enhancement and enrichment of the natural beauty of the earth. Caras Galadhon, the beautiful city of Galadriel and Celeborn, was built in the branches of tall mellyrn, which could reach nearly 200 feet in height. Finally, the elven waybread (lembas) was wrapped in (and kept fresh by) mallorn leaves. We note that mallorn leaves were elongated and ovate to elliptic, with pinnate venation (as supported by Tolkien's descriptions and illustration)—not three-lobed, with palmate venation (as mistakenly shown on many websites).

Description: Tall, tardily deciduous trees, with the pale gold leaves held on the branches during the winter, the bark smooth, thin, and silver, with the branches upswept. Leaves alternate, spirally arranged, simple, ovate to elliptic, pinnately veined, with secondary veins ± parallel, running into the leaf margin, the apex acute to slightly acuminate, the base acute to obtuse, the margin obscurely serrate, with the upper surface of the blade pale green, and the lower surface silvery (due to a dense layer of reflective hairs), and with the petiole somewhat flattened, causing the blades to flutter in the breeze; stipules present, but quickly deciduous. Inflorescences axillary, short, pendulous fascicles. Flowers radially symmetrical, bisexual, the sepals five, distinct, pale green, triangular; the petals five, distinct, elongate-oblong, rounded at the apex, yellow-gold and imbricate in bud; the stamens 10; and the single carpel with a superior ovary and short style. Fruit a nut with a smooth, silver shell containing a single seed (Figure 7.43).

MOSS (VARIOUS GENERA)

(Bryophyta.)

> *At nightfall he brought them to his ent-house: nothing more than a mossy stone set upon turves under a green bank. (LotR 3: IV)*

In the quote we see the ent-house of Quickbeam (Figure 7.44), which was not something that we would even recognize as a house—merely a mossy stone set upon turf blocks, a structure, if it can be called that, of minimal environmental impact, as one would expect of an ent: a race dedicated to the preservation of wild habitats, especially

FIGURE 7.44 *Moss.*

forests. Furthermore, Quickbeam's house was surrounded by a circle of rowan trees (*Sorbus aucuparia*)—living trees, and ones with which he had a special connection. It also contained a bubbling spring. It is not to be doubted that Quickbeam (or any other ent) would view any cutting of trees as wrong—both the wanton and wasteful cutting of Saruman and the orcs, and the careful use (for building material or fuel) by humans such as the Rohirrim or by Aragorn, Legolas, and Gimli. Thus, ent-houses reflect the natural world—only very little modified—containing not much more than the water necessary to life and the living trees of the forest. As in the quote, most of the references to mosses in J. R. R. Tolkien's legendarium are merely descriptive; for example, the "mossy" beard of Fangorn (or Treebeard; see LotR 3: IV) or Lúthien sleeping "upon the moss" (Lays III: line 3229). However, one occurrence stands apart. The Valer Yavanna, the Giver of Fruits and wife of Aulë, through whom all plant and animal life came into existence, is described as "the lover of all things that grow in the earth, and all their countless forms she holds in her mind, from the trees like towers in forests long ago to the moss upon stones or the small and secret things in the mould" (*Valaquenta*). Although Yavanna is especially associated with trees and can in fact adopt tree-form (i.e., "there are some who have seen her standing like a tree under heaven, crowned with the Sun" [*Valaquenta*]), she values *all* of living nature in its full and marvelous diversity, which is expressed here by providing a range of sizes: from the minute ("mosses upon stones") to the very large ("trees like towers"). As she told Manwë, all the life forms of the biosphere "have their worth," and they are interconnected in that "each contributes to the worth of the others." But they can easily be destroyed, especially plants. She thus asked that the "trees might speak on behalf of all things that have roots, and punish those that wrong them!" (SILM 2). Plants have moral significance and can be wronged. We, like Manwë, may think this a strange thought, but Yavanna noted that it was part of the song of creation, in which the branches of great trees were lifted up in praise of Ilúvatar (calling to mind Psalm 148: "Praise the Lord . . . mountains and all hills, fruit trees and all cedars!", see verses 7, 9; see also M. Dickerson and J. Evans, 2006, *Ents, Elves, and Eriador,* and A. Denekamp, 2015, *"Transform Stalwart Trees": Sylvan Biocentrism in The Lord of the Rings*, in *Representations of Nature in Middle-earth*, M. Simonson, ed.). As a result of her prayer, we have the ents, the shepherds of the trees, reminding us of the intrinsic value of the natural world, including even the smallest and seemingly least important of its species.

Etymology: "Moss" is derived from Old English *meos* (and related to *mos*, bog), both from Proto-Germanic **musan* (moss, bog), from Proto-Indo-European **meus-* (damp). The etymological relationship between moss and bog is logical given that mosses are characteristic of damp, boggy habitats.

Distribution and Ecology: Mosses are cosmopolitan in distribution; there are more than 10,000 species within this diverse group of nonvascular plants. They grow in a wide variety of habitats, but they are exceptionally abundant in moist regions. Mosses play an important role in early plant succession (i.e., the establishment of a plant community in an area, either newly available or after disturbance such as agricultural clearing or fire, etc.).

Economic Uses: Mosses are ecologically important in erosion control and are also used to assess of the level of environmental pollutants. Peat mosses (species of *Sphagnum*) are highly acidic and thus have been used as dressing for wounds. Peat is also used as a fuel source.

Description: Tiny, nonvascular plants (i.e., they usually do not have internal water or carbohydrate conducting cells) of a matlike or miniature treelike form, which have a life cycle consisting of a haploid gametophyte plant (i.e., a gamete-producing plant consisting of cells in which the genome has one member of each of the pairs of homologous chromosomes) alternating with a diploid sporophyte plant (i.e., a spore-producing plant consisting of cells in which the chromosomes are in homologous pairs). The gametophytes have erect, horizontal, or pendulous stems that are attached to the substrate by hairlike structures (rhizoids). The leaves are almost microscopic, alternate and spirally arranged, or less commonly two-ranked, simple, variously shaped, and often with a single midvein. The gametophyte produces eggs (each in a flask-shaped structure, the archegonium) and biflagellate sperm (the sperm cells produced within a globose structure, the antheridium). The sperm swim down the neck of the flask-shaped structure to reach the egg, fusing to form a diploid cell, the zygote, which grows into an embryonic sporophyte (surrounded by gametophytic tissues). The sporophyte is attached to (and physiologically dependent on) the gametophyte and grows upward, forming a usually elongate stalk with an often cylindrical, spore-containing structure at its apex. The spore-containing structure (sporangium or capsule) usually opens apically with a cap, which falls off, exposing a ring of small teeth that assists in spore dispersal (by wind). Spores are formed by reduction division (meiosis) and are thus haploid; after dispersal, they grow into gametophytes, continuing the life cycle (Figure 7.44).

MUSHROOMS (*AGARICUS BISPORUS*, AND RELATED SPECIES)

(The Button Mushroom family [Agaricaceae, in the basidiomycetes or club fungi; i.e., Basidiomycota].)

> *There was beer in plenty, and a mighty dish of mushrooms and bacon, besides much other solid farmhouse fare. (LotR 1: IV)*

The importance of mushrooms in the life of hobbits is presented most clearly in *The Lord of the Rings*, in the chapter "A Short Cut to Mushrooms," in which we encounter Frodo, Sam, and Pippin, although hunted by Black Riders, enjoying a brief interval of safety and eating dinner at the home of Farmer Maggot (Figure 7.45). The highlight of this meal was a huge dish of mushrooms and bacon. In the next chapter, we are told that "Hobbits have a passion for mushrooms, surpassing even the greediest likings of Big People" (LotR 1: V), and we begin to understand the reasons behind Bilbo's youthful mushroom-gathering adventures in Farmer Maggot's fields. These mushrooms are not described in detail, but it is likely that they were a species of *Agaricus*. This large genus contains the most commonly cultivated mushroom—the button mushroom (or Portobello; *A. bisporus*), and this species is widespread and common in Europe. However, several species of *Agaricus* are edible, so the mushrooms so enjoyed by Frodo and his companions could also have been species such as the horse mushroom (*A. arvensis*) or the field/meadow mushroom (*A. campestris*). All produce quite similar fruiting bodies (mushrooms). These fungi grow in open areas, especially pastures, so one can easily imagine the mushrooms being gathered around Farmer Maggot's home.

Etymology: The word mushroom comes from late Middle English *muscheron, musseroun,* from Anglo-French *musherun,* Old French, *meisseron,* and perhaps late Latin, *mussirionem,* although this may also have been borrowed from French and is likely derived from a word of pre-Latin origin, used in northern France. The scientific name *Agaricus* is derived from Latin *agaricum,* and Greek *agarikon* (ancient name of a tree-fungus). Quenya names in early use may include *inwetelumbë* and *telumbë.*

Distribution and Ecology: The species of *Agaricus* are widely distributed, and *A. bisporus* (button mushroom, white mushroom, brown mushroom, or Portobello mushroom) occurs in both North America and Europe, where it is commonly found in grasslands and pastures. The tiny spores, produced on gills radiating along the underside of the cap, are dispersed by wind. This fungus is a decomposer, and it lives in the soil as a mycelium (an array of branched filaments) where it breaks down leaf litter and other organic material in temperate communities. *Agaricus* includes both edible and poisonous species. Other edible species are *A. arvensis* (horse mushroom), *A. campestris* (field or meadow mushroom), and *A. silvaticus* (scaly wood mushroom); all are similar to the button mushroom, and all are widely distributed and frequently gathered from the wild. The poisonous varieties, such as *A. xanthodermus* (yellow-staining mushroom), causes gastrointestinal problems in most people. In contrast to the positive references in *The Lord of the Rings,* the only mention of

FIGURE 7.45 *Mushrooms.*

fungi in *The Hobbit* (VIII) is an unpleasant one and likely refers to poisonous mushrooms growing in the deep shade of Mirkwood's trees (shown in the illustration *Mirkwood*; see fig. 47 in W. G. Hammond and C. Scull, 2012, *The Art of the Hobbit by J. R. R. Tolkien*).

Description: For most of their life these saprophytic mushrooms exist as a mass of branching fungal filaments (i.e., a mycelium) growing through and breaking down organic material in the soil; after the filaments of different mating types fuse, the fruiting bodies, mushrooms, form and appear above the ground. These are white to brown when immature, and at maturity form a hemispherical to flattened, fleshy, brown to gray-brown cap, with broad flat scales on a paler background, with radiating gills on the underside, these initially pink, then becoming red brown to dark brown, and bearing basidia; each basidium (microscopic, club-shaped, reproductive structure) producing two chocolate-brown, ellipsoid to nearly round spores. The stalk (or stipe) is cylindrical, with a partial veil, which protects the developing gills; later, once torn by the expanding cap, it forms a ring around the stalk (Figure 7.45).

NASTURTIANS, NASTURTIUMS, INDIAN CRESSES (*TROPAEOLUM MAJUS*, AND RELATED SPECIES)

(In the Nasturtium family [Tropaeolaceae].)

> *The flowers glowed red and golden: snap-dragons and sun-flowers, and nasturtians trailing all over the turf walls and peeping in at the round windows. (LotR 1: I)*

In the first chapter of *The Fellowship of the Ring*, Bilbo and Gandalf sat at the open window of Bilbo's home (Figure 7.46) looking west out at his garden and enjoying the afternoon sunlight on the snapdragons (*Antirrhinum majus*), sunflowers (*Helianthus annuus*), and nasturtians. Gandalf made an admiring comment, and Bilbo replied that he was indeed very fond of his garden, "and of all the dear old Shire" (LotR 1: I). In the Prologue of *The Lord of the Rings*, we are told that hobbits love "good tilled earth: a well-ordered and well-farmed countryside" and that "growing food and eating it occupied most of their time," so we naturally think of them primarily as agriculturists—and it is true that in Farmers Maggot and Cotton, for example, we are provided a positive picture of agrarian life in the Shire. However, it is important to keep in mind that the hobbits of the Third Age also grew plants solely for their beauty, as represented by Bilbo's garden, and thus were also horticulturalists. Like the elves, hobbits sought to live in beautiful surroundings, and, indeed, in Bilbo's and Frodo's heightened appreciation of floral beauty we see an aspect of their character that reveals them as elf-friends. The elves, even more than hobbits, were natural artists; they valued the living world and sought to work with it to highlight and enhance its beauty. Gardening, growing plants purely for aesthetic purposes, was one of the major ways that this was accomplished.

It is also interesting that in this passage J. R. R. Tolkien used the word "nasturtians" instead of the more commonly used "nasturtiums" for the reddish-flowered vine that "peeped in" Bilbo's windows. As Tolkien related in a letter to Katherine Farrer (Letters: No. 148), the proofreader of *The Lord of the Rings* had altered his usage to nasturtiums (along with other changes such as elfin for elven), so Tolkien was "put to the trouble of proving to him his own ignorance." Tolkien was well aware that *Nasturtium* was the scientific name of watercress (i.e., *Nasturtium officinale*), and thus was a source of confusion when used as the common name of *Tropaeolum majus*. He related these facts through a quoted conversation with the college gardener—

> "What do you call these things, gardener?"
> "I calls them *tropaeolum*, sir."
> "But, when you're just talking to dons?"
> "I says *nasturtians*, sir."
> "Not *nasturtium*?"
> "No, sir; that's watercress."

He concluded, as a simple fact of botanical nomenclature, that the usage of the name nasturtiums for these plants was "bogusly botanical, and falsely learned" and preferred to call them nasturtians. We do as well.

Etymology: Nasturtium, the most frequently used common name of *Tropaeolum* species, is derived from Latin; i.e., *nasus* (nose) + *tortus* (twisted), a reference to the pungent smell of these plants, which is due to the presence of mustard oils in all their parts. This name, however, is quite misleading because the same word is the scientific name of the watercresses (i.e., the genus *Nasturtium*, of the Brassicaceae, the mustards, a related family that also contains mustard oils because both families, along with many others, are members of the large order Brassicales). For this reason, as discussed earlier, J. R. R. Tolkien used the less commonly applied name "nasturtians" for these flowers. They are sometimes called "Indian cresses," a name that was in use by the seventeenth century and also used in John Gerard's influential *Great Herball, or Generall Historie of Plantes.* These South American plants, introduced into Europe from Peru (thus the "Indian" element in the name), were called cresses because they have a flavor very similar to that of the European cresses (which are members of the Brassicaceae). The scientific name *Tropaeolum* is derived from the Latin *tropaeum* (trophy, monument), from Greek *tropaion* (monument to an enemy's defeat), from *trope* (to turn) as such monuments in ancient Greece often were erected on the battlefield itself at the very place where the turning point in the conflict occurred, and sometimes the armor

FIGURE 7.46 *Nasturtians, nasturtiums, Indian cresses.*

and weapons of the vanquished were hung there). Carolus Linnaeus imaginatively coined this name because the leaves of nasturtians (which are round, with the petiole attached in the middle) reminded him of shields and their flowers (frequently red or red-blotched) blood-stained helmets.

Distribution and Ecology: *Tropaeolum* is a genus of ca. 90 species and ranges from southern Mexico to southern South America, occurring in a variety of open habitats. *Tropaeolum majus* was introduced into Europe in the seventeenth century from populations growing in the vicinity of Lima, Peru, which likely represent a natural hybrid between *T. minus* and *T. ferreyae,* although there is also some evidence of the genetic influence of *T. peltophorum.* In the context of *The Lord of the Rings,* of course, these flowers must represent an independent introduction to Middle-earth. The flowers have a prominent spur (formed by the calyx) and produce very sweet nectar, which serves as the pollinator reward, and their bright red to orange or yellow (often red-blotched) petals attract hummingbirds, which pollinate the flowers. Their fleshy blue fruits are bird-dispersed.

Economic Uses: Nasturtians (*Tropaeolum major*) are popular garden ornamentals with large red, yellow, or orange flowers. Their flower buds, flowers, and leaves can be eaten, and the related *T. tuberosum* has edible tubers.

Description: Annual to perennial herbs or vines; stems often climbing by the action of twining leaf-petioles, sometimes with tubers. Leaves usually alternate, spirally arranged, simple, orbicular, sometimes lobed, with palmate venation, the apex rounded, the margin entire, usually with the petiole with a peltate attachment (and leaf blade thus shieldlike); stipules present or absent. Flowers solitary and produced in the leaf axils, often each borne on a long stalk, each lower bilaterally symmetrical, bisexual, held more or less horizontally, with five sepals, of which 1 or 3 together form a nectar spur; five colorful (red, orange, or yellow and red-blotched) petals, each with a narrowed base, and the three lower ones usually different from the upper two and often fringed, with eight stamens and three fused carpels, the ovary superior, and a single apically three-branched style. Fruit schizocarpic, separating into three drupe- or nutlike segments, each with a single seed (Figure 7.46).

NETTLES (*URTICA DIOICA*, AND RELATED SPECIES)

(The Nettle family [Urticaceae].)

> *No tree grew there, only rough grass and many tall plants: stalky and faded hemlocks and wood-parsley, fire-weed seeding into fluffy ashes, and rampant nettles and thistles. (LotR 1: VI)*

Frodo, Sam, Merry, and Pippin see nettles in the Bonfire Glade in the Old Forest (Figure 7.47), growing with thistles (*Cirsium* spp., probably *C. vulgare* and/or *C. arvense*), fire-weed (*Chamerion angustifolium*), hemlocks (*Conium maculatum*), wood-parsley (*Anthricus sylvestris*), and unnamed grasses (members of the Poaceae). This clearing is described by J. R. R. Tolkien as a dreary place; it is already late September, and even the showy-flowered plants are past their prime—the hemlocks are "stalky and faded" and the fire-weed "seeding into fluffy ashes" (LotR 1: VI). The nettles and thistles are described as rampant—suggesting that these unpleasant tall herbs, the former with painfully stinging hairs and the latter with sharply spinose leaves, were still flourishing in this large woodland clearing. Certainly, it would be uncomfortable to walk through thickets of either one! Stinging nettles (*Urtica dioica*) are plants of forest edges and waste places, so this species would be perfectly at home in a disturbed habitat such as the Bonfire Glade. Yet, despite the presence of these two well-armed herbs, the bonfire clearing, to the three hobbits, seemed like a "charming and cheerful garden" (LotR 1: VI) after the dark and menacing forest. They were capable of seeing beauty, even in common species that most would consider weeds.

Etymology: The word "nettle" is derived from Old English *netele*, from Proto-Germanic **natilon* (nettle), which was perhaps derived from the Proto-Indo-European root **ned* (to twist, knot), a reference to the use of these plants as a fiber source, especially in fishing nets. The scientific name *Urtica* is the classical Latin name of these plants and is derived from *urere* (to burn) because the plants typically possess stinging hairs.

Distribution and Ecology: Nettles are members of the genus *Urtica*, a group of ca. 45 species that are distributed nearly worldwide but are most diverse in temperate regions. Four species occur in North America (north of Mexico, and of these *Urtica dioica* is the most common), 8 in Europe (with *U. dioica*, again, being the most common), 14 in China, and 4 in Japan. Nettles grow in moist woods and forest-edges, swamps, stream banks, pastures, cultivated fields, fence-rows, hedgebanks, and waste places. The small, greenish, and inconspicuous flowers are wind-pollinated, with the stamens bent downward in bud, but springing elastically upward when mature and ejecting a cloud of pollen.

Economic Uses: Some nettles provide useful fibers (and at one time were especially used in fishing nets, the construction of ropes, and in clothing), and the young shoots are sometimes used as a vegetable, boiled and eaten like spinach. Finally, several species have been used medicinally (in the treatment of a variety of conditions, e.g., fever, colds, rheumatism, coughs, and lung disease).

Description: Annual or perennial, short to tall herbs, sometimes rhizomatous, with simple stinging and nonstinging hairs, the former needlelike and injecting histamine, acetylcholine, and 5-hydroxytryptamine, and other irritating chemicals; stems erect to spreading horizontally, ridged or four-angled. Leaves opposite, borne along the stem, with cystoliths (i.e., microscopic concretions composed of calcium carbonate, nearly filling specialized cells); simple, narrowly to widely ovate to elliptic, with palmate venation, the apex acute or acuminate to rounded, the base cordate to rounded or cuneate, the margin usually dentate to serrate, the leaves petiolate; stipules present. Inflorescences axillary, forming loose to dense clusters, either of all staminate or all carpellate flowers or a mixture of staminate and carpellate flowers. Flowers unisexual, radially symmetrical, the staminate flowers with four greenish tepals and four stamens; the carpellate flowers with four greenish tepals, the outer two smaller than the inner two, and a seemingly single carpel with superior ovary and tufted stigma. Fruit an achene, slightly compressed, loosely enclosed by the two inner tepals (Figure 7.47).

FIGURE 7.47 *Nettles.*

NIPHREDIL (A SPECIES UNIQUE TO J. R. R. TOLKIEN'S LEGENDARIUM, BUT BASED UPON *GALANTHUS NIVALIS*)

(In the Amaryllis or Daffodil family [Amaryllidaceae].)

> *At the feet of the trees, and all about the green hillsides the grass was studded with small golden flowers shaped like stars. Among them, nodding on slender stalks, were other flowers, white and palest green: they glimmered as a mist amid the rich hue of the grass. . . . Here ever bloom the winter flowers in the unfading grass: the yellow elanor, and the pale niphredil. (LotR 2: VI)*

It is told in *The Silmarillion* that at the birth of Lúthien in the forest of Neldoreth (in Doriath), "the white flowers of *niphredil* came forth to greet her as stars from the earth" (SILM 10), and that may have been the first appearance of these flowers in Middle-earth. (Late in his life, J. R. R. Tolkien created for her a beautiful heraldic device based upon this flower; see fig. 194 in W. G. Hammond and C. Scull, 1995, *J. R. R. Tolkien: Artist and Illustrator.*) Lúthien, the elven-daughter of Thingol of Doriath and Melian of the Maiar, grew into a woman of unsurpassed beauty, purity, humility, and power. "As the light upon the leaves of trees, as the voice of clear waters, as the stars above the mists of the world, such was her glory and her loveliness; and in her face was a shining light" (SILM 19), and, with the man Beren, she accomplished much for good in Middle-earth. Niphredil, as late as the Third Age, also grew in Lothlórien, on Cerin Amroth, and it was there that Aragorn and Arwen "walked unshod on the undying grass with *elanor* and *niphredil* about their feet (Figure 7.48). And there upon that hill they looked east to the Shadow and west to the Twilight, and they plighted their troth and were glad" (LotR Appendix A: I (v)). It was then that Arwen clearly stated her conviction that although the Shadow is dark, his valor would destroy it. Aragorn answered "with your hope I will hope," although he could not see how this could be accomplished. Both events, the birth of Lúthien in Doriath and the meeting of Aragorn and Arwen in Lothlórien, surely are points of hope amid the sadness of Middle-earth, and both are associated with niphredil. It seems that a faint memory of these events remains, even today, in that the snowdrop is considered symbolic of hope, purity, and humility. And perhaps power as well, given that such a delicate flower can flourish and bloom in the midst of early spring snows.

Etymology: In Tolkien's etymologies (see HoM-E V) the Sindarin word *nifredil* is identified as meaning "snowdrop" and is related to the word *nifred* (pallor), in reference to the pale color of the flowers. This association also is supported in a letter to Amy Ronald (Letters: No. 312) in which Tolkien noted that niphredil could be considered "a delicate kin of a snowdrop," which is a European wildflower (*Galanthus nivalis*) widely cultivated and naturalized in the British Isles. In fact, Tolkien even used the word snowdrop for these flowers in *The Lay of Leithian* (Lays III: lines 701–703). The name "snowdrop" was first used in the 1633 edition of John Gerard's *Great Herball,* and presumably it refers to the very early flowering of this species, often when snow is still on the ground. In the scientific name *Galanthus nivalis,* the generic name (*Galanthus*) is derived from the Greek *gala* (milk) and *anthos* (flower) in reference to the white flowers, whereas the specific epithet (*nivalis*) means "of the snow," and again, likely is a

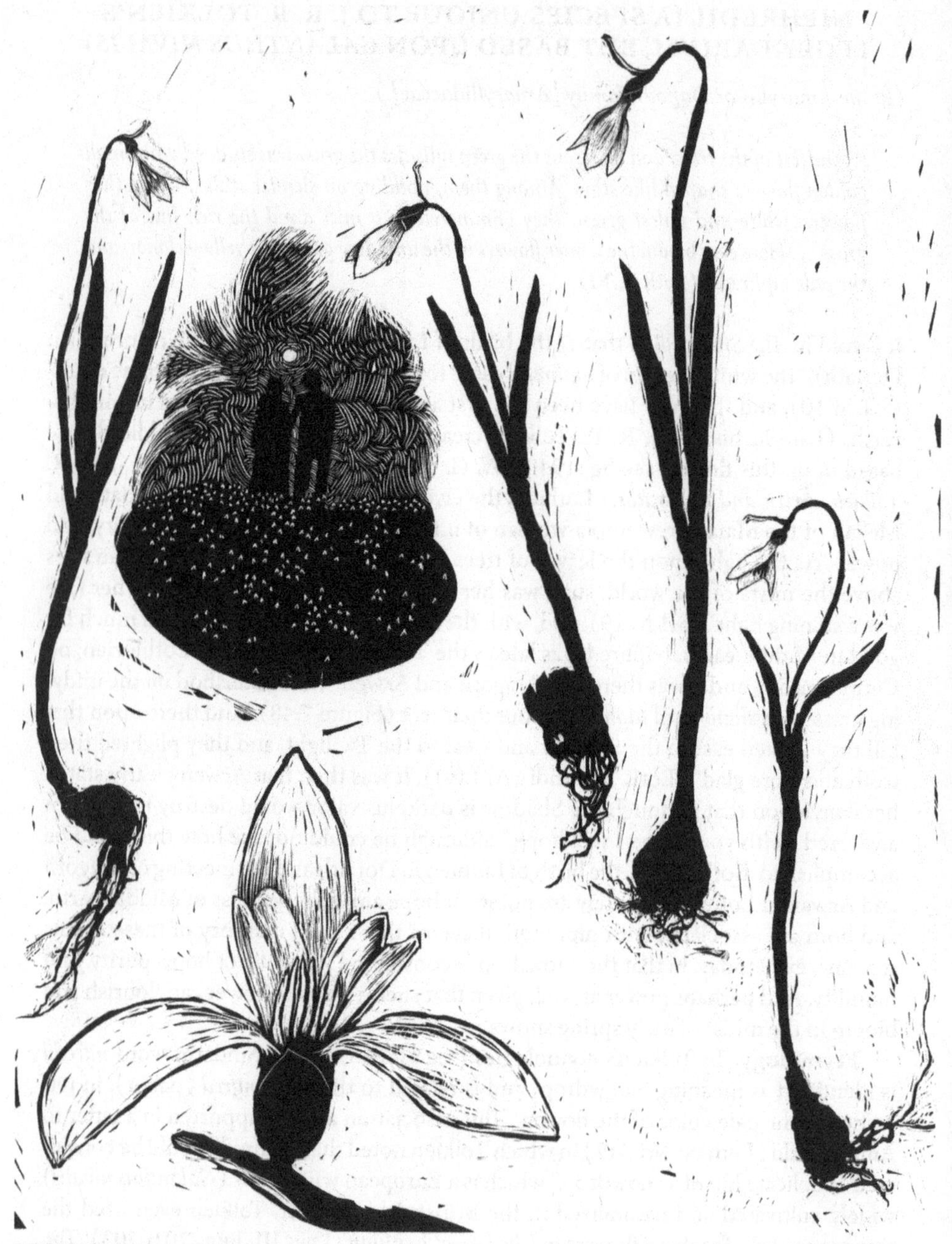

FIGURE 7.48 *Niphredil.*

reference to the growth of this species in the snow of late winter. However, the flowers are also white and can be thought of as snowlike.

Distribution and Ecology: Snowdrops are widespread in Europe from the Pyrenees east to the Ukraine, but are introduced and naturalized in northern Europe (including the British Isles). The species grows in deciduous or coniferous forests, meadows, pastures, and rocky areas, especially in areas of calcareous soil. *Galanthus* is a genus of around eight species, all of which are restricted to the eastern Mediterranean except for *G. nivalis*. The flowers are pollinated by bees, and the seeds are dispersed by ants (which are attracted to the seed appendage). Their ability to bloom so early is developmentally possible because the flower is fully formed inside the bulb during the summer prior to flowering. Growth and flowering thus occur very early in the season, allowing use of light that is blocked later in the spring as the forest canopy closes.

Economic Uses: Snowdrops are popular bulbous ornamentals, grown for their delicate, nodding flowers that mark the end of winter. The plants are poisonous if eaten (containing "amaryllis" alkaloids) and are also used medicinally; for example, they contain the alkaloid galanthamine, which is useful in treating Alzheimer's disease and vascular dementia.

Description (based, in large part, on *G. nivalis*): Perennial herbs, from a bulb having contractile roots, which pull the bulb down into the soil. Leaves alternate, spirally arranged, basal, usually only two, simple, elongate and strap-shaped, glaucous, with parallel venation, the apex ± rounded, the base sheathing, the margin entire; stipules absent. Inflorescence terminal, of a solitary, pendulous flower, on a long scape and associated with a single bract (which is similar in form to the leaves, but much smaller). Flowers radially symmetrical, bisexual, both outer and inner whorls showy (i.e., of tepals), the outer three elliptic to oblong, white, with apex acute to obtuse, the inner three much smaller, obovate, white with a subapical green blotch, with apex emarginate, the stamens six, with orange-yellow elongate-ovoid anthers, these shorter than the inner tepals, and with three fused carpels, the style one, the ovary inferior. Fruit a capsule; each seed with a small, aril-like appendage. The niphredil of *The Silmarillion* and *The Lord of the Rings* differs from our common snowdrop mainly in having a more delicate habit and by its blooming throughout the winter. Note that in the quote heading this section the words "white and palest green" refer to the color of the tepals, the outer ones white and the inner white with a pale green blotch (Figure 7.48).

OAKS (*QUERCUS* SPP.)

(The Beech or Oak family [Fagaceae].)

> *In a great hall with pillars hewn out of the living stone sat the Elvenking [Thranduil] on a chair of carven wood. On his head was a crown of berries and red leaves, for the autumn was come again. In the spring he wore a crown of woodland flowers. In his hand he held a carven staff of oak. (Hobbit IX)*

Oaks are mentioned more times in Tolkien's writings than any other flowering tree—only pines (coniferous trees, which produces cones, not flowers) are referenced more frequently. This frequent mention of oaks reflects their wide distribution and dominance in the forests of Middle-earth (including Beleriand), just as they also dominate the forests of the Northern Hemisphere today. Oaks are repeatedly connected with forests under the control of elves and are abundant, for example, in both northern Mirkwood (controlled by Thranduil in the Third Age) and Doriath (ruled by Thingol in the First Age)—both are kings of "beech and oak and elm" (Lays III: line 72; see also "Of the Rings of Power," in *The Silmarillion*). In fact, Legolas, Thranduil's son, once mentioned that he had "seen many an oak grow from acorn to ruinous age" (LotR 3: VIII). As seen in the initial quote, Thranduil's staff is made of oak and his crown is of berries (possibly the European holly, *Ilex aquifolium* and/or rowan, *Sorbus aucuparia*) and red leaves (probably of oaks, many of which turn orange to red in the autumn, while others turn yellow) (Figure 7.49). Clearly, oak trees were valued in Thranduil's realm both for their beauty and the quality of their wood. There is a hidden connection here between Thranduil and the dwarf Thorin, who in *The Hobbit* was for a time held captive in his halls—both valued the oak. Thorin received his nickname Oakenshield in the great battle with the orcs of Azog at Azanulbizar (near the entrance to Moria), where it was said "that Thorin's shield was cloven and he cast it away and he hewed off with his axe a branch of an oak and held it in his left hand to ward off the strokes of his foes, or to wield as a club" (LotR Appendix A III). The physical characteristics of oaks are mentioned in many passages, although we have space to mention only a few. Several species of oaks are very long lived, and the "huge girth" (LotR 4: VII) of the holm or holly oak (*Quercus ilex*) is mentioned, as are "giant oaks just putting out their brown-green buds" (LotR 4: VII) in the spring, probably a reference to the English oak (*Q. robur*): an indication of the deciduous habit of this species and characteristic of most temperate oaks. Oak trees, especially when growing in open country, can produce very wide-spreading branches, and thus Bilbo was led by Gandalf through "the long green grass and down the lines of the wide-armed oaks" (Hobbit VII). And Gollum, Sam, and Frodo used the crotch of an old holm-oak, formed by its wide-spreading branches, as a protected resting place as they neared the Cross-roads in Ithilien. The spreading branches of the Party Tree described in the first chapter of *The Lord of the Rings,* with lanterns "hung on all its branches" (LotR 1: I) suggests that this proud tree was an open-grown oak. The white oaks (see "Distribution and Ecology") have edible acorns, and this was known by Bilbo. After being rescued by the eagles, and while plodding up slope and down, slowly making his way toward the homestead of Beorn, he was so hungry that he wished that ripe acorns had been on the trees (see Hobbit VII). Finally, bark characteristics are important in oak identification. The cork oak (*Q. suber*) produces especially thick bark, which can be removed and used to make corks. This species of southern Europe must also have been part of the Middle-earth flora (and economy) because it is referenced in the song that Thorin and the other dwarves sang to mock Bilbo Baggins: "Smash the bottles and burn the corks!" (Hobbit I). It is surprising that a tree so common in the forests of every part of Middle-earth (from the Shire to Mirkwood, and south to Gondor and Ithilien) is pictured so sparingly in illustrations relating to Tolkien's legendarium. We have found only three illustrations that clearly show oaks. All are by Alan Lee: two are in *The Two Towers*—the first is in the chapter "The Uruk-Hai" and shows Merry and Pippin walking in Fangorn; the second is in the chapter "The White Rider" showing Aragorn, Gimli, and Legolas in Fangorn); and the third is in *The Hobbit* showing Bilbo in the uppermost branches of a large oak. Oaks were common

FIGURE 7.49 *Oaks.*

in the ancient forest of Fangorn, where Merry and Pippin saw Ents as different from one another as "one tree-kind from another, as birch from beech, oak from fir" (LotR 3: IV). It is curious that Tolkien here compares oaks with firs, the latter a coniferous tree but one bearing a name that descends from the ancestral name for the oak—but he may have been illustrating the etymological connection between these two quite different trees (see "Etymology").

Etymology: The English word "oak" is derived from Old English *ac* (oak tree), from Proto-Germanic **aiks*, of uncertain origin (and with no certain cognates outside Germanic languages). The word "acorn," the name of the distinctive fruits of these trees, is derived from Old English *æcern* (acorn, originally the nuts from any forest tree) and related to the Old English *æcer* (open land, as acorns are the fruits of the open or unfenced lands; related to the word "acre," a unit of land area). The scientific name *Quercus* is the classical Latin name for these trees and is derived from Proto-Indo-European **perkwu-* (oak, or oak forest; but this word has shifted application, evolving into the English word fir.) The early Quenya name of oak is *norno*. Finally, we should also mention here the rather mysterious Gondorian tree, known in Sindarin as *lebethron*. This tree grew in mountainous regions and was valued for its beautiful, hard, and dark wood. It was used to make the walking sticks that Faramir gave to Sam and Frodo (see LotR 4: VII). The first element of this name, *lebeth*, refers to finger, and the second element, *oron* or *orn*, may mean tree. But, it has been suggested by Didier Willis that this second element was derived from the Sindarin name *doron* (= oak; see *Etymologies*, in HoM-E, vol. 5), so the name may mean "finger-oak" although this interpretation is certainly quite speculative. Linking this name with any particular species of oak is even more problematic, but the southern red oak (*Q. falcata*) and mountain chestnut oak (*Q. montana*) have been suggested, and both are native to eastern North America. The former has lobed leaves in which the lobes are somewhat fingerlike, and the latter has its chestnutlike leaves clustered at the tips of the shoots; thus, each leaf of the cluster is like a finger radiating from the twig apex.

Distribution and Ecology: Oaks (members of the large genus *Quercus*, with some 400 species) are widely distributed in the Northern Hemisphere: 90 species are reported from North America (north of Mexico, and many others occur in the mountains of Mexico, with a few extending into Central America and northern South America), 27 reported from Europe, 185 from China, and 14 from Japan. The species belong to two major groups: subgenus *Cyclobalanopsis* (those species with scales on the acorn cup arranged in a series of rings) and subgenus *Quercus* (with scales of the acorn cup overlapping, like the scales of a fish). Members of subgenus *Quercus* are divided into several sections: the largest of these are section *Lobatae* (the red oaks, species having dark bark, leaves with bristle tips, flowers with elongate styles, and nuts maturing in two years) and section *Quercus* (the white oaks, species having lighter bark, leaves without bristle tips, flowers with short styles, and nuts maturing in a single year). The numerous species of oaks are differentiated by variation in the coloration and hairs on their twigs and buds; persistence, shape, and margin of their leaves; and especially the form and distribution of the foliar hairs and form of the nut and associated cuplike structure (cupule). Identification of species can be difficult because the leaves of juvenile plants are often strikingly different from those of mature individuals, and, even in mature trees, the leaves of shaded lower branches differ from those of the upper, sun-exposed branches. Also hybridization is common between red oak species and also between white oak species. Oaks are distinguished from each other by numerous characteristics of leaf shape, lobing, and pubescence, as well as the form of the acorns (and the acorn-cup). They grow in a wide variety of wet to dry habitats, and they often dominate the forests in which they occur. Their inconspicuous flowers are wind-pollinated, and the acorns are dispersed by birds and mammals.

Economic Uses: Oaks are extremely important as a source of timber (used in construction, furniture, barrels for wine, whiskey, etc.). The bark of many species has been used as a dye source, and the thick bark of the cork oak (*Q. suber*), high in suberin, is stripped off the trees (harvested repeatedly, every 8–10 years) and used to make cork. The acorns of some species are very high in tannins and have been used medicinally and in tanning leather, while others have lower tannin content and are used as animal feed (and have also been used as food by humans). Finally, oaks are frequently used in landscaping as stately shade trees.

Description: Deciduous to evergreen trees or shrubs, with the trunks stout, covered with light to dark gray or brown, thick to thin, vertically furrowed bark. Leaves alternate, spirally arranged, simple, unlobed to deeply lobed, ovate to obovate, pinnately veined, forming loops or running into the teeth or lobes, the apex acute to acuminate to rounded, the base various, the margin entire or variously toothed, the teeth (or lobes) with or without bristle-tips, the lower leaf surface showing much variation in the density and form of the hairs, which may be stellate, glandular, or nonglandular, and such hairs may be lacking to densely covering the leaf surface or sometimes restricted to dense hair-tufts (domatia) in the angles of the midvein and secondary veins (and these forming homes for beneficial mites), petiolate; stipules present. Flowers inconspicuous, radially symmetrical, unisexual (with staminate and carpellate flowers on the same plant), with an inconspicuous perianth of six tepals; the staminate inflorescences in the axils of bud scales, at the base on the new shoots, dangling, elongate catkins; the carpellate inflorescences in the leaf axils of the new growth, stiff, a small cluster of flowers or a solitary flower, with each flower associated with a scaly cuplike structure (cupule); staminate flowers each with usually six stamens; carpellate flowers with three fused carpels, with three short to elongate styles, the ovary inferior. Fruits globose, nuts (acorns), each nut borne in a thin- to corky-scaled cupule (the acorn cup) that more or less covers the base of the nut (Figure 7.49).

OLIVE (*OLEA EUROPAEA*)

(In the Olive family [Oleaceae].)

> *Many great trees grew there, planted long ago, falling into untended age amid a riot of careless descendants; and groves and thickets there were of tamarisk and pungent terebinth, of olive and of bay. (LotR 4: IV)*

Olives are only mentioned once in *The Lord of the Rings*. Olive trees were seen by Frodo and Sam in Ithilien, growing among fragrant trees such as tamarisk (*Tamarix* spp.), terebinth (*Pistacia terebinthus*), bay (*Laurus nobilis*), and juniper (*Juniperus* spp.). The presence of olives immediately establishes that the climate of Ithilien is quite different from that of the hobbits' homeland. They have traveled far from the Shire, are in a more southerly region, and one with a Mediterranean climate. The hills and valleys of Ithilien support a lush growth of plants, forming plant communities of greater diversity than anything previously seen by the hobbits. The gardenlike land had recently fallen under the control of Sauron but had not yet been destroyed, although they see signs of environmental damage—"scars of the old wars, and the newer wounds made by the Orcs and other foul servants of the Dark Lord: a pit of uncovered filth and refuse; trees hewn down wantonly and left to die, with evil runes or the fell sign of the Eye cut in rude strokes on their bark" (LotR 4: IV). The two hobbits recognized trees of many kinds, but also see "other kinds unknown in the Shire" (LotR 4: IV)—the hobbits were in a plant community that is beautiful and fragrant but quite outside their past experience. They breathed deeply, enjoying the sweet odors as they walked "through brush and herb," and Sam laughed "for heart's ease not for jest" (LotR 4: IV)—even this damaged environment had the ability to heal, to lift their spirits after the despair and suffering they experienced at the lifeless ridges of the Ered Lithui and Black Gate of Mordor.

Etymology: The English name "olive," like the scientific, generic name *Olea*, is derived from the Latin *oliva* and the ancient Greek *elaia* (olive; and related to the word *elaion* [oil]), which was derived from some Aegean language (perhaps Cretan: *elaiwa*, meaning oil), and may be derived from Proto-Indo-European **loiwom* (oil).

Distribution and Ecology: The genus *Olea* contains just over 30 species and is widely distributed in warm temperate to tropical regions of the Old World. The group's most important species, the olive (*Olea europaea*), is widespread and morphologically variable, occurring from the Mediterranean region, Portugal, and Macaronesia, southward to southern Africa, and eastward through the Middle East into western Asia and in the Himalayan region of India and China. In the Mediterranean region, *O. europaea* subsp. *europaea* has long been cultivated, but wild populations (with thorny branches, smaller leaves and fruits) also occur. Outside of cultivation, olives grow in woods, scrub, or dry rocky places in regions with a Mediterranean climate (i.e., dry summers, with most rainfall occurring in the winter). They are drought- and fire-tolerant. Olive trees are slow growing, develop gnarled and twisted trunks, and can reach a great age (ca. 3,000 years). The small white flowers are wind-pollinated; their drupaceous fruits are dispersed by birds.

Economic Uses: Olives are a very important source of cooking or salad oil. The oil, which is monounsaturated, with high levels of oleic acid, is pressed from the fruit pulp, which contains as much as 40 percent oil. The fruits, either ripe (black) or unripe (green), are very bitter; they can be eaten only after processing (i.e., pickling in brine, for ripe olives, or first treated with lye to neutralize the bitter taste, followed by pickling, for unripe olives). Olives were domesticated nearly 6,000 years ago, probably in the lands just east of the Mediterranean. Additionally, olive trees are occasionally used as ornamentals because of their drought-tolerance, evergreen habit, and attractive leaves that are silvery beneath. In folklore, olives are associated with peace, good will, new life, and hope.

Description: Evergreen trees, with gray, finely fissured bark; branches gray, covered with peltate scales. Leaves opposite and decussate, simple, narrowly elliptic to obovate, pinnately veined, the secondary veins forming a series of loops, the apex acute to slightly mucronate, the base narrowly cuneate, the margin entire, the lower surface light gray or slivery (due to dense covering of peltate scales), shortly petiolate; stipules absent. Inflorescences usually axillary, elongate cymes. Flowers radially symmetrical, bisexual, the sepals four, small, fused, the petals four, fused, forming a white, four-lobed corolla, each lobe only slightly longer than wide; the stamens two, fused to the corolla, with large anthers and short filaments and with two fused carpels; the style short; the ovary superior. Fruit an ellipsoid to nearly globose drupe, black or brownish-green at maturity, with single pit and flesh of high oil content.

PINES (*PINUS* SPP.)

(In the Pine family [Pinaceae].)

> *He [Gandalf] gathered the huge pine-cones from the branches of the tree. Then he set one alight with bright blue fire, and threw it whizzing down among the circle of the wolves. It struck one on the back, and immediately his shaggy coat caught fire, and he was leaping to and fro yelping horribly. (Hobbit VI)*

Pines occurred widely in the Middle-earth of the Third Age, growing mainly in mountains areas, especially in the Misty and White Mountains, the western slopes of the Mountains of Shadow, in northern Ithilien, and on the Lonely Mountain (Figure 7.50). They also had been common in Beleriand and were especially characteristic of the highlands of Dorthonion and the mountains surrounding Hithlum. It is not surprising, therefore, that they are mentioned more times in J. R. R. Tolkien's writings than any other species of tree. Pines, like other members of the Pinaceae, contain resin canals in their wood, leaves, and cones, and these produce sticky and aromatic compounds, especially volatile and fluid terpenes, that are protective in function, defending these trees against attack by insects and other animals but also making these trees quite flammable. Tolkien often stresses these characteristics in his descriptions of pine trees. As Bilbo and the dwarves descend into the valley of Rivendell, "the smell of the pine-trees made him drowsy" (Hobbit III), the memory of the pines on the heights of the Lonely Mountain set ablaze by Smaug and burning like torches (Figure 7.50) is clearly expressed in the dwarves' song in the first chapter of the *Hobbit*, and the dwarves used pine-torches when in the dark halls under the Lonely Mountain. In chapter VI ("Out of the Frying-pan into the Fire"), Gandalf, Bilbo, and the dwarves were trapped by wolves and then goblins in a pine forest on the eastern slopes of the Misty Mountains where they, in order to escape, had climbed various coniferous trees—Fili and Kili end up in a larch (*Larix*; Figure 7.40), Dwalin and Balin in a fir (*Abies*), and the rest in three pines—all of which are members of the Pinaceae. In the opening quote, we see Gandalf using his fire-skills to set alight pine cones and throw them down on the attacking wolves (Figure 7.51). It is not surprising that the dry forest, with its "yellowing bracken, fallen branches, deep-piled pine-needles, and here and there dead trees" (Hobbit VI) was soon in flames. At this point, the arriving goblins used the fire to their advantage; they piled fuel around the trees containing members of the Company, encouraging the flames, which seemed sure soon to set these trees alight. They then began singing a horrible song, the first line of which was "Fifteen birds in five fir trees"; evidently, they could count well enough, but could

FIGURE 7.50 *Desolation of Smaug.*

not properly identify species of conifers! Fortunately, at this point, the Company was rescued by the eagles.

Pines are sun-loving trees characteristic of low-nutrient soils and thus often form open, fire-maintained forests on dry soils, such as well-drained sands, as noticed by Frodo and his companions in the Old Forest (i.e., "the land seemed to be drier and more open, climbing up to slopes where the trees were thinner, and pines and firs replaced the oaks and ashes" [LotR 1: VI]) or in the thin, rocky soils of mountain slopes, as seen by Merry and Pippin when carried by Treebeard, high in the mountains overlooking Isengard (i.e., "bare slopes where only a few gaunt pine-trees grew" (LotR 3: IV]). The pines of such open slopes are often sculpted by the wind, their prominent horizontal branches taking on strange or eerie shapes. On the high plateau of Dorthonion, after its capture by Morgoth, the pine forest became haunted, Taur-nu-Fuin, the "Trackless Forest of Deadly Nightshade" (Lays I: lines 766–767; this forest was illustrated by Tolkien, see W. G. Hammond and C. Scull, 1995, *J. R. R. Tolkien: Artist and Illustrator*, fig. 54, a figure that was later used to represent Mirkwood as well as Fangorn forest). It was there that Beleg, an elf of unsurpassed woodcraft, tracking the Orc band that had taken Túrin captive, was "oppressed by pungent pinewood's odours" and the endless darkness of the huge pines with their intertwined branches obscuring the stars, and he became hopelessly lost. Seemingly alone in the forest, he listened to the wind "moaning in bending boughs; to branches creaking up high over head, where huge pinions of the plumed pine-trees complained darkly in black foreboding" (Lays I: lines 769–773). It is then that he found Gwindor, broken by years in Morgoth's mines, lying on a bed of pine needles. Gwindor told him that he had seen an orc band, perhaps an advance guard, and it was not long before they heard the noise of a great host moving through the pines. Finally, at the northern edge of the forest, they found the orc encampment and saw Túrin, guarded, tied to a withered tree. This meeting of Beleg and Túrin, with its hopeful as well as tragic results, is brought to life by the descriptive detail of *The Lay of the Children of Húrin* (although related more briefly in the prose, *The Children of Húrin*, chapter IX), and the dark and massive pines of the Dorthonion highlands, now transformed by evil, some already dead, with "their lichen-leprous limbs uprooted" (Lays I: line 939), are an important part of Tolkien's vision.

Etymology: The English name "pine" is derived from the Old English, *pin*, borrowed from Old French, *pin*, from Latin, *pinus*, the classical name for the Mediterranean stone pine (*Pinus pinea*; also the scientific name of the entire genus), which may be from Proto-Indo-European **pi-nu-*, from the root **peie-* (to be fat, swell), probably in reference to its sap or pitch. The Sindarin name for pine is *thôn*, as seen in the locality name Dorthonion. The Quenya name may be *sáne* or *aiqairë* (but the latter is also used for firs).

Distribution and Ecology: Pines, a group of around 110 species, are the largest and most widespread genus of conifers in the Northern Hemisphere. The group occurs in North America (south to Nicaragua, including the West Indies) with ca. 70 native or naturalized species (38 of these occur north of Mexico), in Europe and northern Africa with ca. 20 species, and in eastern Asia with ca. 50 species (but many of these introduced). Pines grow in an extremely broad array of habitats, from boreal forests or alpine habitats near tree-line to lowland tropical savannas, and from swampy habitats to deserts. They are often pioneer species that require high light

FIGURE 7.51 *Pines.*

levels and exposed soil for germination, and many require fire for establishment and/or maintenance of their dominance in the plant community. Species are separated by characteristics of the bark and the number of needle-leaves per fascicle, as well as by their length and features of their seed-cones. Pines belong to two major monophyletic subgroups: the hard pines (*Pinus* subgenus *Pinus*), which have hard wood, two vascular strands per needle, the and sheath of scale leaves of the needle-fascicle persistent; and the soft pines (*Pinus* subgenus *Strobus*) with softer wood, a single vascular strand per needle, and the sheath scale leaves of the needle-fascicle deciduous. Both their pollen grains (each with two round air bladders) and winged seeds are dispersed by wind.

Economic Uses: The wood of pines is extremely important and is used in construction, furniture, and pulp (for paper); pines are also the source of resinous products, such as rosin and turpentine. They are also widely used as ornamental trees. A few species have fairly large, edible seeds (pine nuts). Pines are often ecologically important and often dominate the forests in which they occur.

Description: Evergreen trees with a conical crown when young, often rounded or flat-topped with age, with bark with variously formed plates and/or ridges, occasionally smooth. Twigs and leaves dimorphic, some branches elongated with non-photosynthetic scale leaves (i.e., forming long-shoots) and others very short and each with a cluster of (1–) 2–8 photosynthetic, needle leaves surrounded by a sheath of scale leaves (i.e., forming short shoots, called fascicles), the twigs smooth or roughened (due to persistent scale leaves). Leaves alternate, spirally arranged (but appearing fascicled), simple, linear and needlelike, terete to two- or three-angled, usually shaped like pie-pieces in cross-section, rarely flattened, with resin canals, with single or double-stranded midvein, the apex acute, the base slightly narrowed, the angular needle-edges entire; stipules absent. Plants with pollen cones and seed cones on the same tree. Pollen cones ovoid to oblong-cylindric, small, tan to yellow, red, blue, or lavender, with each pollen-bearing structure flat, bearing two sporangia. Seed cones conic to cylindric, erect to pendent, borne in axils of scale leaves; maturing in two (less commonly three) seasons, opening to release the seeds (either immediately after maturation or remaining closed for several years until opened by fire); scales woody or flexible, surface of exposed apical portion thickened and represented by a scar or extended into a pricklelike structure, hook, or claw, each bearing two seeds, free from the associated, much smaller bracts. Seeds usually strongly winged (Figure 7.51).

PIPE-WEED, TOBACCO (*NICOTIANA TABACUM*)

(In the Potato or Nightshade family [Solanaceae].)

> *There is another astonishing thing about Hobbits of old that must be mentioned, an astonishing habit: they imbibed or inhaled, through pipes of clay or wood, the smoke of the burning leaves of a herb, which they called pipe-weed or leaf, a variety probably of Nicotiana. (LotR Prologue: 2, "Concerning Pipe-weed")*

It is significant that none of the characters in *The Lord of the Rings* ever uses the word "tobacco," instead referring to this plant as pipe-weed. Certainly, Tolkien was aware of tobacco's American origin and fairly recent introduction into Europe, and thus the use of this Amerindian word in the distant past of the Third Age of Middle-earth would have been anachronistic and damaging to the Englishness of the Shire, as has been noted by Tom Shippey (1992, in *The Road to Middle-earth*). Thus, the word "tobacco" is only used in the narrative voice in *The Lord of the Rings*. In contrast, it is used more freely in *The Hobbit*, a story quite different in tone, as when Bilbo tells Gandalf that it was "a very fine morning for a pipe of tobacco out of doors" and generously added "If you have a pipe about you, sit down and have a fill of mine!" (Hobbit I). The origin and use of pipe-weed is presented in detail in the prologue "Concerning Pipe-weed," which draws on Merry Brandybuck's *Herblore of the Shire*, in which it is stated that the plant was brought to Middle-earth by the Númenóreans during the Second Age and became naturalized in Gondor (where it was known as "sweet galenas" because of the fragrance of its flowers), was eventually carried north to Bree, and then brought into cultivation in the Shire by Tobold Hornblower (of Longbottom in the Southfarthing). The Hobbits (of Bree and then the Shire) were the first to smoke the dried leaves of the plant. We have, therefore, a fascinating independent introduction of tobacco into ancient Europe, as envisioned in the history underlying the events in *The Lord of the Rings*, thus validating the presence of this American plant in Middle-earth! Smoking pipe-weed was quite popular among hobbits, and in Figure 7.52 we see Merry and Pippin smoking beside Isengard. Finally, it should be mentioned that smoking was taken up by both Gandalf and Saruman, although the latter kept this habit quite secret. Saruman openly criticized Gandalf at the White Council held at Rivendell in the year 2851, arguing that Dol Guldur should not yet be molested, and telling Gandalf that "When weighty matters are in debate, Mithrandir, I wonder a little that you should play with your toys of fire and smoke, while others are in earnest speech." Gandalf laughed, replying that "You would not wonder, if you used this herb yourself. You might find that smoke blown out cleared your mind of shadows within. Anyway, it gives patience, to listen to error without anger." After a scathing and belittling response from Saruman, Gandalf, giving him a sharp look, said nothing, but merely "sent out a great ring of smoke with many smaller rings that followed it" that vanished as he reached out his hand to grasp them (UT 3: IV). The image of Gandalf smoking and blowing smoke rings may be one of the most enduring from *The Hobbit* and *The Lord of the Rings* (see, e.g., painting of Bilbo and Gandalf by Alan Lee in the illustrated *Lord of the Rings*, opposite p. 64, and Figure 7.16), reflecting the pleasure that Tolkien himself received from smoking a pipe.

Etymology: The English word "tobacco" was derived from the Spanish *tabaco*, which may have been taken from a Taino word meaning "a roll of tobacco leaves" (as suggested by Las Casas in 1552) or "a kind of pipe for smoking [or snuffing] tobacco" (as per Oviedo, in 1535). Taino was an Arawakan language and was the principal language in the Caribbean during the time of initial contact between the Spanish and Amerindians. However, it is possible that *tabaco* was instead derived from the Arabic *tabaq* (a type of medicinal herb), and thus could thus represent an Old World name applied to a New World plant. The scientific name *Nicotiana* was coined by C. Linnaeus and honors Jean Nicot, the French Ambassador to Portugal, who in 1559 introduced the plant to France.

FIGURE 7.52 *Pipe-weed.*

Distribution and Ecology: Cultivated tobacco (*Nicotiana tabacum*) is grown as a crop around the world but is native to South America. It is not known from wild populations, being an exclusively cultivated plant, which is likely of fairly recent origin (perhaps within the last 10,000 years). It is a polyploid species, having the full set of chromosomes from each of its parental, diploid ancestors, *N. sylvestris* and *N. tomentosiformis*, both also of South America. Aztec tobacco (*N. rustica*), a related and similar species, is also South American in origin but was brought to North America prior to the time of European contact. It, too, is a polyploid, having evolved through hybridization of *N. paniculata* and *N. undulata* and chromosome doubling. Aztec tobacco occurs in open, disturbed habitats and also is occasionally cultivated. Aztec tobacco was the first to be introduced into Europe in the mid-1500s, but cultivated tobacco was brought to the Old World not much later. The showy flowers are often moth-pollinated (and can produce seed from selfing). Their small seeds are dispersed by wind or rain-wash.

Economic Uses: Tobacco is a major commercial crop; its leaves are smoked, snuffed, or chewed for their mild stimulant effect, and they contain the addictive alkaloid nicotine, among other substances. The plant also is used as an insecticide (nicotine has been shown experimentally to reduce insect herbivory) and medicinally (e.g., treatment of intestinal parasites). Tobacco and related species of *Nicotiana*, a genus of more than 70 species, are used as garden ornamentals because of their showy, trumpet-shaped flowers. Cultivated tobacco and Aztec tobacco are similar in their usage, but the former is much more economically significant. Tobacco was (and is) used for ceremonial and religious purposes in many Amerindian cultures. Finally, the use of tobacco is known to cause heart disease, lung disease, and cancer of the lungs, larynx, mouth, and pancreas, leading to about 4 million deaths per year.

Description (of *N. tabacum*): Usually an annual herb to 2.5 meters, with leaves, stems, sepals, and petals covered with simple, sticky, gland-headed hairs; stems erect, cylindrical. Leaves alternate, spirally arranged, simple, ovate to obovate, pinnately veined, the apex acute to acuminate, the base narrowed, without a petiole, with blade extending onto stem, the margin entire. Inflorescences terminal cymes. Flowers radially symmetrical, bisexual, showy, the sepals five, fused, ovate, green; the petals five, fused into a trumpet-shaped corolla tube, with longitudinal foldlines, and plicate in bud; the tube white and the lobes pink to rose, the apex having the appearance of a five-pointed star; the stamens five, with filaments fused to the corolla, with two fused carpels; the style elongate; the stigma capitate, two-lobed; the ovary superior. Fruit an ovoid capsule, opening to release the tiny seeds (Figure 7.52).

POLE BEANS, BEANS (*PHASEOLUS* SPP.)

(In the Legume family [Fabaceae or Leguminosae].)

> *Actually his [Frodo's] view was screened by a tall line of beans on poles; but above and far beyond them the grey top of the hill loomed up against the sunrise.... The sky spoke of rain to come; but the light was broadening quickly, and the red flowers on the beans began to glow against the wet green leaves. (LotR 1: VII)*

Although the use of species of the genus *Phaseolus* as edible plants (for their seeds and young pods) has now spread around the world, they are native to the Americas and as such should not be expected in Middle-earth. However, they were grown in Tom Bombadil's garden (Figure 7.53), and Frodo saw them when he looked out the eastern window on his first morning in Tom's home instead of "turf all pocked with hoofprints" (LotR 1: VII) of Black Riders, as he half expected based on a misinterpretation of his recent dream (which actually was of Gandalf's rescue from Ornthanc). The arrival of beans in Middle-earth is not explained, but, in the case of Tom Bombadil, one who remembered "the first raindrop and the first acorn" (LotR 1: VII), explanations are unnecessary. The plant described in the introductory quote is clearly *Phaseolus coccineus* (the scarlet runner bean), and it can be differentiated from the common bean (*P. vulgaris*) by its red to orange-red (vs. white to pink or rose-colored) flowers and thick, starchy (vs. slender) roots. It is a perennial vine (although it can be grown as an annual), whereas the common bean is an annual (with both climbing and bushy cultivars); the scarlet runner bean is grown both for its edible pods and seeds as well as its beautiful flowers that attract hummingbirds. Whether Tom also grew the common bean in his kitchen garden is not known, but perhaps common beans were also grown in Middle-earth since Farmer Maggot's homestead is named Bamfurlong, an English place-name probably derived from bean + furlong (i.e., a strip of land reserved for growing beans; see *Nomenclature of the Lord of the Rings*, p. 765, in W. G. Hammond and C. Scull's, 2005, *The Lord of the Rings: A Reader's Companion*).

Etymology: The English word "bean" comes from Old English *bean* (meaning bean, pea, leguminous seed) and is derived from Proto-Germanic **bauno* (bean), and perhaps from Proto-Indo-European **bhabh-* (bean). The name is applied to the large edible seeds of plants of several genera of Fabaceae (the legume family), especially to seeds of members of the genus *Phaseolus*. The scientific name *Phaseolus* was derived from the Latin *phaselus* and the diminutive suffix *-olus* and taken from the Greek *phaseolos* (a name applied to a related legume, the cowpea, *Vigna unquiculata*, which also produces edible seeds).

Distribution and Ecology: Species of *Phaseolus* (beans, although this common name is also applied to some other genera of legumes) occur widely in warm temperate to tropical regions of the Americas, although the domesticated species are now widely grown. The most important species as a food plant is *P. vulgaris* (the common bean); it is distributed widely in Latin America and was domesticated independently in Mexico (small-seeded beans) and Peru (large-seeded beans) about 7,000–8,000 years ago. *Phaseolus lunatus* (the lima bean) is also widely distributed in Latin America; it was also domesticated at least twice, perhaps as early as 8,000 years ago. A less widely known species is *P. coccineus* (the scarlet runner bean); it is native to cool mountainous areas from Mexico to Panama and is thought to have been domesticated in Mexico at least 2,000 years ago. It often grows better than *P. vulgaris* in cool climates. It is cultivated in England and is clearly shown in Tolkien's colored pencil artwork *New Lodge, Stonyhurst* (see fig. 28 in W. G. Hammond and C. Scull's, 1995, *J. R. R. Tolkien Artist and Illustrator*). The colorful flowers of these species are usually pollinated by bees, although *P. coccineus* attracts hummingbirds (as well as some bees), and many cultivars of *P. vulgaris* are selfing. The fruits of wild plants coil when dried, shooting out the seeds, while those of cultivated genotypes dry with the seeds inside and require collection and planting by humans.

Economic Uses: The seeds and young fruits (pods, legumes) of several species of *Phaseolus* are eaten and are highly nutritious, being high in protein, fiber, and complex carbohydrates; the most important domesticated species include *P. vulgaris* (common bean,

FIGURE 7.53 *Pole beans.*

French bean, kidney bean, green bean, pole bean, string bean, wax bean, snap bean, navy or white bean, pinto bean, red bean, and many others; a variable species with numerous cultivars), *P. lunatus* (lima bean), *P. coccineus* (scarlet runner bean), and *P. acutifolius* (tepary bean). It is of interest that beans contain the toxic lectin phytohaemagglutinin (the quantity varying by species and cultivar), but fortunately this compound is destroyed by boiling.

Description: Annual or perennial herbs or vines, with or without simple hairs; stems twining if the plants vines or, in many cultivars, not twining, and the plants then having a bushy habit. Leaves alternate, spirally arranged, pinnately compound, with three leaflets, these ovate, with pinnate venation, the apex acute or acuminate to obtuse or rounded, the base rounded or truncate to obtuse, asymmetric in the two lateral leaflets, the margin entire; stipules present and stipulelike structures also associated with the leaflets. Inflorescences axillary racemes. Flowers bilaterally symmetrical, bisexual, held horizontally, with five fused sepals forming an asymmetric calyx and with uppermost petal differentiated in size and shape, more or less rotund, and forming a banner (or standard), positioned on the outside in bud, two lateral, short-clawed, obovate, wing petals, and the two lowermost petals fused together, elongated, and forming a spirally coiled keel, these usually all the same color (or standard differentiated from the others), white to red, orange-red, violet, or yellow, with 10 stamens, nine of which are fused together by their filaments, and with a single elongate carpel, the ovary superior, and the single style curled. Fruit a legume, elongate, initially green, yellow, or purple, but eventually turning brown and opening to release the variously colored, bean-shaped seeds (Figure 7.53).

POPLARS (*POPULUS* SPP.)

(*The Willow family [Salicaceae].*)

> *Before them a wide grey shadow loomed, and they heard an endless rustle of leaves like poplars in the breeze. "Lothlórien!" cried Legolas. (LotR 2: VI)*

Perhaps poplars (some species of which are also called aspens or cottonwoods) should not be considered plants of J. R. R. Tolkien's Middle-earth, but they are referenced in the passage just quoted, in which Tolkien, as narrator, describes the sound of mallorn leaves rustling in the breeze, which the members of the Fellowship hear as they reach the eaves of the Golden Wood (Figure 7.54). Many poplars have petioles that are flattened near their junction with the leaf blade, causing the leaves to flutter even in the slightest breeze. This fluttering likely allows more light to penetrate into their canopy, allowing greater photosynthetic efficiency (but it also cools the upper canopy leaves and may increase the rate of carbon dioxide uptake), so this tendency to easily flutter is probably advantageous. Tolkien certainly knew and liked poplars, which include species native to England, along with several that are widely planted and naturalized. He mentioned them in a couple of letters to his son Christopher. In a letter sent in October 1943, he told his son that "the poplars are now leafless except for one top spray" (Letters: No. 50), and then in letter sent in April 1944, he stated that the leaves "are out: the white-grey of the quince, the grey-green of young apple, the full green of hawthorn, the tassels of flower even on the sluggard poplars" (Letters: No. 61). Among temperate trees, poplars are distinctive in having easily trembling leaves, although this characteristic is not unique: for example, the tropical tree *Schefflera tremula* (bwa

FIGURE 7.54 *Poplars.*

trembler, palo de viento) has palmately compound leaves with easily trembling leaflets. The sound made by such trembling leaves is relaxing, and, by comparison, we know that mellyrn must have pleased the ears as well as the eyes.

Etymology: The common name "poplar" comes from Anglo-French *popler*, from Old French *poplier* (modern French is *peulplier*), which was derived from Latin *populus* (with a long "o," meaning this group of trees, not *populus*, meaning people, the word from which popular is derived), of unknown origin. The scientific Latin name *Populus* is, as noted, the ancient Roman name for these trees. Perhaps the link between the two meanings of the Latin *populus* relates to the frequent use of poplars in public squares. The Sindarin word for poplar tree may be *tulus* and the Quenya word *tyulussë*.

Distribution and Ecology: Poplars (some species also known as cottonwoods or aspens) include ca. 35 species, which are widely distributed in mainly arctic to warm temperate, moist to wet habitats of the Northern Hemisphere: 8 species are native to (plus 3 introduced to) North America (north of Mexico), 4 are native to Europe (several additional are naturalized or frequently planted), perhaps ca. 20 occur in China (although 71 are recognized in the flora of China, but species limits are very problematic and there may be many fewer species than recognized in that reference), and five are native to Japan. The most widespread species in North America are the balsam poplar (*P. balsamifera*), eastern cottonwood (*P. deltoides*), and trembling aspen (*P. tremuloides*); the most common European species include the silver poplar (*P. alba*), common aspen (*P. tremula*), and black poplar (*P. nigra*). Poplars grow in floodplains, lake and stream margins, aspen parklands, dry to wet, open to dense woodlands and forests. The tiny flowers of poplars are in held dangling catkins and are wind-pollinated; their fruits are capsules, opening to release hair-tufted seeds that are also wind-dispersed.

Economic Uses: Poplars are rapidly growing trees and thus are an important source of inexpensive wood that is much used in plywood, pallets, some musical instruments, snowboards, as a source of pulp, and in tanning leather. The trees are often planted as windbreaks. The buds of some species are used as a source of medicinal resin. Finally, poplars are useful ornamental trees.

Description: Deciduous trees, usually forming clones by root shoots, the bark vertically furrowed to smooth and with horizontal lenticels (similar to that of birches, but not peeling). Winter buds with 3–10 protective scales, often resinous. Leaves alternate, spirally arranged, simple, ovate, palmately to pinnately veined, the veins gradually curving toward margin forming obscure loops, the apex acuminate or acute to obtuse, the base cuneate to acute to cordate, the margin coarsely to finely serrate or crenate-serrate, each tooth with a globose, glandular tip, with hairs various, the leaves petiolate and the petiole round to apically flattened, either in the same plane as the blade or at right angles to the blade, the juvenile and adult often quite different; stipules present, usually minute. Flowers inconspicuous, radially symmetrical, unisexual (with staminate and carpellate flowers on different plants), with a reduced, disk- or cuplike perianth; flowers in the axil of a bract, without nectar glands, densely clustered in axillary or terminal, dangling catkins; staminate flowers with six to numerous stamens, their filaments distinct; carpellate flowers with 3–4 fused carpels, with separate styles, and 2–4, much expanded stigmas; the ovary superior. Fruit a capsule, opening by 2–4 valves; seeds small, with a conspicuous tuft of hairs (Figure 7.54).

POTATOES, TATERS (*SOLANUM TUBEROSUM*)

(The Potato or Nightshade family [Solanaceae].)

> *"What a hobbit needs with coney," he said to himself, "is some herbs and roots, especially taters—not to mention bread." (LotR 4: IV)*

The fact that potatoes were commonly cultivated in the Shire is seen in the chapter "A Long-expected Party," in which Hamfast Gamgee, Sam's father, is described as an expert "in the matter of 'roots,' especially potatoes" (LotR 1: I) (Figure 7.55). Their common usage is also seen in his warning to Sam concerning the dangers of getting mixed up in the affairs of wizards: the Gaffer exclaimed, "Elves and Dragons! Cabbages and potatoes are better for me and you" (LotR 1: I). In this context, both cabbages (*Brassica oleracea* var. *capitata*) and potatoes (*Solanum tuberosum*) stand for the commonplace vegetables that could be found in nearly every hobbit's garden, in contrast to the exotic and faraway. Such a connection is certainly appropriate for cabbages because that vegetable is native to—and was domesticated in—Europe. Potatoes, however, are a New World domesticated vegetable, derived from wild species growing in the Andes (South America) and only known in Europe after the time of Columbus. Even the use of the common name "potato" hints at this since the name is derived from an Arawakan language! How could potatoes possibly be part of the culture of hobbits in Middle-earth? Tolkien was aware of this problem, and perhaps this is one reason that the word "taters" (instead of potatoes) is most frequently used for this important starchy vegetable. One could argue that potatoes and taters, as used in *The Lord of the Rings*, refer to some other root crop—although Sam's later suggestion that he would like to make "fish and chips" (LotR 4: IV) for Gollum make this unlikely. Thus, as has been noted by others (see, e.g., the entry for "potato" in the *Encyclopedia of Arda*, Fisher, 2016), we consider it likely that potatoes (*S. tuberosum*) had been introduced into Middle-earth by the Númenóreans and then later adopted as a food plant by hobbits. After all, the Shire is located in territory that was once part of the Númenórean kingdom of Arnor. We see a similar connection with Númenor in the geographical distribution of athelas and pipe-weed (tobacco).

The quote highlighted at the beginning of this entry, however, is perhaps the most widely known and comes from the chapter "Of Herbs and Stewed Rabbit." The humorous incident occurred in Ithilien when Sam wished that he had herbs and root vegetables (such as carrots, turnips, or potatoes) to add to the rabbit stew that he was cooking. He asked Gollum to bring him some herbs, but he replied "Sméagol doesn't . . . eat grasses or roots" and exclaimed that he would not "grub for roots and carrotses and—taters" (LotR 4: IV). He didn't even know what potatoes were (so perhaps they were not widely eaten outside the Shire), and Sam responded by providing their full name—potatoes—and explained that they were "rare good ballast for an empty belly" (LotR 4: IV).

Etymology: The English common name "potato" is taken from the Spanish *patata*, which is derived from Taino (a now extinct Arawakan language once spoken in the Caribbean) and originally referred to the sweet potato (*Ipomoea batatas*, which is widely cultivated in and native to tropical America). The name was applied to *Solanum*

FIGURE 7.55 *Potatoes, taters.*

tuberosum, an Andean species, in the late sixteenth century. "Tater," a colloquial pronunciation, is first recorded in mid-eighteenth-century England. The derivation of the scientific name *Solanum* is obscure, and it is an ancient Latin name of some unknown plant. The Sindarin name may have been *ceforf*.

Distribution and Ecology: The cultivated potato (*Solanum tuberosum*) is native to the Andes of South America and was one of the staple crops of the Incas, along with quinoa (*Chenopodium quinoa*), maize (*Zea mays* subsp. *mays*), and oca (*Oxalis tuberosa*). It ancestors originally grew (and wild potatoes still grow) in open montane habitats, but it is now also widely cultivated throughout the temperate zone. Potatoes are members of the large genus *Solanum*, a group of ca. 1,250 species that is very broadly distributed. The showy flowers of some species of *Solanum* are pollinated by bees (which grab the stamens and shake out the pollen), and their berry fruits are usually bird dispersed. Although *Solanum tuberosum* does form berries, cultivated plants are often propagated by using small tubers or tuber pieces.

Economic Uses: Potatoes are the fourth most important food crop in the world (after corn, rice, and wheat), and the starchy and nutritious tubers are boiled, baked, fried, or stewed. The plant is now grown throughout the temperate zone and also in tropical highlands. *Solanum tuberosum* was domesticated by Amerindians around 8,000 years ago in the Andes from wild species in the *S. brevicaule* complex, and the species includes diploid to pentaploid genetic entities. Potatoes were first grown in Europe in the mid to late sixteenth century, but took a long time to be widely accepted as a food.

Description (of *S. tuberosum*): Perennial or annual herbs, producing starchy tubers, these variously shaped, and yellow, brown, red, or purple; stems erect or sprawling, terete to angled, sometimes slightly winged, with simple hairs, these with or without glandular-hairs (and similar hairs also on the leaves). Leaves alternate and spirally arranged, pinnately compound, with both large and small leaflets more or less alternating, the leaflets ovate to elliptic, with pinnate venation, the apex acute to acuminate, the base obtuse to slightly cordate, often slightly asymmetrical, the margin entire; stipules absent. Inflorescences terminal cymes. Flowers bisexual, radially symmetrical, held horizontally to pendulous, with the five sepals fused, the five petals strongly fused, white to pink or pale violet, spreading to slightly reflexed, forming a broad, starlike corolla with broadly triangular lobes, and five stamens, with very short filaments and elongate, bright yellow anthers that stick together, forming a ring around the style, opening by apical slitlike pores; the carpels two, fused, the ovary superior, and with a single style and a slightly expanded, slightly two-lobed stigma. Fruit a globose berry, greenish, with numerous flattened seeds (Figure 7.55).

REEDS (*PHRAGMITES AUSTRALIS*)

(The Grass family [Poaceae or Gramineae].)

> *This remnant sailed with Círdan south to the Isle of Balar, and they made a refuge for all that could come thither, for they kept a foothold also at the Mouths of Sirion, and there many light and swift ships lay hid in the creeks and waters where the reeds were dense as a forest. (SILM 20)*

It is clear that reeds were widespread in the wetlands of Middle-earth of the First through the Third Ages, as they are in Eurasia and North America today. They occur in the plant communities associated with many rivers, marshes, and lakes. In Beleriand, they are mentioned from the marshes of Nervast (surrounding Lake Linaewen), Lake Ivrin (the source of the River Narog in West Beleriand), and Lake Aeluin (in Dorthonion) south to the mouths of the River Sirion. In the Third Age, they were nearly cosmopolitan and are especially characteristic of marshes and swampy river margins, such as along the Withywindle and Brandywine; the Midgewater marshes (east of Bree); Swanfleet marsh (where the Glanduin meets the Greyflood River); along the Forest River and Esgaroth (Long-lake); the Gladden Fields (marshes where the Gladden River meets the Anduin); in wetlands along the Entwash, especially in the marshes where it joins the Anduin; the adjacent Wetwang and north-eastward in the Dead Marshes; and, finally, in the wetlands of Lebennin near the mouth of the Anduin.

Tolkien stresses the tall stature of this common wetland grass and its tendency to form dense stands. Thus, after the *Nirnaeth Arnoediad* (Battle of Unnumbered Tears), when the elves were forced out of most of Beleriand, Círdan established a refuge on the Isle of Balar and the nearby mouths of the Sirion River, where he kept many swift ships, hidden from the orcs of Morgoth by the dense and tall "forest" of reeds (see quote and Figure 7.56). Reeds are also distinctive in having large, plumose inflorescences, and these are noted by Sam and Frodo in their passage through the Dead Marshes, where "dead grasses and rotting reeds loomed up in the mists like ragged shadows of long-forgotten summers" (LotR 4: II). As they rested in the "reed-thicket," hiding like "little hunted animals," the reeds "hissed and rattled," and they heard "the faint quiver of empty seed-plumes, and broken grass-blades trembling in small air-movements that they could not feel" (LotR 4: II). Certainly, these tall grasses are an essential component of our mental image of the Dead Marshes, a wetland that has been distorted by the evil of Sauron. However, it would be incorrect for us to consider the dismal picture of the Dead Marshes provided by Tolkien as characteristic of his feelings for wetlands in general—even though we may find walking through such habitats unpleasant. Tolkien loved wetlands, and most marshes in his legendarium are pictured as places of beauty and ecological significance. Thus, Linaewen is said to be the home of multitudes of birds "of such as love tall reeds and shallow pools" (SILM 14), the Swanfleet is described as a place of "countless swans housed in a land of reeds" (LotR 6: VI), and the Gladden Fields are described as "a wilderness of islets, and wide beds of reed and rush, and armies of yellow iris that grew taller than a man" (UT 3: I, footnote on Gladden Fields; see also Figure 7.25). More significantly, the reeds surrounding Lake Aeluin are said to bewail the death of Barahir (the father of Beren; see Lays IV: lines 550–554), and "their rustling plumes" in the marshes bordering Lake Ivrin contribute to Túrin's release from grief resulting from his accidental killing of Beleg—indeed, it is said that "On Ivrin's lake is endless laughter" (Lays I: lines 1519–1526). Thus, Tolkien pictures the sound of the wind in the reeds as one of beauty and comfort. But in the Dead Marshes even that sound has been twisted by Sauron: it becomes merely a hiss or a rattle, resulting from air-movements that the tired hobbits cannot even feel and so providing no relief—the opposite of the breezes of Lake Ivrin. Finally, if additional evidence is needed, Tolkien's positive view of reeds is shown in Frodo's song of praise to Goldberry, the daughter of the Withywindle and beautiful wife of Tom Bombadil. She is said to be as "slender as a willow-wand"

FIGURE 7.56 *Reeds.*

and "clearer than clear water"; she is the "wind on the waterfall, . . . the leaves' laughter", and a "reed by the living pool!" (LotR 1: VII). All these comparisons relate to water and the plants growing in watery habitats, and all express the beauty and joy of Goldberry, her interactions with Tom, and their joint stewardship of the beautiful wetlands of the Withywindle valley.

Etymology: The English name "reed" is derived from the Middle English *rede*, and the Old English *hreod* (both meaning reed and referring to this large wetland grass), and these are related to the German *Ried*. These names are hypothesized to be derived from the Proto-Germanic **kreut-* (again, referring to reeds); no cognates are known outside the Germanic languages. The Latin scientific name *Phragmites* is derived from the Greek *phragmites* (growing in hedges), derived from *phragma* (hedge or fence), apparently from its hedgelike growth along ditches. The Sindarin name for reed bed is *esgar*, as seen in the locality name Esgaroth (= Reedlake, because of the reed-dominated swamp located on its western shores). The Quenya name perhaps is *linquë*, but this also may apply to grasses in general; the words *liskë* and *fen* may also apply.

Distribution and Ecology: Reeds have an almost cosmopolitan distribution, occurring in temperate and tropical wetlands, both freshwater and brackish, and often forming dense stands. These plants are members of the genus *Phragmites* and are usually considered to represent a single widespread and variable species, *Phragmites australis*, although sometimes these plants are divided among a few poorly characterized species. In North America, both native populations (*P. australis* subsp. *americanus*) and introduced and invasive populations (*P. australis* subsp. *australis*) occur. The native populations have drooping inflorescence branches, are green in color, with usually widely scattered stems, and have leaves with ligules 0.4–1.7 mm long, whereas the introduced populations have ascending inflorescence branches, are bluish gray-green, form dense stands, and have shorter ligules (0.1–0.4 mm). The invasive genotype was early brought from Europe, perhaps in ship ballast. The plants are wind-pollinated, and the small fruits (grains) are wind- or water-dispersed (Figure 7.56).

Economic Uses: Reeds are used as thatch, in lattices, woven into mats and baskets, used to make brooms and flutelike musical instruments, and the pulp can be used for paper. The grains, young shoots, and rhizomes can be eaten. Reeds are also used in phytoremediation wastewater treatment. Arrow shafts have been made from their stems, and they are called arrow-reeds in the poem *The Sea-bell* (TATB 15).

Description: Perennial, rhizomatous herbs; stems erect, jointed, round in cross-section, hollow in the internodal regions, to 4 meters tall. Leaves alternate, two-ranked, distributed along the stem, simple and differentiated into a spreading blade, a sheath that closely encircles the stem, with its margins overlapping but not fused, and with a distinct line of hairs (i.e., a ligule) at the junction of blade and sheath; the blade linear, flat or folded, 15–40 cm long and 2–4 cm wide, the venation parallel, with apex long-acuminate, the margin entire, and blade usually breaking off from the sheath in the late fall; stipules absent. Inflorescences terminal, plumose panicles, to 35 cm long and 20 cm wide, often purplish when young, but pale yellow-brown at maturity. Flowers very tiny, arranged in hairy spikelets of 2–10 flowers, with each flower surrounded by two bracts (a lemma and palea), with 1–3 anthers and two fused carpels, the ovary superior, the two stigmas plumose. Fruit a grain (i.e., small, dry, with a single seed, fused to the fruit-wall).

ROCKROSES (*HELIANTHEMUM NUMMULARIUM*, AND RELATIVES)

(The Rockrose family [Cistaceae].)

> *The bushes, and the long grasses between the boulders, the patches of rabbit-cropped turf, the thyme and the sage and the marjoram, and the yellow rockroses all vanished, and they found themselves at the top of a wide steep slope of fallen stones, the remains of a landslide. (Hobbit VI)*

Bilbo, Gandalf, and the dwarves, escaping from the goblins, found themselves on the eastern slopes of the Misty Mountains (Figure 7.57). Gandalf led them along a rough path while Bilbo looked for something to eat, and they passed (as described in the quote) through open rocky habitats with scattered shrubs (hawthorns, *Crataegus*) and perhaps also low pines (*Pinus*) but dominated by sun-loving herbs such as blackberries (*Rubus*), thyme (*Thymus*), sage (*Salvia*), marjoram (*Origanum majorana*), and rockroses. It was then that the path ended, and they came to a steep scree slope. In attempting to hike down through the loose rocks, they started a landslide and became part of the mass of sliding stones. They escaped from harm only by catching hold of (or in the case of Bilbo and a few others, by hiding behind) the trunks of pines, which grew at the base of the slope. In this action-packed scene, we actually do not "see" the rockroses: they have "vanished" (and are just a memory) as Bilbo and the dwarves slide down the scree slope. The habitat is certainly appropriate here, as rockroses are plants of open rocky habitats. The species that J. R. R. Tolkien likely had in mind is the common rockrose (*Helianthemum nummularium*), a low subshrub with usually bright yellow flowers, which is widely distributed in Europe and grows commonly in the British Isles. This scree-slope passage recalls a similar experience that Tolkien had on a walking tour in Switzerland in 1911 (for more details, see D. A. Anderson's, *The Annotated Hobbit*). *Helianthemum nummularium* is a common and conspicuous plant in the Alps, so perhaps its occurrence in the open rocky meadows above the scree slope is also based on Tolkien's Switzerland hiking experience.

Etymology: The name "rockrose" is derived from rock + rose, in which "rock" is derived from Old English *rocc,* which in turn comes through Old North French *roque,* from Latin *rocca,* of uncertain origin, and "rose" is derived from Old English *rose,* from Latin *rosa,* and Greek *rhodon* (rose), again, of uncertain origin. The Latin scientific name *Helianthemum* is derived from Greek *helios* (sun) and *anthemon* (flower) because the flowers are often yellow and open in the sunshine.

Distribution and Ecology: The ca. 80 species of rockroses occur from Western Europe and northeastern Africa to central Asia. The group is most diverse in the Mediterranean region, and 31 species occur in Europe; only a single species is native to China. Twenty-four American species often are included, but these should be segregated in the genus *Crocanthemum,* and they are more closely related to species of *Hudsonia* than they are to *Helianthemum.* (The species of *Helianthemum* have opposite and decussate leaves, while those of *Crocanthemum* are alternate and spirally arranged; in addition, the style of *Helianthemum* is tapered from a broad apex to a narrower base while that of *Crocanthemum* is broader at the base and tapers toward the apex.) Rockroses grow in open habitats (e.g., cliffs, scrub, grasslands, dry and open forests),

FIGURE 7.57 *Rockroses.*

on sandy, chalky, or calcareous-rocky soils. Species of rockroses are differentiated by characteristics of habit (perennial vs. annual plants), leaf shape, presence or absence of stipules, presence and form of the multicellular hairs (on various plant parts), inflorescence structure, the length of the floral pedicels, sepal size, and color of the petals (white, pink, yellow). The showy flowers are visited by various bees, flies, and beetles; the flowers open during bright sunlight and usually remain open for only a few hours. The small seeds are dispersed by wind or rain-wash.

Economic Uses: Rockroses are occasionally used as ornamentals (especially in rock gardens) because of their showy, often yellow flowers.

Description (of *Helianthemum* spp.): Dwarf shrubs to herbs, perennial or annual, with stems erect to spreading; hairs solitary to stellate (occasionally peltate scales), usually on both stems and leaves. Leaves opposite and decussate, simple, ovate to obovate, and sometimes quite narrow, pinnately veined, the secondary veins usually obscure, the apex rounded to obtuse, the base cuneate to cordate, the margins entire, often revolute, sometimes fringed with hairs; the lower surface with sparse to dense hairs, petiolate; stipules usually present. Inflorescences terminal racemelike cymes. Flowers radially symmetrical, bisexual, the sepals five, distinct; three inner ones large, the two outer much smaller, the petals five, distinct, usually yellow, less commonly pink, orange, or white (and occasionally red-spotted basally), obovate and apically rounded, crumpled in bud and thus slightly wrinkled at maturity; the stamens numerous, distinct, usually yellow, and three fused carpels; the style single, slender at base and broadening toward the stigma; the ovary superior. Fruit a three-valved capsule, with numerous small seeds (Figure 7.57).

ROOTS (PROBABLY *DIOSCOREA* SPP.)

(The Yam family [Dioscoreaceae].)

> *One day in the waning of the year he [Mîm] told the men in Bar-en-Danwedh that he was going with his son Ibun to search for roots for their winter store; but his true purpose was to seek out the servants of Morgoth, and to lead them to Turin's hiding-place. (CoH VIII; see also SILM 21)*

Túrin and his outlaw band first encountered Mîm, the petty-dwarf, near the isolated high hill, Amon Rûdh. Mîm and his two sons were out gathering roots (Figure 7.58), and this rather mysterious plant only appears in J. R. R. Tolkien's writings in association with the story of this dwarf. The outlaws would have liked to kill Mîm and certainly were disappointed to find only "roots and small stones" (CoH VII) in his sack, but Túrin forbade it: Mîm's life was ransomed through the offer of his home atop Amon Rûdh. Later, Túrin was joined at Amon Rûdh by his friend Beleg, an elf of Doriath, and they gathered men around them, driving out the orcs of Morgoth from the region between the Teiglin River and the west march of Doriath. Túrin and Beleg were eventually betrayed by Mîm, who again had left the hidden refuge, ostensibly to gather roots (see quote), but actually to seek out the servants of Morgoth in order to betray Túrin and Beleg. The reason for Mîm's betrayal, however, may not be as clear as suggested in the preceding quote: at least he did not want Túrin killed. And, as the event is related in *The Silmarillion* (see chapter 21) Mîm's encounter with orcs was perhaps accidental, and it was only their threat to torture his son that led Mîm to treachery (see also CoH VIII).

FIGURE 7.58 *Roots.*

In the *Children of Húrin*, we discover that the flesh of these wild roots was white and that "when boiled they were good to eat, somewhat like bread" and that they could be "hoarded like the nuts of a squirrel" (CoH VII). Mim told the outlaws that these starchy roots were unknown to the elves and men. He would not tell them their real name, calling them only "earth-bread," and he would not show them how to find them because, as he said, men are "greedy and thriftless, and would not spare till all the plants had perished" (CoH VII). All of these characteristics match the starchy, tuberlike rhizomes of species of *Dioscorea* (yams). They can be white in color, are very starchy, can be stored for long periods of time, and are frequently cooked by boiling. Additionally, yams are widely distributed, grow natively in Europe, and thus they are a likely candidate for earth-bread. Hopefully, Mîm's opinion of humans is less accurate than his plant knowledge! Sweet potatoes (*Ipomoea batatas*, in the morning-glory family, Convolvulaceae) are quite similar to yams and are an alternative possible identification of earth-bread, yet they do not occur natively in Europe, originating in the tropical regions of the Americas, so we consider them an unlikely choice.

Etymology: The English word "root" is derived from Old English *rot*, through Old Norse to Proto-Germanic **wrot* (root, herb, plant). The Latin scientific name, *Dioscorea*, honors the Greek physician and botanist Dioscorides, the author of *De Materia Medica*, an encyclopedia of herbal medicine. The common name "yam," used for members of this genus, is derived from Portuguese *inhame* and Spanish *igname*, which in turn are derived from a West African language (where many of these important edible tuber crops are native).

Distribution and Ecology: The yams (i.e., species of *Dioscorea*) are a group of more than 600 species with a nearly worldwide distribution, occurring in both tropical and temperate regions although most diverse in the tropics. Four species occur in Europe, 6 are native or naturalized in North America (north of Mexico), and 52 occur in China (of which only two are introduced). These twining vines occur in a wide variety of habitats. Their inconspicuous flowers are pollinated by insects (mainly flies), and their usually flattened or winged seeds (borne in three-winged capsules) are dispersed by wind. The species are not well studied, and their limits and differentiating characteristics are often unclear.

Economic Uses: Several species have been independently domesticated in various parts of the world, and these plants are important crops because of their large, starchy rhizomes or tubers. The most important of these are the white yam (*D. rotundata*) and the closely related yellow yam (*D. cayenensis*; both native to Africa, with many different cultivars), winged yam (*D. alata*; native to southeastern Asia and also widely cultivated), Chinese yam (*D. polystachya*; native to China and quite cold tolerant), cush-cush yam (*D. trifida*; native to South America), and air potato (*D. bulbifera*; native to both Africa and Asia, widely cultivated, and naturalized in Florida). The starchy tubers or rhizomes are boiled, baked, or fried, like potatoes. Some require more processing to remove bitter compounds than others, and some species contain steroidal saponins and are used in the synthesis of cortisone and human sex hormones.

Description: Twining vines from starchy rhizomes or tubers (and sometimes also producing tubers in the leaf axils); stems branched or unbranched, sometimes winged; hairs simple, T-shaped to stellate, and sometimes with prickles. Leaves alternate to nearly opposite, usually spirally arranged, usually simple, but occasionally palmately lobed (or rarely compound), palmately veined, the major veins converging and connected by a network of smaller veins, ovate to elliptic, the apex acute to acuminate, the base usually cordate, the margin entire, with a petiole; stipules absent. Inflorescences axillary, variously cymose (but often appearing racemose), sometimes reduced to a solitary flower. Flowers radially symmetrical, usually unisexual (with the staminate flowers and carpellate flowers on different plants), the perianth parts all alike (i.e., of tepals), these six, small, distinct to slightly fused, white to greenish, pale yellow, or purple; the stamens usually six, distinct, and with three fused carpels; the style three-branched; the ovary inferior. Fruit usually a triangular and three-winged capsule, sometimes a samara or berry, with usually few flattened or winged seeds (Figure 7.58).

ROSES, WILD ROSES, EGLANTINE (*ROSA* SPP., BUT ESPECIALLY *R. RUBIGINOSA* AND *R. ARVENSIS*)

(The Rose family [Rosaceae].)

> *The swift growth of the wild with briar and eglantine and trailing clematis was already drawing a veil over this place of dreadful feast and slaughter. (LotR 4: IV)*

Gollum, Frodo, and Sam had been walking all night, and at sunrise they found themselves in northern Ithilien—they were surrounded by woods of unfamiliar resinous trees and sweet-smelling herbs. It seemed good to the hobbits "to walk in a land that had only been for a few years under the domination of the Dark Lord" (LotR 4: IV). They left the road looking for a place to camp, and as Sam scrambled through the brush, "smelling and touching the unfamiliar plants and trees," he stumbled upon an area that had been cleared and burned. There "he found a pile of charred and broken bones and skulls" (LotR 4: IV). But already the sprawling shrubs and vines, eglantine (*Rosa rubiginosa,* also called sweet briar), clematis (*Clematis vitalba*), and briars (probably *Rubus* sp., although this common name is also applied to *Smilax* spp. and occasionally also to especially prickly species of *Rosa*) were covering this place of slaughter (see introductory quote). Eglantine, a wild rose broadly distributed across Europe and also commonly naturalized in North America, is a compact shrub with fairly large pink flowers, although the five petals of each flower shade to white at the base and the tuft of stamens in its center are yellow. Its stems are prickly and have glandular hairs; the foliage has a pleasant ripe apple scent. Their edible hips are ovoid or ellipsoid to nearly globose and bright red. It was early spring, so the eglantine may not yet have been in bloom, although our mental image of this scene has always included a few flower buds. Some of the colorful hips may also have persisted. Sam was reminded afresh that Ithilien was in the control of the enemy—and he noticed other signs that orcs had been in the area fairly recently: piles of filth and trash, trees cut down and left to die, with their bark cut and defaced by the sign of the eye of Sauron. The eglantine rose, in this situation, may represent both pain (as represented by its stout, curved prickles, reflecting the environmental pain and destruction brought about by Sauron) and pleasure (as represented by its beautiful flowers and fragrant leaves, reflecting the amazing power of the natural world to overcome such damage). Finally, we note that especially "thorny" roses are sometimes also called brambles (although this common name more commonly applied to species of blackberry, i.e., *Rubus*). Some of the thorny plants that Sam and Frodo encountered in

Mordor may actually have been such roses (growing intermixed with blackberries; see LotR 6: II; Figure 7.59).

Another rose plays an equally important role in Tolkien's legendarium, this one is a wild rose with white flowers (probably representing *R. arvensis*). We read in *The Lay of Leithian*, that Lúthien, while dancing in the forest of Neldoreth, was alerted that a stranger, Beren, was in the woods, and she leapt "among the hemlocks tall" (*Conium maculatum*) and hid

> under a mighty plant with leaves
> all long and dark, whose stems in sheaves
> upheld an hundred umbels fair;
> and her white arms and shoulders bare
> her raiment pale, and in her hair
> the wild white roses glimmered there. (Lays III, lines 621–626)

Lúthien dashed into the woods. Beren stared but could not call out; he blindly groped his way across the clearing. He was enchanted and forlorn, bruised and torn from his long journey, and was left far behind. Later, in his seemingly hopeless quest to recover a Silmaril from the crown of Morgoth, Beren asked Finrod Felagund for help and briefly told him of his first encounter with Lúthien—and words failed him as he remembered her dancing "with wild white roses in her hair" (Lays III, line 1789). In folklore, the rose has many often conflicting connections, but here it seems likely that the wild white roses represent Lúthien's beauty and purity and Beren's amazing love for her.

Etymology: The English name "rose" is derived from Old English *rose,* which is derived from the Latin *rosa* (which is also the source of the scientific name, *Rosa.* The word "eglantine" (one of the common names of *Rosa rubiginosa*) is from French *églantine,* from Old French *aiglent,* from Latin **aquilentus* (rich in prickles), from *aculeus* (prickle). The Sindarin name of this flower is *meril,* while the Quenya name is *lossë.*

Distribution and Ecology: Roses (*Rosa* spp., about 140 species) are widely distributed in the cooler parts of the Northern Hemisphere; 33 species are recognized in the flora of North America, 47 in the flora of Europe, and 95 species in the flora of China, and 8 in Japan. They grow in a wide variety of habitats. Recognition of species often is difficult because of hybridization, asexual reproduction, and more or less continuous morphological variation. Species are distinguished by characters such as habit (e.g., climbing or not, form of the rhizomes); variation in abundance of the hairs; whether glandular or nonglandular; the frequency, shape, and distribution of prickles on the stem; leaf characters (e.g., their size, number of leaflets, leaflet shape, apex, and marginal condition, stipule form, and whether or not they are fused to the petiole); the number of flowers per inflorescence; number and shape of the floral bracts; floral characters (e.g., shape of the hypanthium and receptacle, persistence and form of the sepals, size and color of petals, whether or not the styles are fused); and size and shape of the hip. The showy flowers are pollinated by insects, especially various bees, and the colorful hips (accessory fruits with the hard, dry achenes surrounded by a colorful, more or less globose and fleshy floral-cup) are bird-dispersed.

Economic Uses: Many beautiful rose hybrids and cultivars are extremely important as cultivated ornamentals; roses are also used as an important source of essential oils in the scent-making industry. The fleshy and colorful matured floral cup (hip)

FIGURE 7.59 *Roses, wild roses, eglantine.*

surrounding the achene fruits is eaten and is high in vitamin C. Some species are also used medicinally (especially in teas).

Description: Perennial, often rhizomatous shrubs or vines; stems erect, arching, or creeping, with long and short shoots; hairs variably developed, some nonglandular and others gland-headed; both stems and leaves usually with sharp-pointed modified hairs (i.e., prickles) and these sparse to dense, erect to curved, weak to stout. Leaves usually deciduous, alternate, spirally arranged, pinnately compound or with three leaflets (rarely simple); the leaflets variously shaped, with pinnate venation, the apex acute or acuminate to rounded, the base (cordate to) rounded to acute or narrowly cuneate, the margin dentate or doubly so, to crenate, the lower surface without prickles or with prickles on midvein similar to those of the stem, without hairs or with unicellular nonglandular or multicellular gland-headed hairs; stipules present, free or adnate to the petiole. Inflorescences terminal, the flowers numerous to few, in variously shaped cymes or reduced to a solitary flower. Flowers usually bisexual, radially symmetrical, erect, with an urn-shaped to globose floral cup (hypanthium) bearing five erect to reflexed sepals, five (but often numerous in horticultural cultivars) distinct white to yellow or red petals, and numerous stamens; the carpels usually numerous, distinct but sometimes fused by the styles, with or without hairs, the ovaries superior, each with a slender style. Fruit a cluster of achenes, surrounded by the fleshy red to purple-black floral cup (or hip) (Figure 7.59).

ROWAN (*SORBUS AUCUPARIA*, AND RELATED SPECIES)

(In the Rose family [Rosaceae].)

> *"There were rowan-trees in my home," said Bregalad, softly and sadly, "rowan-trees that took root when I was an Enting, many many years ago in the quiet of the world. The oldest were planted by the Ents to try to please the Entwives; but they looked at them and smiled and said that they knew where whiter blossom and richer fruit were growing. Yet there are no trees of all that race, the people of the Rose, that are so beautiful to me. And these trees grew and grew, till the shadow of each was like a green hall, and their red berries in the autumn were a burden, and a beauty and a wonder." (LotR 3: IV)*

In European folklore and mythology, rowans are consistently viewed positively—they are often considered magical or protective against witchcraft (because the small apple-like fruits are reddish and terminate in five triangular sepals, evoking a five-pointed star; see also Watts, 2007). Rowan trees also receive positive treatment in Tolkien's writings. Rowans grew in a circle around the home of the ent Bregalad, and "whenever he saw a rowan-tree he halted a while with his arms stretched out, and sang, and swayed as he sang" (LotR 3: IV) (Figure 7.60). It is certainly fitting that his name in Westron is Quickbeam, which is derived from the Old English name of this species (i.e., *Sorbus aucuparia*, the only compound-leaved species of *Sorbus* native to the British Isles). In fact, his name is doubly suitable since he is also quick in making up his mind! He tells of the ecological disequilibrium brought about by Saruman: "But the birds became unfriendly and greedy and tore at the trees, and threw the fruit down and did not eat it" (LotR 3: IV) and of the eventual destruction of his beloved rowans by Saruman's orcs. Bregalad is aware, of course, that rowans belong to the "people of the rose," which plant systematists call the family Rosaceae. Many Rosaceae are economically important (as food plants; e.g., apples, pears, plums, cherries, peaches, almonds, blackberries, raspberries, and strawberries, or as ornamentals; e.g., roses, cinquefoils, shadbushes, flowering quinces, crabapples, and flowering cherries). Rowan trees also grew around the open place on Amon Hen where Frodo pondered "everything he could remember of Gandalf's words" (LotR 2: X), seeking direction regarding how best to continue the quest. Perhaps they provided him some protection because he suddenly felt that "unfriendly eyes were upon him" and turned to see Boromir—who soon after tried to take the Ring. Rowans are widely distributed in Middle-earth and often occur in open habitats such as rocky bluffs, and they grew among the birches upon Amon Rûdh—"Upon the eastern side a broken land climbed slowly up the high ridges among knots of birch and rowan" (CoH VII)—which for a while was the home of Túrin and his outlaw band.

Etymology: The English name "rowan" related to the compound "rowan-tree" was derived through Old Norse *reynir* from the Germanic **raud-inan* (to redden), in reference to the colorful fruits of this tree. The Old English name was *cwic-beám*, which survives in the names "quickbeam" and "quicken-tree." The Latin name, *Sorbus*, is the classical name for this tree and may have been derived from Proto-Indo-European **sor-/*ser-* (red, reddish brown), again in reference to the colorful fruits. In North America, this species (and its close relatives: *Sorbus* subgenus *Sorbus*) are known as mountain- ashes because they have pinnately compound leaves similar to those of the ash (*Fraxinus*). Quickbeam's Sindarin name, *Bregalad*, means "quick (or lively) tree."

Distribution and Ecology: Rowans or mountain-ashes belong to the genus *Sorbus*, a group of more than 100 species, although species limits are quite problematic because the sexual species are known to hybridize, producing offspring that are morphologically intermediate, have increased chromosome numbers, and that reproduce asexually, thus blurring species distinctions. The species of *Sorbus* are distributed broadly in cool temperate regions of the Northern Hemisphere and are most diverse in eastern Asia, especially China, where 67 species occur (and many of these are restricted to this region). Around 18 species occur in Europe and ca. 10 grow in North America. The species with pinnately compound leaves such as rowan (*Sorbus aucuparia*), mountain-ash (*S. americana*), Huangshan rowan (*S. amabilis*), Japanese rowan (*S. commixta*), showy mountain-ash (*S. decora*), Chinese rowan (*S. discolor*), service tree (*S. domestica*), Siberian mountain-ash (*S. sambucifolia*), Green's mountain-ash (*S. scopulina*), Sitka mountain-ash (*S. sitchensis*), and relatives, belong to subgenus *Sorbus*; while the simple-leaved species mostly are placed in subgenus *Aria*. The showy flowers are pollinated by a wide variety of insects, and the colorful fruits are eaten and dispersed by birds. Rowans occur in a wide variety of deciduous, mixed, or coniferous forests and are especially common in open areas or on forest borders.

Economic Uses: *Sorbus aucuparia* and related species are widely used as ornamentals because of their showy white flowers and bright orange or red fruits, which are held on the tree into the winter, providing food for various birds. The fruits are bitter but can be used in jellies, jams, and preserves or in flavoring alcoholic beverages; however, the seeds should not be eaten because they contain cyanide. The wood is used in furniture, tool handles, walking sticks, and for carving.

FIGURE 7.60 *Rowan.*

Description (as applied to *Sorbus* spp.): Deciduous trees and shrubs, the bark rough to smooth and then with horizontally elliptic lenticels. Leaves alternate, spirally arranged, odd-pinnately compound (in subg. *Sorbus*, i.e., *S. aucuparia* and its relatives), with the leaflets 7–43, or simple (in subg. *Aria*), with the leaves or leaflets narrowly to broadly ovate to elliptic (or less commonly even obovate) or ± oblong, pinnately veined, with secondary veins forming loops; the apex acute or obtuse to acuminate; the base slightly asymmetrical (leaflets) or symmetrical and cuneate or acute to slightly cordate (leaves); the margin clearly to obscurely serrate, sometimes also lobed, with simple, nonglandular hairs, petiolate; stipules present but quickly deciduous. Inflorescences terminal, rounded to pyramidal cymes (with nodes near stem apices contributing axillary cymes). Flowers radially symmetrical, bisexual, the sepals five, triangular, the petals five, basally narrowed, white or pink, and the stamens numerous, with anthers globose, yellow or pink, all borne on a nectar-producing floral cup (= hypanthium) with usually 2–5 fused carpels, the styles 2–5, sometimes basally connate, the ovary inferior or at least partly so. Fruit small or less commonly large pomes, ovoid or globose to ellipsoid or oblong, bright orange or red, to pink, yellow, or white, with a cartilaginous core (Figure 7.60).

RUSHES (*JUNCUS* SPP.)

(In the Rush family [Juncaceae]).

> *They waded the stream, and hurried over a wide open space, rush-grown and treeless, on the further side. Beyond they came again to a belt of trees: tall oaks, for the most part, with here and there an elm tree or an ash. (LotR 1: IV)*

Rushes were widespread and characteristic elements of the various wetland plant communities in Middle-earth during the Third Age, and they are almost always mentioned in connection with such habitats—the exceptions relate to their traditional uses as weaving material (e.g., in the construction of mats and chair-seats, as seen in the homes of Tom Bombadil and Beorn; see "Economic Uses"). The introductory quote from *The Fellowship of the Ring* is typical in that rushes are mentioned as a component of the wet, open, herb-dominated vegetation near the Stock-brook, which was waded by Pippin, Sam, and Frodo on their way to Crickhollow in Buckland (Figure 7.61). Another typical association is seen in *The Hobbit*, where the wood elves' song describing the passage of wine barrels down the Forest River includes the lines "Past the rushes, past the reeds, Past the marsh's waving weeds," (Hobbit IX). Rushes were common in the marshes of the Gladden Fields, where they grew with reeds (*Phragmites australis*) and yellow-flags (*Iris pseudacorus*). In the "Disaster of the Gladden Fields" (in *Unfinished Tales*), it is said that Isildur came up out of the waters of the Anduin River, "struggling among great rushes and clinging weeds" and there suddenly knew that the One Ring was gone—it had slipped off his finger and "gone where he could never hope to find it again" (UT: 3: I). In a few moments he was dead—killed by orcs lurking along the shore. Rushes also grew in Beleriand during the First Age. For example, we are told that "rush and reed" (Lays I: line 1519) grew in the marshes along the shores of Lake Ivrin where Túrin was healed from grief.

FIGURE 7.61 *Rushes.*

Etymology: The common name "rush" is derived from Old English *resc* or *risc* (rush), from Proto-Germanic **rusk-*, from Proto-Indo-European **rezg-* (to plait, weave), a reference to the stems and leaves of such plants, which were (and occasionally still are) used in making mats, baskets, or the seats of chairs, and the like. The scientific name *Juncus* is the classical name for the plant and perhaps is derived from *jungere* (to join), another reference to the use of these plants in weaving and rope-making.

Distribution and Ecology: Rushes (i.e., members of the genus *Juncus*) grow nearly everywhere (except Antarctica) and include some 300 species. They occur in a variety of habitats but are especially common in swamps; marshes; bogs; wet habitats near springs, streams, or lakes; wet ditches; low woods; moist meadows; or coastal wetlands. In North America (north of Mexico) 95 species can be found, whereas 53 species occur in Europe, 76 in China, and 26 in Japan. Important characteristics in their classification include whether or not the leaves have a blade (in addition to the sheath), the form of the blade (flat or terete, and, if terete, whether or not it is segmented), inflorescence structure, and features of the capsules and seeds. Their rather inconspicuous but lilylike flowers are wind-pollinated, and the capsules release tiny seeds that are dispersed by wind, water, or external transport on animals (sticking to fur or feathers).

Economic Uses: Large species, such as *J. effusus,* are used to make mats, or woven and used in baskets or chair seats. The last use is evident in *The Hobbit,* in which it is mentioned that Beorn had "low-seated benches with wide rush-bottoms and little short thick legs" (Hobbit VII) in his home. The first usage may be attested in *The Fellowship of the Ring,* in which it is mentioned that, in the home of Tom Bombadil, the floor was flagged and "strewn with fresh green rushes" (LotR 1: VII); its walls were also of stone and "mostly covered with green hanging mats" (LotR 1: VII) and these could also have been made of rushes. Many rushes are ecologically significant components of wetland habitats, and the use of rushes in Goldberry and Tom's home is certainly appropriate given their connection to the wetlands of the Withywindle River valley.

Description: Usually perennial, rhizomatous or clumped herbs; stems erect, usually round in cross-section, usually solid, quite short to just over 1 m tall. Leaves alternate, three-ranked, basal to distributed along the stem, simple and differentiated into a spreading blade, a sheath that closely encircles the stem, with its margins unfused and without a ligule at the junction of blade and sheath; the blade linear, flat, folded, or terete, sometimes septate, variable in length and width and sometimes very reduced; the venation parallel, with apex acute to acuminate, the margin entire and without hairs; stipules absent. Inflorescences terminal but sometimes appearing lateral, variable in form, with flowers well-spaced to densely clustered. Flowers radially symmetrical, bisexual, tiny, not in spikelets, and each flower with a perianth of six tepals, three or six stamens, and three fused carpels, the ovary superior, the three stigmas elongate. Fruit a capsule, associated with the persistent tepals and releasing numerous tiny seeds (Figure 7.61).

SAGES (*SALVIA* SPP.)

(The Mint family [Lamiaceae, or Labiatae].)

> *"A few bay-leaves, some thyme and sage, will do—before the water boils," said Sam (LotR 4: IV)*

Several species of sage make brief appearances in *The Hobbit* and *The Lord of the Rings*. Sage (likely the culinary sage, *Salvia officinalis*, a Mediterranean species) was one of the fragrant herbs that Sam and Frodo encountered in Ithilien. Sam gathered culinary sage and bay leaves (*Laurus nobilis*) for the rabbit stew that he made for Frodo (see quote; Figure 7.62). But other sages also grew in the forest openings and meadows of this former Garden of Gondor; we are told that there were "sages of many kinds putting forth blue flowers, or red, or pale green" (LotR 4: IV), which is not surprising considering that *Salvia* contains species with an exceptionally wide range of flower colors: white, cream, yellow, orange, red, blue, purple, and even brown. For example, the flowers of *S. coccinea* (a widespread tropical American species) are bright red, as are those of *S. splendens* (native to Brazil). Both are widely cultivated, and J. R. R. Tolkien may have had them in mind when writing this passage. *Salvia pratensis* and *S. verbenaca* both have blue to purple flowers, are widespread in Europe, grow natively in England, and are cultivated as ornamentals. They are likely the species referenced as the blue-flowered sages of Ithilien, although other species of *Salvia* have blue flowers (e.g., the beautiful sky-blue flowers of

S. azurea, a species of the southeastern United States). Interestingly, no species of *Salvia* is known to have green flowers, although *S. mexicana* 'Limelight' has persistent yellow-green calyces so at first glance may seem to have greenish flowers (the corollas of this species are actually deep purple). Culinary sage is yet another of the species that Tolkien used to highlight the Mediterranean aspect of the flora of Ithilien, along with plants such as olives (*Olea europaea*), cedar (*Cedrus libani*), terebinth (*Pistacia terebinthus*), and bay (*Laurus nobililis*). But other sages, even today, occur commonly in northern Europe, and this was true also in the Third Age. For example, Bilbo saw "thyme, sage, and marjoram" (Hobbit VI) on the western slopes of the Misty Mountains and "Salvia" was even used as a name among hobbits (LotR Appendix C, genealogy of the Brandybucks), so sages must have been grown as ornamentals or medicinal herbs in hobbit gardens. Finally, sages are members of the mint family (Lamiaceae), an economically important group of aromatic herbs and shrubs, as a result of the presence of an amazing diversity of ethereal oils in their tissues. This can be experienced and appreciated by comparing the flavors or odors of peppermint, spearmint, horehound, lavender, basil, catnip, oregano, rosemary, sage, and thyme, to name only a few! Other mints growing with the sages in Ithilien were thyme (*Thymus vulgaris*) and marjoram (*Origanum majorana*).

Etymology: The English and scientific names are related. The common name "sage" is derived from the Old French *sauge*, from the Latin *salvia* (the ancient Roman name for culinary sage), from *salvus* (healthy), in reference to its healing and preserving qualities.

Distribution and Ecology: Sages (the genus *Salvia*, containing ca. 900 species) are broadly distributed (with centers of diversity in Central and South America, the Mediterranean region, and Eastern Asia); 84 native or naturalized species grow in China, 65 in North America (north of Mexico), 36 in Europe, and 9 in Japan. They grow in a wide variety of more or less open habitats. The species are distinguished by numerous characteristics in almost any part of the plant, but especially in the form and density of the hairs on the stems and leaves; the shape and size of the leaves, whether lobed or not, or even compound; and type of margin; the inflorescence structure; and various floral characteristics (e.g., calyx length; corolla size, shape, or color; and form of the staminal connective). Their colorful flowers are pollinated by bees or birds. Pollination is accomplished by an unusual lever mechanism. The two stamens are unusual in shape, with the two halves of the anther widely separated by an elongated connective that forms a lever. Phylogenetic studies have indicated that the species of *Salvia* do not constitute a clade (i.e., they are not each other's closest relatives). It has been suggested, therefore, that the circumscription of *Salvia* be expanded to include some other genera (e.g., *Rosmarinus, Perovskia*). When a pollinator enters a young flower (with mature stamens and the stigma not yet receptive), it causes the stamens to move, depositing pollen on the pollinator's body, whereas in older flowers (with a receptive stigma), the style bends down, picking up pollen from the pollinator's body. The small fruits are merely shaken from the persistent calyx by the action of wind or disturbance of the plant.

Economic Uses: *Salvia officinalis* (culinary sage; native to the Mediterranean region) and several other species have long been used as medicinal plants (for a very wide range of aliments). *Salvia officinalis* is also used as a spice in cooking. *Salvia divinorum*, a Mexican species, is a psychoactive plant, causing hallucinations because it contains a unique diterpenoid (salvinorin A), which is the psychoactive compound. Many species are attractive garden ornamentals.

FIGURE 7.62 *Sages.*

Description: Aromatic, annual or perennial herbs (to occasionally low shrubs); stems erect to spreading, with various sparse to dense simple hairs, and these with or without glandular-heads. Leaves opposite and decussate, simple, unlobed to lobed, and sometimes even compound, narrowly to broadly triangular or ovate to obovate, with pinnate venation, the apex acute or acuminate to rounded, the base acute to cordate or hastate, with or without a petiole, and reduced in size toward the inflorescence; stipules absent. Inflorescences terminal, with flowers appearing to be in whorls along an elongate axis, but actually in paired reduced cymes, each in the axil of a bract. Flowers bisexual, bilaterally symmetrical, held horizontally, with the five sepals fused, forming a tubular to cup-shaped calyx, two-lipped; the five petals strongly fused, white, cream, yellow, orange, red, blue, purple, or brown, two-lipped, the upper lip two-lobed, straight to arched upward, and the lower lip three-lobed, spreading, and two stamens, each with short filament and a prolonged T-shaped connective dividing the anther into two widely separated halves (one of which may be sterile); the carpels two, fused; the ovary superior, four-lobed, and with a single style arising from the middle of the four ovary-lobes and apically forked, each style branch ending in a minute stigma. Fruit a schizocarp of four nutlets, glabrous and smooth (Figure 7.62).

SALT MARSH GRASSES, CORDGRASSES (*SPARTINA* SPP.)

(The Grass family [Poaceae].)

> *To the salt marshes*
> *where snipe and seamew and the sea-breezes*
> *first pipe and play they press together*
> *sweeping soundless to the seats of Ylmir [Ulmo],*
> *where the waters of Sirion and the waves of the sea*
> *murmurous mingle.* (*Lays I: lines 1550–1555*)

Saltmarshes, frequently dominated by cordgrasses (i.e., members of the genus *Spartina*), are a characteristic feature of the muddy flats that experience daily tidal flooding along the coasts of eastern North America and western Europe. In the quote from the *Lay of the Children of Húrin* (in the *Lays of Beleriand*), Tolkien beautifully described the meeting of the waters of the Sirion River, sacred to Ulmo, with those of Belegaer, the great western sea (Figure 7.63). Such tidal marshes are ecologically significant and are home to a large number of bird species; it is difficult to envision such habitats without recalling the calls of seagulls and the feeding of many cryptically colored wading birds such as snipe (Figure 7.63) and sandpipers. The marshes of the Sirion delta are also mentioned in *The Silmarillion*, where they are described as "empty of all living things save birds of the sea" (SILM 14). Such passages reinforce the easily forgotten fact that much of Beleriand in the First Age was unpopulated and consisted of wild lands: forests, swamps, grasslands, and marshes that were only rarely visited by elves, dwarves, or humans. It is good to remember that the "trees and the grasses and all things growing or living in the land belong each to themselves" (LotR 1: VII)—as Goldberry told Frodo after he had asked whether the Old Forest belonged to Tom Bombadil—yet Tom still is the Master, using his powers appropriately as an environmental steward. We are called to a similar stewardship of the natural areas in our world, and, like the elf Voronwë, we can stand "knee-deep in grass" and listen, enchanted by the "singing of the birds" (UT 1: I). Such wild lands are beautiful—and much of the beauty of Beleriand (and the reason that the region so attracts us) is a result of the dynamic interplay of cultivated/settled areas with such wilderness habitats.

Etymology: The English common name "cordgrass" is a compound word; "cord" is from Old French *corde* (rope, string, twist) from Latin *chorda*, and Greek *khorde* (string of a musical instrument, catgut), which is derived from the Proto-Indo-European root **ghere-* (intestine), whereas "grass" is derived from Old English *græs* (herb, plant, grass), from Proto-Germanic *grasan*, derived from Proto-Indo-European **ghros-* (young shoot, sprout), from the root **ghre-* (to grow, become green). The Latin scientific name *Spartina* is derived from the Greek *spartine* (rope, traditionally made from *Spartium junceum*).

Distribution and Ecology: The 16 species of cordgrass mainly grow in wet, saline habitats, and many species form dense coastal meadows that are flooded daily by tidal action. They occur natively in North and South America, Europe, and perhaps northern Africa; introduced species that have become invasive occur not only within the native range of the genus, but also in Tasmania, New Zealand, and China. Nine native and two introduced species occur in North America; of these, the most widespread coastal species are smooth cordgrass (*S. alterniflora*) and saltmeadow cordgrass (*S. patens*). Four native species and three introduced species grow in Europe. One of these, *S. anglica* (common cordgrass) is a tetraploid derivative of the diploid *S.* × *townsendii*, which is a sterile hybrid between small cordgrass (*S. maritima*), a native species, and the introduced *S. alterniflora*, which was brought from eastern North America to the coast of England in the 1820s. The allotetraploid, which originated in southern England in the 1870s, is fertile and is rapidly eliminating small cordgrass from many coastal habitats. Identification of the species is often difficult, relying on characteristics such as whether or not rhizomes are produced (and their color), the length of the leaf blades and ligules, the form of the leaf margins, inflorescence structure, and the form of the spikelets and flowers (e.g., spikelet size, number of veins on the glumes, awn length, anther length). The small flowers of cordgrasses are wind-pollinated, and the seeds float and are dispersed by water (often in ocean currents), although dispersal may also occur by fragmentation of the rhizome system.

Economic Uses: Species of *Spartina*, abundant salt marsh grasses, are salt-tolerant (with salt-excreting glands on their leaves) and ecologically important in coastal habitats. Several non-native species have become invasive. Cordgrasses have been used as a source of thatch and as forage, and they have been widely planted for reclamation of coastal mud flats (although with harmful consequences for native species when such plantings are outside the native range). Saltmarshes are ecologically significant and are currently imperiled by sea level rise.

Description: Perennial, rhizomatous herbs; stems erect, jointed, round in cross-section, hollow in the internodal regions, from 10 cm to 3.5 m tall. Leaves alternate, two-ranked, distributed along the stem, simple and differentiated into a spreading blade, a sheath that closely encircles the stem, with its margins overlapping but not fused, and with a membranous ligule that is lined with hairs (at junction of blade and sheath); the blade linear, flat or inrolled, 6–96 cm long and 0.5–25 mm wide, the venation parallel, with apex narrowly acute, the margin entire, often roughened; stipules absent. Inflorescences terminal, of few to numerous, erect to spreading spikelike branches, to

FIGURE 7.63 *Salt marsh grasses, cordgrasses.*

3–95 cm long, usually narrow, green to pale purple, yellow-brown at maturity. Flowers very tiny, arranged in flattened spikelets, these reduced to a single flower and with outer pair of bracts (glumes) strongly keeled and each flower surrounded by two bracts (a lemma and palea), with three anthers and two fused carpels; the ovary superior; the two stigmas plumose. Fruit a grain (i.e., small, dry, with a single seed; Figure 7.63).

SAXIFRAGES (*SAXIFRAGA* SPP.)

(In the Saxifrage family [Saxifragaceae].)

> *The grots and rocky walls were already starred with saxifrages and stonecrops. (LotR 4: IV)*

Saxifrages are part of the botanical bounty of the forests of Ithilien (Figure 7.64). Most species love moist rocky habitats, growing out of crevices, and, as their name implies, seeming to crack the rocks out of which they grow. And this characteristic habitat—"grots and rocky walls"—is noted by J. R. R. Tolkien. Additionally, the rocky walls are said to be "starred" with saxifrages—a reference to the showy, starlike, five-petaled flowers of these plants. The key characteristics of these diminutive plants are thus deftly presented in this sentence. The sentence is part of a paragraph (in the early part of the chapter "Of Herbs and Stewed Rabbit") in which no fewer than 17 different plants are mentioned; 34 different plants are referenced in this chapter as a whole, more than in any other in *The Lord of the Rings*. Clearly, Tolkien wants the reader to focus on the botanical richness of the forests, meadows, and ponds of Ithilien—the garden of Gondor. He even states that Sam and Frodo saw kinds of trees unknown in the Shire and likewise notes that growing in Ithilien were "many herbs of forms and scents beyond the garden-lore of Sam" (LotR 4: IV). The botanical diversity of the region not only enhances its beauty, but illustrates its intrinsic value, which was then threatened by Mordor.

Etymology: The English common name "saxifrage" comes from Old French *saxifrage*, from Late Latin *saxifrage herba*, literally rock-breaking herb, from *saxum* (stone, rock) and *frangere* (to break) because of the common habitat of these herbs that frequently grow in the crevices of rocks.

Distribution and Ecology: The saxifrages are a large group, including around 400 species, which are for the most part widely distributed across the Northern Hemisphere, with 25 reported as native or naturalized in North America (north of Mexico; although others are grown as ornamentals), 123 recorded from Europe, 203 from China, and 15 from Japan. Saxifrages are mainly cold temperate, arctic, or alpine plants, growing in a variety of habitats such as rocky seepage areas; wet talus slopes; screes; open mountain slopes; wet to moist, open to shaded rocky ledges, cliffs, and crevices; gravelly or rocky areas; stream margins; arctic and alpine tundra or meadows; and arctic marshes. The species are distinguished by numerous characteristics, especially those relating to their life span and habit, form of their leaves (e.g., shape, size, extent of lobing, margin, pubescence, etc.), and flowers (e.g., symmetry, hypanthium, petal shape and size, coloration, ovary position, etc.). Their showy flowers are pollinated by small short-tongued insects (mainly flies and bees), and their numerous small seeds are scattered by wind or passing animals.

FIGURE 7.64 *Saxifrages.*

Economic Uses: Several species and hybrids are used as ornamentals, especially in rock gardens.

Description: Usually perennial herbs, sometimes rhizomatous, with or without hairs, these often simple, glandular or not; stems more or less erect, usually fairly short. Leaves alternate, in a basal rosette and/or distributed along stems, simple, quite variable in shape, unlobed to shallowly or deeply lobed, with pinnate to palmate venation; the apex acute to obtuse or rounded; the base cuneate to cordate; the margin entire, crenate, serrate, or dentate, petiolate or sessile; stipules absent. Inflorescences terminal, variable in form and sometimes reduced to a solitary flower. Flowers radially to occasionally bilaterally symmetrical, bisexual, usually erect, with a floral cup (= hypanthium) bearing usually five green sepals, usually five (occasionally fewer) petals, these variously colored, 10 distinct stamens; carpels two, more or less connate, with two styles, and the ovary superior to inferior. Fruit a capsule, releasing numerous small seeds (Figure 7.64).

SEAFIRE

(Various bioluminescent dinoflagellates, such as Gonyaulax spp. and Lingulodinium polyedrum of the Gonyaulacaceae, Noctiluca scintillans of the Noctilucaceae, Pyrocystis spp. of the Pyrocystaceae, and Pyrodinium bahamense of the Dinophyaceae; of the Class Dinophyceae, Division Pyrrhophyta.)

> *Then Tuor bowed in reverence, for it seemed to him that he beheld a mighty king. A tall crown he wore like silver, from which his long hair fell down as foam glimmering in the dusk; and as he cast back the grey mantle that hung about him like a mist, behold! he was clad in a gleaming coat, close-fitted as the mail of a mighty fish, and in a kirtle of deep green that flashed and flickered with sea-fire as he strode slowly towards the land. In this manner the Dweller of the Deep, whom the Noldor name Ulmo, Lord of Waters, showed himself to Tuor son of Huor of the House of Hador beneath Vinyamar. (UT 1: I)*

This is one of the most dramatic scenes in the entire legendarium—the meeting of Tuor, a human being, with Ulmo, one of the Valar (Figure 7.65), the angelic powers who have assisted in the creation of the world and had assumed the function of guarding and governing it—and this meeting set in motion a series of events that changed the history of Middle-earth, leading in time to Tuor's marriage with Idril Celebrindal, Turgon's daughter, and the sailing of Eärendil, their son, with his wife Elwing to Valinor, bearing a Silmaril and the errand of elves and men: pardon for the Noldor, mercy upon men and elves, and help in their struggle with the evil forces of Morgoth. Eärendil stood before the assembled Valar, "And his prayer was granted" (SILM 24). Morgoth's realm in Middle-earth would be destroyed. The meeting of Tuor with Ulmo also is beautifully illustrated by Ted Nasmith (in the illustrated *Silmarillion*, opposite p. 247), and his painting shows clearly the flashing of seafire at the Vala's feet. Many have experienced the flashing of dinoflagellates, lighting up marine waters when they are disturbed, and it is always magical. These distinctive unicellular aquatic algae contribute to the beauty and significance of this scene, which unites diverse aspects of Ulmo's oceanic realm: his mantle and hair reflect physical characteristics of the marine shoreline—mist and foam; his coat the scaly surface of fish, vertebrate residents of the marine realm; and his deep green kirtle flashed and flickered with seafire, reflecting the color of chlorophyll and the "plants" living in the marine environment.

Etymology: The common name "dinoflagellate" comes from the Greek *dinos* (whirling) and the Latin *flagellum* (whip), a reference to the two whiplike structures, or flagella, on the "body" of this single-celled organism, causing it to move forward and spin at the same time.The name "seafire" comes from sea + fire, where "sea" is derived from Old English *sæ* (sheet of water, sea, lake, pool), from Proto-Germanic **saiwaz*, of unknown origin, and "fire" is from Middle English *fier*, Old English *fyr* (fire), which is derived from Proto-Germanic **fur-i-*, from Proto-Indo-European **perjos*, from the root **paewr-* (fire, as an inanimate substance).

Distribution and Ecology: Dinoflagellates are unicellular photosynthetic algae and are abundant in marine waters, where they are planktonic (i.e., tiny organisms floating in ocean currents). They are ecologically significant in being important primary producers, occupying a comparable position to plants in terrestrial environments. However, some species occur in freshwater, living in rivers, ponds, and lakes. Others are symbiotic, living beneficially within the bodies of various marine animals, such as sponges, cnidarians (jellyfish, corals, sea anemones), or mollusks. Bioluminescence is present in some and is a defensive mechanism, reducing predation by startling potential predators. The bioluminescent response is initiated most frequently by mechanical stimulation (e.g., movement of the water), but it can also be induced by chemical or temperature changes in the water. Those dinoflagellates showing bioluminescence contain scintillons, minute cytoplasmic bodies that contain dinoflagellate luciferase, an enzyme involved in the biochemistry of bioluminescence, and luciferin, a chlorophyll-derived tetrapyrrhole ring that acts as the substrate for the light-producing reaction. The light produced is usually blue because it travels the farthest through water and can be seen by the greatest number of aquatic organisms. Certain dinoflagellates produce toxic compounds, and, when their population levels increase, they form algal blooms called red tide.

Economic Uses: Dinoflagellates are economically important as primary producers (i.e., organisms that because they are photosynthetic; they can take the energy of sunlight and use it to fix carbon dioxide, producing glucose and, subsequently, other organic compounds). They are eaten by other organisms, providing a key component in oceanic food webs. Certain species produce powerful neurotoxins. They can cause economic damage when, as a result of population increase, a red tide results, leading to the poisoning of fish and accumulation of the toxins in shellfish (which if eaten cause poisoning). Finally, fossil dinoflagellates are used to date geologic sediments (useful in petroleum exploration).

Description: The body of a dinoflagellate consists of a single cell, and they are usually photosynthetic (i.e., contain a chloroplast surrounded by three membranes and representing a reduced, endosymbiotic red alga, with chlorophyll and accessory pigments giving them a golden brown color), although some lack chlorophyll and are parasites. The single cell is quite distinctive—it usually is entirely covered by variously ornamented cellulosic plates and has two flagella, one in an equatorial groove around the cell (the vibration of which causes a spinning forward movement) and the second attached near the first but passing down a longitudinal groove and extending behind

FIGURE 7.65 *Seafire.*

the cell like a tail (and it functions more or less like a rudder). The cell can be more or less globose or with various projections, and it usually contains a light-sensitive organelle (eye-spot). Because dinoflagellates contain a chloroplast, they are often thought of as plants. However, the chloroplast of the true plants (i.e., clade containing green plants, red algae, and glaucophytes) is an endosymbiotic cyanobacterium that is surrounded by only two membranes, while (as mentioned) the dinoflagellates usually have an endosymbiotic red algal cell as their chloroplast—which represents a separate acquisition of photosynthetic ability. Dinoflagellates are more closely related to nonphotosynthetic protists such as apicomplexans (the group containing the malarial parasite) and ciliates (a group of unicellular organisms with the cell covered by numerous cilia, such as *Paramecium*) than they are to the green plants—the large group containing various green algae, mosses, liverworts, ferns, conifers, and flowering plants (Figure 7.65).

SEAWEEDS

(Various large marine algae, such as Macrocystis, Fucus, Laminaria, Saccharina, Alaria, Undaria, and Sargassum; of the Brown Algae [Phaeophyceae].)

> *[W]hen the long-haired riders on their lathered horses*
> *with bit and bridle of blowing foam,*
> *in wrack wreathed and ropes of seaweed,*
> *to the thunder gallop of the thudding of the surf.*
>
> (Lays I: lines 1567–1570).

In these verses from the poem *The Lay of the Children of Húrin*, we have an imaginative description of the waves of Belegaer, the Great-Sea, driven by the "Dweller in the Deep" (i.e., Ulmo) and crashing ashore near the mouths of the Sirion River (Figure 7.66). Various species of large brown algae, frequently called *kelps*, are characteristic of such environments because they are abundant in the deep near-shore waters extending upward into the intertidal zone, and they form the "wrack" and "ropes of seaweed" pictured. It is not surprising that these impressive photosynthetic organisms, so characteristic of the shores of Middle-earth, are especially loved by Uinen, the spouse of Ossë, who loves the coasts and islands. The various genera of large brown algae are pictured in *The Silmarillion* as among the oldest living things. In the twilight under the stars, before the awakening of the elves, we find "in the seas the great weeds, and on earth the shadow of great trees" (SILM 3). Just as the great trees dominate plant life in the terrestrial environment, towering over other land plants, these large brown algae, in genera such as *Macrocystis, Fucus, Nereocystis, Laminaria, Saccharina, Undaria*, and *Alaria*, form a marine forest, towering over the other algal members of these marine communities (as shown in Tolkien's beautiful *The Hills of the Morning* (see fig. 1 in W. G. Hammond and C. Scull's, 1995, *J. R. R. Tolkien: Artist and Illustrator*). Finally, we should not think of these large brown algae as purely organisms of cool to cold waters—in the tropical Atlantic, the floating brown alga, *Sargassum*, is abundant and also accumulates in the wrack zone.

Etymology: "Seaweed" is from sea + weed, where "sea" is derived from Old English *sæ* (sheet of water, sea, lake, pool), from Proto-Germanic **saiwaz*, of unknown origin, and "weed" is derived from Old English *weod* (grass, herb, weed), from Proto-Germanic

FIGURE 7.66 *Seaweeds.*

*weud-, again, of unknown origin. The Sindarin word for seaweed may be *gaeruil* or *uil*, and the Quenya name may be *felpa*.

Distribution and Ecology: Brown algae, a group of 300 genera and ca. 2,000 species, are widely distributed in marine or brackish waters from the tropics to the arctic, in deep near-shore waters, the intertidal zone, and tidal and/or rock pools, and these photosynthetic organisms are ecologically significant and conspicuous especially in cool to cold Northern Hemisphere waters.

Economic Uses: Several species are important food plants (used in soups or salads). Many species contain the viscus polysaccharide alginic acid, which is extracted commercially and used in foods and pharmaceuticals.

Description: The multicellular body (or thallus) yellow-green to brownish in color, small to large, filamentous, or structurally complex, with rootlike structures (*holdfasts*, these branched or cuplike), stemlike structures (*stipes*, unbranched or branched, short to elongate, sometimes hollow), and leaflike structures (*blades* or lamina; i.e., the flattened portion or portions of the thallus, unlobed or lobed, and thallus with a single blade or several blades), and often with air bladders or floats. The cells photosynthetic, with a chloroplast surrounded by four membranes, containing chlorophyll a and c, and with carotenoid pigments, especially fucoxanthin; the chloroplast derived from an endosymbiotic red alga (thus these organisms cannot be considered to be plants, which have chloroplasts surrounded by only two membranes, containing chlorophyll a and b, which are derived from an endosymbiotic cyanobacterium). The life cycles of brown algae vary from an alternation of generations (haploid vs. diploid) to a diploid-dominated cycle. Their motile cells (spores or gametes) have two flagella, one lacking hairs (whiplash flagellum) and the other covered with numerous tubular hairs (tinsel flagellum) (Figure 7.66).

SEDGES (*CAREX* SPP.)

(The Sedge family [Cyperaceae].)

> *They came upon many hidden pools, and broad acres of sedge waving above wet and treacherous bogs; but Shadowfax found the way, and the other horses followed in his swath. (LotR 3: V)*

Sedges are characteristic of wetland habitats such as those in the West Emnet (e.g., fens, bogs, marshes, and pools) through which Gandalf, Legolas, and Gimli passed as they traveled southward from the borders of Fangorn toward Edoras. These plants are quite similar to grasses but have usually triangular stems (not terete stems, as in grasses) with their leaves arranged in three (instead of two) ranks. Their leaves are composed of a spreading blade and a sheath that is close-fitting around the stem, as in grasses, but the sheath margins are fused (not overlapping, but free). Although sedges are only mentioned in this passage, it is certain that these abundant yet inconspicuous wetland plants were also common in the wetlands of Middle-earth. Their presence in such habitats provides forage for grazing animals such as horses, and, in this way, their use is similar to that of grasses. Sedges (i.e., species *Carex*), however, are unimportant as a human food source, unlike the grasses, which provide many extremely important grains (e.g., wheat, barley, rice, and corn). In fact, it is surprising that such similar plants have fruits that are so different in their food value.

Etymology: The word "sedge" (in its general meaning of a coarse grasslike herb growing in wetland habitats) is derived from Old English *secg*, from Proto-Germanic **sagjoz*, and probably from Proto-Indo-European **sek-* (to cut), alluding to the sharply serrate-margined (and thus "cutting") leaves of some species. The common name is used for members of the genus *Carex*, but sometimes it is generalized and used for any member of the family Cyperaceae. The scientific Latin name *Carex* is the Roman name for these plants (or similar grasslike plants) and is probably derived from the Greek *charaktos*, meaning toothed, and *keirin* (to cut), again referring to the sharp teeth found along midvein and margins to the leaves of some species. The Sindarin name is possibly *esg* (although this name seems also to be connected to reeds).

Distribution and Ecology: The sedges (i.e., species of *Carex*, a genus of ca. 2,000 species) are nearly cosmopolitan in distribution; however, they are poorly represented in lowland tropical regions, especially sub-Saharan Africa. About 180 species occur in Europe, 480 in North America (north of Mexico), ca. 500 in China, and ca. 200 in Japan. They grow in a wide variety of habitats, but most are species of wetlands. Their inconspicuous flowers are wind-pollinated, and their achene fruits are associated with a flask-shaped bract that often assists in water dispersal.

Economic Uses: Considering the size of the group, its species are of little economic importance. Some species provide forage for livestock (especially those grazing in wetlands), and a few are used as fiber plants or are minor ornamentals.

Description: Usually perennial herbs, often with rhizomes; stems erect, usually triangular and solid in cross-section, short to tall. Leaves alternate, three-ranked, simple and differentiated into a spreading blade, a sheath that closely encircles the stem, with its margins fused, and with a distinct flap of tissue (i.e., the ligule) at the junction of the blade and sheath; the blade linear, with parallel venation, flat or folded. Inflorescences terminal, various. Flowers very tiny, unisexual (and the staminate and carpellate on the same or different plants), arranged in one-flowered spikelets, and these usually gathered into spikes or racemes, each flower associated with a bract; the carpellate flower with two bracts, the second flask-shaped, the perigynium, surrounding the gynoecium, the staminate flowers with 1–3 nonsagittate anthers; and the carpellate flowers with 2–3 fused carpels; the ovary superior; the two or three stigmas elongated, papillose. Fruit an achene.

SEREGON AND STONECROPS (*SEDUM* SPP.)

(In the Stonecrop family [Crassulaceae].)

> *Now that hill [Amon Rûdh] stood upon the edge of the moorlands that rose between the vales of Sirion and Narog, and high above the stony heath it reared its crown; but its steep grey head was bare, save for the red seregon that mantled the stone. And as the men of Turin's band drew near, the sun westering broke through the clouds, and fell upon the crown; and the seregon was all in flower. Then one among them said: "There is blood on the hill-top." (SILM 21; see also CoH VII)*

Amon Rûdh, a key geographical locality in the story of Túrin Turambar, is described in detail in *The Silmarillion* and in the *Children of Húrin*; it is also shown in two beautiful paintings (one by Ted Nasmith, opposite p. 204, published in an illustrated edition of *The Silmarillion*, and the other by Alan Lee, opposite p. 128, published in the *Children of Húrin*), in which the red top of the mountain, covered in seregon, is clearly seen. A footnote by Christopher Tolkien (in the "Narn i Hîn Húrin," of the *Unfinished Tales*) indicates that seregon was a plant of the kind called stonecrop in English, and the habitats, rocky cliffs, and the flat rocky top of the mountain Amon Rûdh are characteristic of stonecrops. The flowers of stonecrops show a wide range of colors (see description), but those of seregon were dark red. The quote given here is slightly expanded in the *Children of Húrin*, in which it is indicated that Túrin's reply to the statement about "blood on hill-top" was "Not yet." The statement, made by Andróg, one of Túrin's outlaw companions, proved to be prophetic. Blood was spilled on the hill-top when Túrin and his outlaw followers, along with the elf Beleg, were betrayed by the petty-dwarf Mîm. Andróg and many other outlaws were killed on the summit of Amon Rûdh in a fight with a band of Orcs, Beleg was bound, and Túrin was captured (Figure 7.67) and then taken by the Orc band toward Morgoth's stronghold of Thangorodrim. Seregon is only mentioned in connection with Amon Rûdh, but other species of stonecrop also occur in Middle-earth. They occurred in Ithilien, where Sam and Frodo saw that the "grots and rocky walls were already starred with saxifrages and stonecrops" (LotR 4: IV). Later they saw them again at the Cross-Roads, where they encountered a huge sitting figure, a carving in stone, perhaps of Isildur, Anárion, or some Gondorian king, which had been defaced by the maggot-folk of Mordor. The head was broken off and replaced by a "round rough-hewn stone, rudely painted . . . in the likeness of a grinning face with one large red eye in the midst of its forehead." Then Frodo spotted the old king's head, lying by the roadside. "The eyes were hollow and the carven beard was broken, but about the high stern forehead there was a coronal of silver and gold. A trailing plant with flowers like small white stars had bound itself across the brows as if in reverence for the fallen king, and in the crevices of his stony hair yellow stonecrop gleamed" (LotR 4: VII). The king again had a crown of silver and gold—provided by two of the plants of Ithilien—the color of their flowers perhaps providing a subtle reminder of the two trees of Valinor, one silver (Telperion) and the other gold (Laurelin). A stonecrop and an unknown twining vine (perhaps a morning-glory, *Ipomoea,* of the Convolvulaceae), as representatives of the world of flora, gave Frodo and Sam hope in a pretty dark place shortly before they had to face the Morgul-vale and the stairs of Cirith Ungol.

Etymology: The English common name "stonecrop" is a combination of the words "stone" and "crop" and is derived from the Old English *stancropp*; it refers to the apparent ability of these succulent plants to grow out of bare rocks and stone. The scientific Latin name, *Sedum*, coined by C. Linnaeus, was taken from the Roman naturalist Pliny, who had used the name for these plants; the name is derived from the Latin *sedare* (to alleviate, calm) and is based on the plant's medicinal uses (as outlined by Pliny in his *Naturalis Historia*), although some have suggested that it is derived from *sedeo* (to sit), alluding to its low habit. Finally, *seregon* is the Sindarin name of a particular species of *Sedum* with deep red flowers that grew on Amon Rûdh, and the name means blood of stone, derived from the root *sereg* (blood) and "gon" (stone).

FIGURE 7.67 *Seregon and stonecrops.*

Distribution and Ecology: Stonecrops are a large group of succulent herbs comprising around 450 species. The group (i.e., the genus *Sedum*) is widely distributed in the Northern Hemisphere, occurring in North America (including Mexico), Central America, Europe, Asia, and northern and eastern Africa. In North America (north of Mexico) 41 species occur, while 57 grow in Europe, 121 in China, and 34 in Japan, according to the published floras of those regions. They are plants of various plant communities, but usually grow on rocky, gravelly, or talus slopes; bluffs; cliffs; ledges; shaded to sunny rock outcrops; in flat rocky or gravelly areas; and in rocky crevices. They sometimes occur in living mats (over rocks) along with mosses and lichens, but they also are known from stream banks and gullies. The numerous species are distinguished by characters such as their habit, leaf arrangement; extent of succulence; size, shape, coloration, and pubescence; and floral characters, such as petal size, shape, and color. Their flowers have small nectar glands (associated with each of the carpels) and are pollinated by various insects. The tiny seeds of *Sedum* are probably dispersed by wind.

Economic Uses: Many species of stonecrop are cultivated as garden ornamentals because of the form of their succulent leaves and their showy flowers; they are interesting plants in rock gardens. Some are occasionally eaten as salad herbs or used medicinally. They are often used as a roof covering (or a component in the mix of species) in green roofs. They are especially useful in this regard because they are drought-resistant (storing water in their leaves; they also have crassulacean acid metabolism, a modified photosynthetic system that allows them to keep their stomata closed during the day and thus save water, but open them at night to take in and fix carbon dioxide).

Description: Annual to perennial herbs (rarely subshrubs), with or without hairs; stems erect to creeping, sometimes succulent. Leaves usually alternate, but sometimes opposite or whorled, often forming a basal rosette and additionally distributed along stems, simple, more or less succulent, linear to orbicular, ovate to obovate, with pinnate venation but veins obscure, the apex acute to rounded, the base narrowed, spurred or not, the margin entire; stipules absent. Inflorescences terminal cymes. Flowers radially symmetrical, bisexual, erect, with usually five distinct or slightly fused sepals; usually five distinct, spreading, or erect petals, these yellow, white to red, or purple (but deep red in *seregon*), and usually 10; stamens, these free or slightly fused to petals; carpels five, usually distinct, each associated with a nectar gland; and the ovaries superior. Fruits a cluster of follicles, these erect or spreading and each with numerous seeds (Figure 7.67).

SLOE, PLUMS, AND CHERRIES (*PRUNUS SPINOSA*, *P. DOMESTICA*, *P. CERASUS*, *P. AVIUM*, AND RELATIVES)

(The Rose family [Rosaceae].)

> *But the Entwives gave their minds to the lesser trees ... and they saw the sloe in the thicket, and the wild apple and the cherry blossoming in spring, and the green herbs in the waterlands in summer, and the seeding grasses in the autumn fields. (LotR 3: IV)*

Given the hobbits love of food and the importance of these trees in providing flavorful fruits, it is not surprising that various species of the genus *Prunus* (plums, cherries, and their relatives) are mentioned in *The Lord of the Rings*. After the war, Sam worked hard in restoring the environmental damage to the Shire caused by Saruman, including the wanton felling of many trees, using his gift from Galadriel. Sam thought that only his great-grandchildren would see the Shire as he remembered it, but the gift of Galadriel had, as Frodo suspected, tremendous power. The following year, 1420, turned out to be marvelous, with bountiful harvests. We are told that hobbit children "sat on the lawns under the plum-trees and ate, until they had made piles of stones like small pyramids or the heaped skulls of a conqueror" (LotR 6: IX). Note that they sat on the lawns—these are not wild plums, but cultivated trees—illustrating that hobbits were proficient in the cultivation of fruits, such as plums, apples, cherries, strawberries, and raspberries, all members of the rose family. Gardening and horticulture, it is clear, were valued in the Shire, as Frodo told Faramir when they met in Ithilien. Where did hobbits get a domesticated tree, the European plum (*Prunus domestica*), which is much improved in fruit size and flavor over its wild relatives, such as the sloe? Obviously, generations of selection must have occurred—planting the seeds of trees with the largest and most flavorful fruits and doing this over and over again. The European plum may be derived from a hybrid of the cherry plum (*P. cerasifera*) and the sloe (*P. spinosa*) that was then improved by selection, or perhaps it was derived merely by selecting the best-fruited plants from polyploid individuals of the cherry plum. Either way, much genetic change is involved—change such as that existing between domesticated dogs and their wolf ancestors. In the context of Tolkien's legendarium, the race that was most focused on this process (i.e., plant domestication) is the entwives (Figure 7.68). With their interest in agronomy and horticulture, they certainly noticed (as indicated in the opening quote) both the sloe (*P. spinosa*), a wild Eurasian species that produces fruits used in wine-making, and—because of its thorns—is a common component of hedges, and the various wild cherry species, from which have been domesticated both the sweet cherry (from *P. avium*) and the sour cherry (from *P. cerasus*). Like the sloe and European plum, both the sweet and the sour cherry are Eurasian in distribution. The entwives may have begun this process of domestication and freely shared their horticultural knowledge, thus enriching the diet of the other races of Middle-earth, including hobbits. Finally, we note that sloe, as a wild species, is mentioned descriptively in Tolkien's works. In fact, this common species was so part of the landscape of the Shire that it is included in Bilbo's walking song: "Apple, thorn, and nut and sloe" (LotR 1: III), sung by Frodo, Sam, and Pippin on their first full day of walking through the Shire on their way to Crickhollow. Sloe probably was quite widespread in Middle-earth because the Fellowship also observed it, along with hawthorn (*Crataegus*), near the shores of the Anduin as they approached the Emyl Muil.

Etymology: The English "sloe" (the common name of *Prunus spinosa*) is derived from Old English *slah* (pl. *slan*), from Proto-Germanic **slaikhwon*, from the Proto-Indo-European **sleie-* (blue, bluish, blue-black), in reference to the color of the fruits of this species. The word "plum" (for the cultivated *P. domestica*) is from Old English *plume*, from an early Germanic borrowing from Latin **pruna*, from the Latin *prunum* (plum), from Greek *prounon* (plum) and *proumne* (plum tree), of unknown origin and perhaps from an Asiatic language. The scientific name *Prunus* is also from the Latin *prunum*. The English "cherry" is derived from Anglo-French *cherise*, from Latin *ceresia*, based on Greek *kerasos* (cherry tree). The Old English word for cherry was

FIGURE 7.68 *Sloe, plums, and cherries.*

ciris, but this died out after the Norman invasion and was replaced by the French word. The Sindarin word for cherry is *aep*, and *aeborn* for cherry-tree, while the Quenya word for cherry may be *pio*, and for cherry-tree *aipio* (but these words could also be applied to plums).

Distribution and Ecology: The genus *Prunus* (cherries, plums, and their relatives), with more than 200 species, is nearly cosmopolitan in distribution, growing in both temperate and tropical regions; however, the group is most abundant in the north-temperate zone. Forty-four species are native or naturalized in North America (north of Mexico), 17 in Europe, about 85 in China, and 24 in Japan. Plums, cherries, and their relatives grow in a wide array of wet to very dry, open to wooded habitats. The sloe (*P. spinosa*) is a shrub of open woods and hedges and is widespread across Europe, Asia, and northern Africa. The European plum (*P. domestica*) probably originated as a cultivated tree in Eastern Europe or adjacent southwestern Asia but was early carried into Western Europe. It also occasionally escapes from cultivation, growing in disturbed habitats as a naturalized species. The European plum can be distinguished from sloe by its fruits: pendent, globose to oblong, 20–75 mm long, and nonastringent in the plum versus more or less erect, globose, 10–15 mm long, and astringent in the sloe. Sweet cherry (*P. avium*) is a Eurasian species, and it was early domesticated in either southeastern Europe or Asia Minor; the sour cherry (*P. cerasus*) may also have originated in southeastern Europe and/or western Asia, as a hybrid between the sweet cherry and a wild cherry (*P. fruticosa*). Both sweet and sour cherries were carried to many new regions by the Romans. The sweet cherry has leaves that are dull above, flowers with an urn-shaped floral cup (or hypanthium), and red to dark purple or black fruits, while the sour cherry has glossy leaves, flowers with a broadly bell-shaped hypanthium, and red to dark red fruits. The flowers of plums, cherries, and their relatives are pollinated by various insects, especially bees, and the colorful fruits are dispersed mainly by birds.

Economic Uses: Species of the genus *Prunus* provide several important edible fruits: for example, the European plum (a cultivated species, widely used for prunes, prune juice, etc.), Japanese or Chinese plum (*P. salicina*; and various cultivars that are hybrids of this species and various North American species; widely used as a fresh fruit), sweet cherry, sour cherry, peach and nectarine (*P. persica*, a cultivated species derived from the Chinese wild peach, *P. davidiana*), apricot (*P. armeniaca*), and almond (*P. dulcis*). These and several other species are grown as ornamental trees or shrubs because of their beautiful flowers, often produced en mass in the spring before the leaves emerge. Some, such as the black cherry (*P. serotina*), produce valuable timber. Finally, several have medicinal uses, and all species contain hydrogen cyanide in their vegetative parts, which are thus poisonous.

Description: Deciduous or evergreen trees or shrubs, the bark smooth and with horizontal lenticels, sometimes peeling, to roughened or furrowed, or forming irregular plates or scales, the branches usually forming long and short shoots, sometimes with thorns, with or without simple hairs. Leaves alternate, spirally arranged, simple, ovate to obovate, pinnately veined, with secondary veins forming loops, the apex acuminate or acute to rounded, the base narrowly cuneate or acute to cordate, the margin entire to variously toothed, and the teeth sometimes glandular, the surfaces with or without simple hairs, often with glands at base of blade or on petiole, and with or without a petiole; stipules present, usually deciduous. Inflorescences terminal or axillary, few to several-flowered, racemes or various clusters, sometimes reduced to a solitary flower. Flowers radially symmetrical, usually bisexual, the sepals five, variously shaped, the petals five (but more in some ornamentals), basally narrowed, white to dark pink, and the stamens numerous, with anthers globose, pale yellow or yellow to reddish, all borne on a nectar-producing cup (= hypanthium) with a single carpel, the style elongate, the ovary superior. Fruit a small to large drupe, these globose or ovoid to obovoid, yellow to orange or red, purple to blue or blue-black, with a single pit (Figure 7.68).

SNAPDRAGONS (*ANTIRRHINUM MAJUS*, AND RELATED SPECIES)

(In the Snapdragon family [Plantaginaceae].)

> *They used to go up like great lilies and snapdragons and laburnums of fire and hang in the twilight all evening! (Hobbit I)*

We get the first glimpse in *The Hobbit* that Bilbo Baggins was not quite the respectable hobbit that he appeared to be, when, in talking with Gandalf, he recalled the wizard's wonderful tales about dragons, goblins, and visiting elves and also his spectacular fireworks (see Figure 7.69). J. R. R. Tolkien (as narrator) notes that Bilbo was not so prosaic as he liked to believe and "also that he was very fond of flowers" (Hobbit I)—after all he had compared these fireworks with three different flowers—lilies (*Lilium* spp.), snapdragons (*Antirrhinum majus*), and laburnums (*Laburnum anagyroides, L. alpinum*, and their hybrid, *L.* × *watereri*). And Bilbo's love of flowers also is evident in the attention he paid to his garden at Bag End, which contained snapdragons, sunflowers (*Helianthus annuus*, and possibly also related species), and nasturtians (*Tropaeolum majus*). Although he did not at all realize it, Bilbo was ready for an adventure, which as Gandalf noted would be very good for him. Quite soon, he would be on that adventure: would visit elves, be captured by goblins, and have to figure out exactly how to deal with a dragon! This readiness, at least in part, is seen in his vivid imagination and appreciation of beauty, both clear indications that he was ready to meet with elves and would profit from the experience.

Etymology: "Snapdragon," a name used since the late sixteenth century, is based on an imaginative resemblance of the flowers to a dragon's mouth, which can be made to open and close by squeezing the flowers. The scientific Latin name *Antirrhinum* is based on Greek *antirrhinon*, from *anti* (in the sense of like) and *rhis* or *rhin-* (nose) from the resemblance of the flower (and possibly also the fruit) to an animal's face and snout.

Distribution and Ecology: The limits of the genus *Antirrhinum* are problematic, with some botanists restricting the genus to a Mediterranean clade of ca. 20 more or less perennial species (including *A. majus* and its close relatives, with a chromosome number of 16) whereas others recognize a more inclusive monophyletic group, also including ca. 15 annual species (*Antirrhinum* sect. *Saerorhinum*) of western North America (which have 30 or 32 chromosomes and are excluded as the genus *Sairocarpus* by those who circumscribe *Antirrhinum* restrictively). Most are plants of rocky and often calcareous habitats, and they frequently grow on rock walls. *Antirrhinum majus* was taken to England during the Roman period and has naturalized

FIGURE 7.69 *Snapdragons.*

in the British Isles; it also has become widely naturalized in North America. The species of *Antirrhinum* are distinguished by characters such as the presence or absence of hairs on the stem and leaves or the form of these hairs (glandular or eglandular), orientation of the stems, shape and arrangement of the leaves, and size and color of the flowers, but identification is often difficult as a result of hybridization. The flowers of snapdragons and close relatives are pollinated by solitary bees; their small seeds are wind-dispersed.

Economic Uses: The snapdragon (*A. majus*) is a commonly cultivated ornamental herb with white to red or yellow flowers; this species also is used in scientific studies of genetic control of floral development. *Antirrhinum* species are occasionally used medicinally (as an anti-inflammatory), and edible oil can be extracted from the seeds.

Description: Low shrubs or perennial herbs (but sometimes grown as annuals); stems erect to spreading, with or without simple hairs, and these (when present) with or without glandular heads. Leaves opposite and decussate and/or alternate and spiral, simple, linear or narrowly to broadly ovate to obovate, with pinnate venation, the apex acute to rounded, the base narrowed, the margin entire; stipules absent. Inflorescences terminal racemes. Flowers bisexual, bilaterally symmetrical, held horizontally, with the five sepals slightly fused forming a deeply lobed calyx; the five petals strongly fused, white to red, purple, or yellow and often bicolored, two-lipped, the upper lip two-lobed and the lower lip three-lobed, with a prominent projection that closes the mouth of the corolla tube; four stamens, two long and two short, included within the corolla; the carpels two, fused; the ovary superior, and with a single style and small, slightly two-lobed stigma. Fruit an asymmetrical capsule with two locules, one opening by a single apical pore, and the other opening by two apical pores; seeds numerous, small (Figure 7.69).

SORREL (*RUMEX ACETOSA*, AND RELATED SPECIES OF *RUMEX* WITH SOUR LEAVES; I.E., *RUMEX* SUBGENUS *ACETOSA*)

(The Knotweed family [Polygonaceae].)

> *He [Bilbo] nibbled a bit of sorrel, and he drank from a small mountain-stream that crossed the path, and he ate three wild strawberries that he found on its bank, but it was not much good. (Hobbit VI)*

Sorrel is only mentioned once in J. R. R. Tolkien's writings—in *The Hobbit*—and (as we see in this quote) it is one of the plants that a famished Bilbo ate soon after emerging from the goblin tunnels (Figure 7.70). The plant that Tolkien likely had in mind here is the common sorrel (*Rumex acetosa*), a species that is widely distributed across Europe and Asia, occurs in northwestern Africa, and has been introduced into North America. Its young leaves are edible and pleasantly sour; they can be eaten raw or cooked (and used in soups, stews, or as greens). However, there are many similar sorrels, such as sheep sorrel (*R. acetosella*) or heartwing sorrel (*R. hastatulus*), and these also are edible. Common sorrel occurs in alpine meadows and perhaps Tolkien even remembered it from his 1911 walking tour in Switzerland (although, of course, it is also common in

England). He commented in a letter to his son Michael that they "went on foot carrying great packs," slept wherever they could find shelter, and often "after a meagre breakfast fed ourselves in the open" (Letters: No. 306). We can imagine Tolkien, like Bilbo, hiking along a mountain trail, stopping to nibble on sorrel and wild berries.

The true sorrels (members of the genus *Rumex*) are not to be confused with wood-sorrels, which are species of *Oxalis* (of the plant family Oxalidaceae). A common European species that also grows in the Alps (and is widespread in Europe and Asia) is *Oxalis acetosella*. Its leaves, as implied by its name, are high in oxalates and have a similar sour flavor. The streamside habitat, perhaps implied in the passage quoted here, suggests that the plant here was most likely *Rumex acetosa*—not *Oxalis acetosella*—and this is also suggested by Tolkien's usage of the name "sorrel" instead of "wood-sorrel."

Etymology: The common name "sorrel" comes from Middle English *sorel*, which was borrowed from Old French *surele*, from *sur* (sour), derived from Proto-Germanic **sura-* (sour) and the Proto-Indo-European root **suro-* (sour) because of the taste of its leaves, resulting from their high oxalic acid content. The scientific Latin name *Rumex* is the classical name for the sorrels, and the name is probably derived from *rumo* (to suck), alluding to the ancient practice of sucking its sour leaves to alleviate thirst.

Distribution and Ecology: The ca. 200 species of *Rumex* (usually called "docks," or in the case of the most sour species, "sorrels") are very widely distributed in temperate regions (of both hemispheres). Sixty-three species of *Rumex* occur in North America (north of Mexico), 50 species grow in Europe, 27 species in China, and 14 in Japan. The species of *Rumex* subgenus *Acetosa* (ca. 50 species, including subg. *Acetosella*) have unisexual flowers, and the staminate and carpellate flowers are borne on separate plants; in contrast, in members of *Rumex* subgenus *Rumex*, the flowers are usually bisexual, or bisexual and unisexual (and then with staminate and

FIGURE 7.70 *Sorrel.*

carpellate flowers on the same plant). In addition, members of subg. *Acetosa* usually have leaves that are more or less sagittate at the base, while in the species of subg. *Rumex* the leaf-base can be cuneate, truncate, rounded, or cordate, but never sagittate. Finally, the species of subg. *Acetosa* have leaves that contain high levels of oxalic acid, and they are called sorrels, whereas those of subg. *Rumex* have lower acidity and are called docks. The sorrels occur in a variety of open habitats, especially montane meadows, grasslands, heaths, rocky or sandy shores or slopes, banks of rivers or streams, waste places, disturbed habitats, and cultivated fields. Their tiny, more or less pendulous flowers are wind-pollinated; the achenes are associated with persistent tepals that assist in dispersal by wind and/or water.

Economic Uses: Many species of *Rumex* (the sorrels or docks) are edible, and their leaves are used fresh (in salads) or can be cooked (in soups or sauces); the leaves can also be dried and used in teas. The sorrels are also occasionally used medicinally (as diuretics).

Description (of species in *Rumex* subg. *Acetosa*): Perennial to annual herbs, sometimes with taproot and/or rhizomes; stems usually erect to ascending; hairs (when present) simple; tissues sour due to high levels of oxalic acid. Leaves alternate, spirally arranged, basal and/or distributed along the stem, simple, ovate to obovate, narrow to broad, with pinnate venation, the apex acute to obtuse, the base usually sagittate with pointed lobes oriented either downward or upward but sometimes merely cuneate, the margin entire, petiolate (at least on basal leaves); stipule present, a membranous tubular structure surrounding the stem (i.e., an ocrea). Inflorescences terminal, paniclelike, with the tiny flowers ± arranged in a series of whorls. Flowers radially symmetrical, unisexual; the staminate flowers on one plant and the carpellate flowers on another, with perianth of green to reddish tepals, with three in the outer whorl and three in the inner whorl, the inner ones ± broader and differentiated from the outer, with six stamens and three fused carpels, the three styles distinct, with well-branched stigmas; the ovary superior. Fruit an achene, three-sided, surrounded by the persistent perianth (Figure 7.70).

SUNFLOWER (*HELIANTHUS ANNUUS*)

(The Aster or Composite family [Asteraceae or Compositae].)

> *Inside Bag End, Bilbo and Gandalf were sitting at the open window of a small room looking out west on to the garden. The late afternoon was bright and peaceful. The flowers glowed red and golden: snap-dragons and sun-flowers. (LotR 1: I)*

In the first chapter of *The Fellowship of the Ring*, we see Bilbo Baggins and Gandalf sitting beside the open window of Bilbo's home, which gave them a beautiful view of Bilbo's flower garden (see Figure 7.46 and Figure 7.81). As is discussed under the entry for nasturtians, this seemingly casual reference to Bilbo's garden indicates that he, like many hobbits, appreciated the beauty of flowers and sought to enhance the beautiful environment surrounding his home. Since the floral heads of cultivated sunflowers usually face eastward, they would have been especially attractive from Bilbo's window! The inclusion of sunflowers in his garden would also have attracted various songbirds, which value its nutritious seeds, and these avian visitors also could have been observed from his window, adding to his pleasure. How sunflowers found their way to the Middle-earth of the Third Age is a bit of a mystery because the common sunflower (*Helianthus annuus*), like all sunflowers, is native to North America; it was restricted to this continent until fairly recently, when its cultivation spread worldwide as it became an important edible oil and seed crop. It is possible that sunflowers were introduced into Middle-earth by the Númenóreans—as certainly were kingsfoil and pipeweed and probably also potatoes.

Etymology: The name "sunflower" comes from sun + flower and refers to species of the genus *Helianthus*, especially *H. annuus*, which was introduced from North America into Europe in the early 1500s; the name is based on the appearance of the floral head, which has a central zone of disk flowers surrounded by radiating corollas of the ray flowers. The English word "sun" is from Old English *sunne*, derived from Proto-Germanic **sunnon*, and from Proto-Indo-European *s(*u*)*wen-* (sun); the word "flower" was taken from Old French *flor*, from Latin *florem* (flower). The scientific Latin name *Helianthus* is derived from Greek *helios* (sun) + *anthos* (flower).

Distribution and Ecology: The sunflowers comprise a group of 52 species that are restricted to North America, but a few species have naturalized in the Old World and *H. annuus* (which is widely distributed in North America) is widely grown as an oil-crop in temperate regions of the world. The agronomic cultivars of this species are unbranched, with a single large head, whereas individuals in wild populations are well branched, with numerous smaller heads. The species was domesticated about 2,500 years ago (although many wild populations still exist). Sunflowers grow in a wide variety of open and disturbed habitats. Their flowers are pollinated by various bees. The fruits are valuable wildlife food; they are dispersed by birds and mammals but can also be blown by wind or moved in rain-wash.

Economic Uses: The common sunflower (*H. annuus*) is an important crop plant; it is the source of sunflower oil (cooking oil, also used in margarine, salad dressings, soaps, and paints). The salted seeds are eaten as a snack. The fruits are also used as bird-seed, and the plant is used as livestock forage (the fruits, as meal, or the vegetative parts, as a silage plant). Finally, the various species of *Helianthus* are also beautiful garden ornamentals, grown because of their showy, sunlike floral heads.

Description (of *H. annuus*): Annual herbs, with simple glandular and nonglandular hairs; stems erect, branched distally or unbranched. Leaves mostly alternate, spirally arranged, distributed along the stem, simple, more or less palmately veined, the apex acute to acuminate, the base cuneate to cordate, the margin serrate, the lower surface with stiff nonglandular hairs and usually also gland-headed hairs, with well-developed petiole; stipules absent. Inflorescences terminal, or terminating each branch, the flowers densely clustered in heads, each head with expanded, flat receptacle, bearing ca. 150 to more than 1,000 flowers, the peripheral ones numbering ca. 17 to more than 100, sterile, usually yellow, bilateral, with an elongated corolla, and these called ray flowers; the central ones, numbering 150 to more than 1,000, fertile, yellow to reddish, radially symmetrical, with a tubular corolla formed by five fused petals and with five triangular lobes, and these called disk flowers. Both the disk and ray flowers surrounded by numerous ovate or narrowly ovate bracts, each with usually acuminate apex. Each disk flower with sepals modified, forming two elongate scales (and sometimes also a few smaller ones), with five fused petals, five stamens, these arising from the corolla tube, and with their anthers forming a cylinder around the style, and with two fused carpels, the style elongate, with two style branches, and on each of these the stigmatic region restricted to two lines; the ovary inferior. Fruit an obovoid-compressed, black or black-and-white striped achene.

TEA (*CAMELLIA SINENSIS*)

(The Tea family [Theaceae].)

What's that? Tea! No, thank you! (Hobbit I)

Although tea was introduced into England later than coffee, it quickly became the more popular beverage and is currently the most popular caffeine-containing drink in the British Isles. Thus, in *The Hobbit*, we are not surprised to find a rather frazzled Bilbo asking Gandalf if he would like a cup of tea—an action quite in keeping with the Englishness of the Shire. Interestingly, tea is not mentioned elsewhere in *The Hobbit*, and this popular beverage is also absent from *The Silmarillion* and only mentioned a few times in *The Lord of the Rings*. This may be because tea was not introduced into England until the seventeenth century, so would not fit in with the more archaic cultures depicted in these works—but the Shire may be the exception. It is pictured as rural England in the late nineteenth century, and, as such, it clearly contrasts with the surrounding "Wilderland" containing dwarves, elves, trolls, and dragons. Additionally, such anachronisms add comic flavor—one can hardly imagine someone like Bilbo facing such wild adventures! Therefore, J. R. R. Tolkien must have considered the inclusion of tea in the chapter "An Unexpected Party" appropriate. He did not remove it when he had the chance: either in the 1960 extensive rewrites (see John Rateliff, 2007, *The History of The Hobbit, Part Two: Return to Bag-End*, p. 775) or in the 1966 revisions in connection with the Third Edition (see Douglas Anderson, 2002, *The Annotated Hobbit*, p. 41). Tea is also mentioned in the humorous poem "Perry-the-Winkle" (in *The Adventures of Tom Bombadil*), which was reported to have been written by Sam Gamgee and thus must, at least in some degree, reflect the customs of the Shire. In this poem, Perry-the-Winkle is said to have tea and cramsome bread every Thursday. We can safely say that tea was part of the culture of the Shire, although clearly it was not as popular as beer (at least among Bilbo, Frodo, and their friends).

Etymology: The English word "tea" is derived from Dutch, *thee*, from Malay, *teh*, and ultimately from Chinese (Amoy dialect), *t'e*, which corresponds to Mandarin *ch'a*. The scientific Latin name *Camellia* was coined by C. Linnaeus in honor of Rev. George Kamel (1661–06), a pharmacist and missionary to the Philippines, to mark his contributions to botany.

Distribution and Ecology: Tea (*C. sinensis*) is native to the region from India eastward to Indochina, southern China, to southern Korea and Japan (although its exact range is difficult to interpret because the species has been cultivated for such a long period and has naturalized in many areas where it has been grown). It is a member of the genus *Camellia*, which contains ca. 120 species that are distributed widely—from Nepal and India eastward throughout eastern and southern Asia. Tea grows in moist, evergreen, broad-leaved forests (as do most species of *Camellia*). In the wild, the showy flowers of *Camellia* are usually pollinated by insects (mainly bees and wasps), but *C. japonica* is pollinated by the Japanese white eye bird (*Zosterops japonica*). Their capsules release seeds with a high oil content, which are eaten, cached, and dispersed by various rodents.

Economic Uses: Tea (made from the leaves of *C. sinensis*) is one of the world's most important caffeine-containing beverages. It originated in southwestern China but was of widespread use in China by the eighth century and taken to Europe by the Portuguese in the sixteenth century. Tea drinking became popular in England in the seventeenth century. In addition to caffeine, tea contains theobromine and theophylline (useful in treating respiratory diseases and asthma), as well as many beneficial antioxidants (polyphenols) and essential oils. *Camellia sinensis* is occasionally used as an ornamental shrub, but it is not as important in this regard as the related species *C. japonica* and *C. sasanqua*. *Camellia oleifera* is an important source of cooking oil (extracted from its seeds).

Description (of *C. sinensis*): Evergreen shrubs or small trees, with simple hairs, the branches erect. Leaves alternate and spiral, simple, oblong to elliptic or obovate, pinnately veined, with the secondary veins forming loops, the apex acute to acuminate, the base acute or cuneate, the margin serrate, and each tooth with a deciduous glandular apex, with short petioles; stipules lacking. Inflorescences axillary, the flowers solitary or in two- or three-flowered clusters. Flowers radially symmetrical, bisexual, the sepals usually five, overlapping, the petals 6–8, the outermost 1–3 slightly smaller than the others, white, slightly connate basally, but appearing ± distinct, overlapping, obovate and cupped, rounded apically, the stamens numerous, distinct (or the outermost very slightly fused), with long filaments and globose yellow anthers, and usually three fused carpels; the style single, elongated, apically three-branched, and the ovary superior. Fruit a ± lobate capsule with usually a single large seed per locule.

THISTLES (*CIRSIUM* SPP.)

(In the Sunflower family [Asteraceae or Compositae].)

> *[H]e made her soft pavilions*
> *of lilies, and a bridal bed*
> *of flowers and of thistle-down*
> *to nestle down and rest her in.* (*TATB 3, "Errantry": lines 34–37*)

Thistles are common herbs of open, sunny habitats and are immediately recognizable by their spiny often lobed leaves and showy floral heads, each of which contains many small, narrowly trumpet-shaped flowers surrounded by numerous overlapping, spiny bracts. Several species are quite weedy, and they are especially characteristic of disturbed areas, such as fields, clearings, or roadsides. It is thus not surprising that the sole mention of thistles (discounting the Bree surname, Thistlewool) in *The Lord of the Rings* occurs when Frodo, Sam, Merry, and Pippin, while in the Old Forest, see these spiny plants in the Bonfire Glade, growing in the disturbed opening along with "rough grass and many tall plants: stalky and faded hemlocks [*Conium maculatum*] and wood-parsley [*Anthricus sylvestris*], fire weed [*Chamerion angustifolium*] seeding into fluffy ashes, and rampant nettles [*Urtica dioica*] and thistles." (LotR 1: VI). The species of thistle that Tolkien probably had in mind are spear thistle (*Cirsium vulgare*) and field thistle (*C. arvense*), and both are abundant in waste places in England (and Europe), where they are native. Both species have been introduced into North America and Eastern Asia where they are considered noxious weeds. Another visually striking feature of these plants, easily seen in the late summer and autumn, is their distinctive fruits. Each fruit is associated with an elongated and plumose fringe of persistent sepals, which are fluffy, catch the wind, and carry away the small dry fruit (containing only a single seed). The plants are thus efficiently dispersed across open habitats by wind. Tolkien picked up on this noteworthy characteristic of thistles and many of their relatives in the sunflower family in the poem "Errantry" (see initial quote) in which the wandering mariner, in an attempt to woo a pretty butterfly, made her a bridal bed of flowers and thistledown. This poem by Bilbo Baggins (see the introduction to *The Adventures of Tom Bombadil*) indicates that hobbits were quite familiar with thistles, which were undoubtedly found throughout the shire, where roadsides, clearings, and agricultural lands were common—all typical thistle habitats. Interestingly, Bilbo chanted another version of this poem in the Hall of Fire in Rivendell (Figure 7.71).

Etymology: The name "thistle" is derived from Old English *thistel*, and Proto-Germanic **thistilaz*, which possibly comes from the Proto-Indo-European **steig-* (to prick, stick, or pierce), a reference to the spiny leaves of these plants. The scientific Latin name *Cirsium* is derived from the Greek *kirsion* (a thistle), which is probably derived from *kirsos* (swollen vein), based on the plant's medicinal use in antiquity in the treatment of swollen veins.

Distribution and Ecology: *Cirsium*, a genus of about 200 species, is widely distributed in the Northern Hemisphere, with 62 species reported from North America (north of Mexico), ca. 60 from Europe, 46 in China, and 55 in Japan. The plants range from sea level to alpine regions, growing from tropical to arctic regions in a wide variety of usually open habitats such as swamps, meadows, prairies, forest openings and edges, dunes, and deserts. Hybridization is common in thistles, and many species are quite variable comprising several intergrading, geographical, or ecological races, thus making their identification often quite difficult. Their showy flower heads surrounded by spiny bracts are pollinated by various insects. Interestingly, the anthers are fused into a cylinder and open toward the hairy style, which grows through the middle of the fused anthers, picks up the pollen, and presents it to the visiting insects. The small, dry fruits are associated with numerous, persistent, highly modified sepals (called a pappus), each of which is linear and plumose, and the plumose pappus forms a parachute-like structure, leading to dispersal of the fruits by wind.

FIGURE 7.71 *Thistles.*

Economic Uses: The stems and roots of thistles (i.e., species of *Cirsium*) are occasionally eaten, although the roots (which contain the indigestible oligosaccharide inulin) cause flatulence. Thistles are of minor use as cultivated ornamentals, although many are strikingly attractive plants that occur in a wide array of open habitats. Although many species are not weedy, a few are widely distributed, noxious weeds, such as bull or spear thistle (*C. vulgare*), Canada or field thistle (*C. arvense*), and European swamp thistle (*C. palustre*). Thistles provide food for wildlife (e.g., goldfinch, linnet, and other finches). They are of minor medicinal use. Finally, the "fluff" or thistledown can be used as tinder.

Description: Annual, biennial, or perennial herbs, with or without simple hairs; when present, these sparse to dense; stems erect, sometimes narrowly spiny-winged. Leaves alternate, spirally arranged, both in a basal cluster and also distributed along stem; simple, but often shallowly to deeply, pinnately lobed, oblong or ovate to obovate, pinnately veined, the apex acute and spine-tipped, the base narrowed, the margin bristly to spinose-toothed, glabrous to densely pubescent, the upper leaves usually progressively smaller; stipules absent. Inflorescences terminal, the flowers densely clustered in heads, each head with 25 to more than 200 flowers that are surrounded by several series of overlapping, spreading to erect, usually spine-tipped bracts. Flowers radially to slightly bilaterally symmetrical, usually bisexual (but unisexual, and with staminate and carpellate heads on different individual plants in *C. arvense*), with the sepals modified, forming numerous plumose bristles or plumose narrow-scales (i.e., the pappus); five fused petals, white to pink, red, purple, or yellow, forming a long, slender, straight to distally bent corolla tube that abruptly expands at its apex and forms five narrow and radiating lobes; five stamens, these arising from the corolla tube and, with their five anthers, forming a cylinder around the style; with two fused carpels, the style elongate, single; the ovary inferior. Fruit an ovoid-compressed achene with persistent plumed pappus at its apex (Figure 7.71).

THRIFT (*ARMERIA MARITIMA*)

(Leadwort family [Plumbaginaceae].)

> *Then Tuor went up the wide stairs, now half-hidden in thrift and campion, and he passed under the mighty lintel and entered the shadows of the house of Turgon. (UT 1: I)*

Thrift (or sea-thrift, sea-pink) is mentioned only once in J. R. R. Tolkien's writings, occurring along with sea campion (*Silene uniflora*) in the preceding quote. As discussed under the campion entry, Tuor had been forced from his homeland, Dor-lómin, and eventually reached the Great Sea and ruins of Vinyamar, in which he found the shield, mail-shirt, and sword that years earlier had been left for him by Turgon (in obedience to the command of Ulmo). It is certainly ecologically appropriate that the weathered stairs leading up to Turgon's palace, which had long been abandoned, are covered by thrift because this species is characteristic of rocky habitats and cliffs along the coasts of northern Europe, where it often grows with sea campion. But these two maritime species also provide a subtle signal that Tuor has approached the domain of Ulmo—and bearing the shield and sword, he is ready to meet the Lord of Waters, who will give Tuor a task that will impact the entire history of Middle-earth.

Etymology: The word "thrift" comes from Middle English *thriven* (to thrive), from Old Norse *thrift*, from *thrifask* (to thrive), of unknown origin. The scientific name *Armeria* is based on the Celtic *ar mor* (at seaside), an allusion to its coastal habitat.

Distribution and Ecology: Thrift (*Armeria maritima*) is widely distributed in Eurasia and North America, typically occurring in coastal habitats such as maritime rocks and cliffs, salt marshes, and sand dunes, but it also grows in gravelly tundra and river floodplains or alpine meadows. The showy white to pink or red flowers are pollinated by various insects; the small dry fruits are associated with a persistent papery calyx and dispersed by wind. The species (one of perhaps 50 in the genus) is morphologically quite variable and is divided into several geographically and ecologically differentiated varieties.

Economic Uses: Several cultivars of thrift are grown as ornamental herbs. The species is rarely used medicinally, but was at one time used in treating obesity.

Description: Perennial herbs; the hairs simple (when present); stems erect. Leaves alternate and spirally arranged, in basal rosettes, simple, linear, with only midvein evident, the apex acute, the base broad, the margin entire; petiole and stipules lacking. Inflorescences terminal, dense clusters of flowers composed of several scorpioid cymes associated with papery bracts and atop a long scape. Flowers bisexual, radially symmetrical, with five fused sepals, the calyx 10-ribbed and funnel-shaped, with five white to pink or red petals, these slightly fused at the base but with well-developed lobes, with five stamens, each fused to the corolla; the carpels five, these fused, but with five separate and elongate styles; the ovary superior. Fruit a small capsule, opening transversely and single-seeded, and enclosed by the persistent, papery calyx.

TURNIPS (*BRASSICA RAPA*, WITH AN EMPHASIS ON TURNIPS AND THEIR WILD RELATIVES)

(In the Mustard family [Brassicaceae].)

They passed along the edge of a huge turnip-field, and came to a stout gate. (LotR 1: IV)

The connection of turnips with the traditional farming culture of the Shire, as seen when Frodo, Sam, and Pippin encounter a large turnip field owned by Farmer Maggot, is not surprising. And, only a little earlier, Farmer Maggot and his dog had encountered a Black Rider in this very field (Figure 7.72). Turnips are one of the oldest European root crops and were domesticated (from their wild European ancestors) well before the time of the ancient Greeks and Romans. They provide food for humans as well as fodder for livestock and are still a popular cold weather root crop in the British Isles and elsewhere in Europe. Farmer Maggot exemplified the best of the agricultural ethos of the Shire, which represents (at least in part) an idealized version of the rural England of J. R. R. Tolkien's childhood. The Shire, as Tolkien noted, "is in fact more or less a Warwickshire village of about the period of the Diamond Jubilee" (Letters: No. 178). As Tom Bombadil later told Frodo, speaking of Farmer Maggot, "There's earth under his old feet, and clay on his fingers; wisdom in his bones, and both his eyes are open" (LotR 1: VII). The link between Maggot and this traditional yet valuable root crop is solid—this Shire farmer is intimately connected to the soil, both in practical terms through his cultivation of turnips (a root crop requiring him to get "clay on his fingers"

FIGURE 7.72 *Turnips.*

in its harvesting) and, more generally, through his good stewardship of the land and knowledge of the local environment (including the things that threaten it).

Etymology: The word "turnip" is from the English word *turnepe*, used ca. 1500 AD and probably is derived from *turn* (i.e., turned or rounded, based on its globose shape, as though turned on a lathe) + Middle English *nepe* (turnip), from Old English *næp*, from Latin *napus* (turnip). The scientific name *Brassica* is from the ancient Latin name for cabbage.

Distribution and Ecology: *Brassica rapa* is widely distributed in Europe and Asia and also occurs in northern Africa; the species is widely naturalized in the Americas and Australia. It occurs in a variety of open and disturbed habitats (roadsides, waste places, cultivated fields, orchards, gardens, etc.) The species is a member of the large genus *Brassica*, which contains some 35 species and occurs natively in Europe, Asia, and northern Africa, but is also widely naturalized. The genus contains several important crop plants (e.g., cabbage, collard greens, broccoli, cauliflower, Brussels sprouts, kohlrabi, curly kale, brown mustard, mustard greens, black mustard, and rapeseed or canola) in addition to the cultivars of *B. rapa* discussed here. These are easily grown from seed, and several were very early domesticated. Their usually yellow flowers are pollinated by various generalized insects, and the small seeds are dispersed by wind (or secondarily by rain wash). Of course, the seeds of cultivated entities such as turnips are transported by humans.

Economic Uses: Turnips (*B. rapa* var. *rapa*) are one of the oldest root crops. They were a well-established crop before the time of the ancient Greeks and Romans, and they are still a popular root vegetable. The spherical red and white roots can be eaten raw or cooked and are especially popular in soups and stews; the leaves (turnip greens) are also eaten. They are also used as livestock feed. Wild *B. rapa* is perhaps of central European origin (and recognized as var. *sylvestris*) but is now a very widespread weed. The species is morphologically variable and has been domesticated more than once; in addition to the turnip (*B. rapa* var. *rapa*), important cultivars include the pe tsai, celery cabbage, heading Chinese cabbage or Peking cabbage (*B. rapa* var. *pekinensis*), pak choi, bok choi, or Chinese white cabbage (*B. rapa* var. *chinensis*), and pak choi sum or Chinese flowering cabbage (*B. rapa* var. *parachinensis*). These greens are usually boiled, stir-fried, or pickled.

Description (of *B. rapa*): Annual or biennial herbs, the tissues with the pungent odor and taste of mustard oils; stems erect, unbranched to distally branched, with hairs, absent or sparse and simple; roots slender or with a fleshy taproot, and then usually purple near the leaves and white below; the flesh varies from white, to yellow, orange, or red. Leaves alternate and spirally arranged, those occurring basally, pinnately compound to deeply pinnately lobed, oblong or obovate, with pinnate venation, the apex rounded, the base decurrent, the margin dentate to serrate with a well-developed petiole, while those borne distally on the elongating stem, smaller, unlobed, oblong to ovate, the base more or less cordate and clasping stem, the margin entire or sinuate to sparsely dentate or serrate and lacking a petiole; stipules absent. Inflorescences terminal, in racemes. Flowers bisexual, radially symmetrical, held erectly, with the four sepals distinct; four petals, yellow, clawed, rounded apically; six stamens, with four of these longer than the other two; the carpels two, fused; the ovary superior, elongated, with a very short style. Fruit an elongated silique (i.e., with the two valves falling away from a persistent, rimmed partition and releasing the seeds); seeds several, small, black to brown or reddish (Figure 7.72).

UNNAMED HOLLYLIKE TREE (*BANKSIA* SP., PERHAPS RELATED TO *B. INTEGRIFOLIA*)

(The Macadamia family [Proteaceae].)

> *In this way they came at least to what looked like an impenetrable wall of dark evergreen trees, trees of a kind that the hobbits had never seen before: they branched out right from the roots, and were densely clad in dark glossy leaves like thornless holly, and they bore many stiff upright flower-spikes with large shining olive-coloured buds. (LotR 3: IV)*

These hollylike trees enclosed the bowl-like depression that was the traditional place where the ents gathered for discussion of important matters (i.e., held their Entmoots) and were seen by Merry and Pippin as they were carried by Treebeard. They had dark, glossy, evergreen and somewhat leathery leaves; thus, they were somewhat hollylike, but lacked the spinose-toothed margins of European holly (*Ilex aquifolium*). And, interestingly, they also had stiff, upright flower spikes. Striking, uprightly held inflorescences are quite unusual among trees, but immediately call to mind many species in the Southern Hemisphere family Proteaceae, a group that also contains many plants with hollylike leaves. The plants described in this passage may have been a species of *Banksia*, especially one of the arborescent species such as *B. integrifolia*, which grows today in coastal forests of eastern Australia. This is an extremely unusual distribution for a tree of Middle-earth—nearly all the trees of J. R. R. Tolkien's legendarium have north-temperate distributions (occurring in Eurasia and North America), whereas *Banksia*, in contrast, is restricted to the southern Hemisphere: Australia and New Zealand, with relatives in southern Africa. Perhaps the ents in their long travels encountered this species far in the South, appreciated its beauty, and brought it to Fangorn Forest, just as species of *Banksia* are appreciated by many today and are used horticulturally in ecologically suitable landscapes.

Etymology: *Banksia* was named by C. Linnaeus in honor of Sir Joseph Banks who collected the plant during Captain Cook's first expedition.

Distribution and Ecology: The 77 species of *Banksia* are restricted to Australia (and New Zealand) and mainly occur in coastal regions of that continent. They occur in various kinds of arid to humid forests and scrublands (but are absent from deserts). The species of *Banksia* are quite variable in the form of their leaves and color of their flowers. Their flowers produce nectar and are pollinated by birds or mammals, while the follicular fruits open to release winged seeds that are dispersed by wind. Many species are quite adapted to fires, being protected by thick bark and sprouting from the underground woody tubers, whereas others are killed but have woody fruits that open and shed seeds, in response to fire.

Economic Uses: Species of *Banksia* are popular as ornamental trees or shrubs; some are also used in the cut flower industry.

Description: Trees with erect branches, with alternate and spirally arranged simple, leathery, pinnately veined leaves, with base narrowed and apex acute to rounded, the margin entire to clearly serrate or dentate, sometimes the teeth so large as to be considered leaf-lobes, the lower surface with dense hairs, and stipules absent; inflorescences terminal, erect, densely flowered, spikelike clusters, with the flowers in pairs. Flowers bisexual, bilaterally symmetrical, with four fused colorful, greenish yellow to yellow, orange, or red tepals, with four stamens fused to the tepals and a single carpel, with a superior ovary and elongate style. Fruits woody, horizontally opening follicles, clustered in dense, conelike clusters opening to release winged seeds.

WATERLILIES (*NYMPHAEA ALBA*, AND RELATED SPECIES)

(The Waterlily family [Nymphaeaceae].)

> *About her [Goldberry's] feet in wide vessels of green and brown earthenware, white water-lilies were floating, so that she seemed to be enthroned in the midst of a pool. (LotR 1: VII)*

White waterlilies (certainly the European *Nymphaea alba,* although the North American *N. odorata* is quite similar) figure prominently in the encounter of Frodo and his companions with Tom Bombadil along the River Withywindle. Soon after Merry and Pippin are trapped by Old Man Willow, Frodo and Sam meet Tom, singing to himself, walking back along the riverbank; in his hands "he carried on a large leaf as on a tray a small pile of white water-lilies" (LotR 1: VI), which he was bringing to his wife Goldberry. In short order, he rescued Merry and Pippin, and the hobbits were told to follow him home as quickly as they could. Upon entering the house of Tom Bombadil, they saw Goldberry—her yellow hair rippled down her shoulders, and her gown was "green as young leaves, shot with silver like beads of dew"—she seemed to be "enthroned in the midst of a pool" (LotR 1: VII) (Figure 7.26). She is the daughter of the Withywindle, as she told the hobbits, as is also related in the poem "The Adventures of Tom Bombadil." In that poem, Old Tom is sitting on the riverbank in the summer, and "up came Goldberry, the River-woman's daughter; pulled Tom's hanging hair. In he went a-wallowing under the water-lilies, bubbling and a-swallowing" (lines 12–14) (Figure 7.73). She took his hat, but Tom replied "You bring it back again, there's a pretty maiden! . . . I do not care for wading. Go down! Sleep again where the pools are shady far below willow-roots, little water-lady!" (lines 19–22). Clearly, she is a minor nature deity, a water nymph, and what flower could be more suitable to her than the waterlily, a beautiful aquatic plant, commonly of pools and streams, whose Latin name, *Nymphaea,* is derived from the Greek, *nymphe,* a water nymph or goddess of springs, streams, and watery meadows (but which, in its original meaning had referred to a bride or young wife).

Certainly, waterlilies were as widely distributed in the wetlands of Middle-earth as they are in our world today. Sam and Frodo also encountered them in a small clear lake in Ithilien, along with iris swords (*Iris pseudacorus*), where they stopped and quenched their thirst. They also occurred along the Sirion and Narog Rivers in the willow forest of Nan-Tathren, in Beleriand, during the First Age.

Etymology: "Waterlily" is a compound word, in which the English word "water" is derived from Old English, *waeter,* from Proto-Germanic **watar,* going back to Proto-Indo-European **wod-or* (water), from the root **wod* or perhaps **wed* (meaning something wet). The English "lily" is derived from Old English *lilie,* which in turn comes from the Latin *lilia,* plural of *lilium* (meaning a lily), and is a cognate with the Greek

FIGURE 7.73 *Waterlilies.*

leirion. These Latin and Greek names perhaps were borrowed from an Egyptian word. The scientific Latin name *Nymphaea* is derived from the Greek *nymphaia* (waterlily), from *nymphe* (goddess of springs, a water nymph).

Distribution and Ecology: Waterlilies belong to the genus *Nymphaea*, are distributed nearly worldwide, and include about 40 species. They occur in aquatic and wetland habits of cold temperate to tropical regions. Nine species occur in North America (north of Mexico) and, of these, the white waterlily (*N. odorata*) is the most widely distributed. Four species grow in Europe, where the European white waterlily (*N. alba*) is the most common and the only species native to the British Isles. Five species occur in China, 1 in Japan, and 17 in Australia. The showy flowers are pollinated by insects (especially beetles, flies, and bees), while the seeds, along with their mucilaginous arils, float (at least initially) and are dispersed by water currents or birds.

Economic Uses: Waterlilies are ecologically important in wetland habitats, providing food for wildlife, and their floating leaves and beautiful flowers add to the aesthetic quality of such habitats. Waterlilies are popular ornamentals (used in pond gardens) and many cultivars and hybrids have been developed.

Description (of *Nymphaea* spp.): Rhizomatous, perennial aquatic herb, the rhizomes usually in the mud under the water. The hairs simple, usually producing mucilage (slime). Leaves alternate and spirally arranged, simple, round to widely ovate or elliptic, with usually more or less palmate veins, the apex rounded, the base deeply cut, with the lobes overlapping to divergent, the margin entire (to toothed), glabrous and waxy above, the blade with the petiole elongate; the blade thus usually floating on the surface of the water. Flowers solitary, with long stalks and floating or raised above the surface of the water, radially symmetrical, bisexual, erect, large, often fragrant, white (in *N. alba, N. odorata*, and many others), but also pink, blue, or yellow, with four sepals, mostly greenish, eight to numerous petals (actually petaloid staminodes, i.e., sterile stamens), these colorful and showy, ovate to obovate, grading into stamens. The stamens numerous, cream-colored to yellow, attached along the lateral surface of the ovary. Carpels five to numerous, fused, the stigmatic disk with prominent, distinct, upwardly incurved appendages around the margin; ovary superior in relation to the perianth parts but inferior in relation to the stamens. Fruit a berrylike, irregularly opening capsule, on curved or coiled stalks and thus maturing under water; the seeds with mucilaginous arils (Figure 7.73).

WHEAT (*TRITICUM AESTIVUM*, AND CLOSE RELATIVES)

(The Grass family [Poaceae, Gramineae].)

> *Below them lay the woods of Oromë, and westward shimmered the fields and pastures of Yavanna, gold beneath the tall wheat of the gods. (SILM 8)*

Nearly all civilizations have been (and still are) based on the cultivation of one or more species of grain. Thus, soon after the end of the last glacial period, we find evidence of the domestication of several grass species: wheat and barley in the Middle East (shortly before 10,000 years ago), rice in China (around 10,000 years ago), maize in Central America (about 7,500 years ago), and oats in the eastern Mediterranean (more than 3,000 years ago). The cultures of J. R. R. Tolkien's legendarium are no different and even the angelic beings and elves living in Valinor depended upon "the tall wheat of the gods" (SILM 8). In the quote we see the forests, pastures, and wheat fields of Valinor as viewed by Melkor and Ungoliant from the summit of Hyarmentir, a tall southern peak of the Pelóri mountain range. Agricultural knowledge, such as how to grow wheat and process its grain for food, therefore, was likely provided to the elves by the Valar, especially Yavanna. In fact, as related in the document "Of Lembas" (see *History of Middle-earth*, Vol. 12, p. 403–405), this special wheat of the Valar was given to the elves and supported them on their long journey to Valinor. The elves grew it in Middle-earth and found that "no worm or gnawing beast would touch" it. In contrast, humans, when they arose, encountered no Vala—they received no request to journey to Valinor. Instead, they dwelt in Middle-earth and may have been taught about wheat and its uses by the entwives. Treebeard told Merry and Pippin that the "land of the Entwives blossomed richly, and their fields were full of corn" (LotR 3: IV) and that men had learned their crafts. In the song of the ents and entwives, the entwives sing of spring coming to "garth and field" and corn "in the blade," followed by the time of harvest, "when straw is gold, and ear is white" (LotR 3: IV). All these are clear references to the cultivation of wheat (*Triticum aestivum* and related species) and not to corn (i.e., Indian corn or maize, *Zea mays* subsp. *mays*) as often assumed by American readers (see "Etymology"). Of course, the cultivation of wheat is important in an agrarian society such as the Shire, and, in the poem "Bombadil Goes Boating" we find Tom and Farmer Maggot talking of "wheat-ear and barley-corn, of sowing and of reaping" (TATB 2). Hobbits may have acquired such agricultural knowledge very early from the surrounding human cultures. Thus, in Middle-earth, we see wheat flour used in making foods as common as bread, cakes and biscuits—as when Merry and Pippin, at the ruined gates of Isengard, offered Aragorn, Gimli, and Legolas butter and honey for their bread—or as sacred as lembas, wrapped in mallorn-leaves, which were provided by Galadriel and sustained Frodo and Sam as they traveled across the tortured plain of Gorgoroth (Figure 7.74).

Etymology: "Wheat" is derived from Old English *hwæte*, from Proto-Germanic **hwaitjaz* (wheat, literally white, referring to the color of the grain once it is ground), from Proto-Indo-European **kwoid-yo-*, from the root **kweid-* or **kweit-* (to shine). The word "corn" (meaning grain and usually used of the dominant grain grown in a particular region) is derived from Old English *corn*, from Proto-Germanic **kurnam* (small reed), from the Proto-Indo-European root **gre-no-* (grain), from which also is derived the Latin *granum*. The restricted application of the word, as referring to maize (*Zea mays* subsp. *mays*), occurred in North America, first as "Indian corn" (around 1600 AD) but later shortened by dropping the adjective "Indian." The word "corn" usually refers to wheat in England, but is used for oats (*Avena sativa*) in Scotland and Ireland. The scientific name *Triticum*, the classical Latin name for wheat, comes from the Latin word *tritus* (well-trodden).

Distribution and Ecology: Wheat originated in the Middle East, where its wild relatives still occur, but bread wheat (*Triticum aestivum*) is only known as a cultivated crop. It is now very widely grown in cool to warm temperate regions. The flowers are tiny and wind-pollinated. The fruits are harvested and dispersed by human action; their wild relatives had fruiting clusters that easily broke apart, causing the seeds (and

FIGURE 7.74 *Wheat.*

associated bracts) to be shed and transported externally by animals (on their fur or feathers), by wind action, or rain-wash on the soil surface.

Economic Uses: Wheat is one of the most important crop plants in the world; among the grains, it is second only to rice in annual production for human consumption, providing 20 percent of all food calories. It is ground into flour, which is used to make bread (and related products, such as biscuits, cookies, pastries, etc.), with the sticky gluten allowing the dough to rise (because the carbon dioxide produced by yeast during fermentation is trapped). Wheat straw has been used as a thatch and is important animal bedding. Finally, wheat (usually mixed with other grains, especially barley) is used in making beer and other alcoholic beverages. Wheat was domesticated more than 10,000 years ago in the Middle East, in the Fertile Crescent region, probably first in a region that is now part of southeastern Turkey. Its biological history is complex, involving hybridization and polyploidy (genome doubling). Bread wheat (*T. aestivum*) is a hexaploid, having a genome compiled from three different ancestral diploid species. In the distant past, *Aegilops speltoides* (a wild goat grass) hybridized with *T. urartu* (wild Einkorn wheat) to produce a tetraploid wheat, *T. turgidum* subsp. *dicoccoides* (wild Emmer wheat), from which Emmer wheat (*T. turgidum* subsp. *dicoccon*), also tetraploid, was selected. Emmer wheat then crossed with *Aegilops tauschii* to produce hexaploid wheat (i.e., the various cultivars of *T. aestivum* or bread wheat). *Triticum urartu* was also domesticated, and that plant is known as *T. monococcum* (Einkorn wheat). There are numerous different groups of cultivars of *T. aestivum*; there are also several cultivar groups among the tetraploid wheat plants, the most important of which is *T. turgidum* subsp. *durum* (macaroni wheat).

Description (of *T. aestivum* and close relatives): Annual herbs, without rhizomes; stems erect, jointed, round in cross-section, hollow in internodal regions. Leaves alternate, two-ranked, simple and differentiated into a spreading blade, a sheath that closely encircles the stem, with its margins not fused and with a membranaceous flap of tissue (the ligule) at the junction of the blade and sheath; the blade linear, flat, with or without hairs. Inflorescence terminal, spikelike, with one spikelet per node, not breaking apart in cultivated plants, but falling apart in the wild relatives. Flowers very tiny, arranged in spikelets, with 2–9 flowers, with spikelets usually arranged in two rows along axis and each flower surrounded by two bracts (a lemma and palea) and some of these (the lemmas) often with elongated awns, with three sagittate anthers and two fused carpels; the ovary superior; the two stigmas plumose. Fruit a large plump grain, only loosely enclosed by the bracts; endosperm hard or mealy (Figure 7.74).

WHITE FLOWERS OF THE MORGUL VALE (BASED ON *ARUM* SPP.)

(The Arum or Aroid family [Araceae].)

> *Wide flats lay on either bank, shadowy meads filled with pale white flowers. Luminous these were too, beautiful and yet horrible of shape, like the demented forms in an uneasy dream; and they gave forth a faint sickening charnel-smell; an odour of rottenness filled the air. (LotR 4: VIII)*

In the context of J. R. R. Tolkien's legendarium these luminous white flowers are restricted to the Morgul Vale, growing only in the meadows along the polluted Morgulduin River (Figure 7.75). Like Minas Morgul itself, they shine with a pale light, and their form is said to be beautiful and yet horrible. It is very likely that the inspiration for this plant is some member of the Araceae, probably a species of the genus *Arum*, which is a largely Mediterranean genus that is quite diverse in Europe (i.e., having 10 species native there). Some of these species have white "flowers"—actually inflorescences—that are of an unusual form, with a showy, more or less ovate, leaflike bract, the *spathe*, that attracts the pollinators to the tiny flowers, which are densely packed on an associated, thickened axis, the *spadix*. But members of the genus *Arum* have more similarities to the description in this quote (from "The Stairs of Cirith Ungol") than just their frequently white spathes. The inner surface of these spathes is covered with downward pointed and oil-covered papillae (minute projections), which form a slippery surface, causing visiting flies to fall into a trap formed by the lower part of the spathe. The minute papillae reflect light and the spathe, therefore, seems to glow—the luminous quality referenced by Tolkien. The inflorescences of *Arum* (and many other genera of Araceae) can be strangely shaped—a strangeness that in this passage is heightened—and described as demented and horrible. *Arum* species produce the unpleasant odors of rotting flesh because their inflorescences are pollinated by small flies that are also attracted to carrion, so this aspect of their biology also closely matches Tolkien's description. Finally, some species of *Arum* grow along rivers and streams, as pictured in the Morgul Vale. These similarities, when taken together, strongly suggest that Tolkien based his conception of these demented white flowers on the strange inflorescences of members of the diverse European genus *Arum*.

It is likely that the white arum flowers of the Morgul Vale were once solely beautiful, just as Minas Morgul had once been fair and radiant. But now Minas Ithil (the Tower of the Moon) had become Minas Morgul, and, as seen by Sam and Frodo, it gave off a pale and ailing light—"wavering and blowing like a noisome exhalation of decay, a corpse-light, a light that illuminated nothing" (LotR 4: VIII). The tower, under the power of the Lord of the Nazgûl, had become corrupt and horrible, a place of death and sorcery. Arum flowers, like most of their relatives in the Araceae, are poisonous, and they are often associated with death, as are the species of the similar genera *Zantedeschia* (arum lilies) and *Amorphophallus* (voodoo-lilies, corpse-flowers, and giant arums)—yet another connection with Minas Morgul. These flowers may have a history similar to that of the orcs, which (if elven tradition can be trusted) may have been descended from elves who long before had been abducted by Morgoth, then twisted and ruined. As Frodo told Sam, "the Shadow that bred them can only mock, it cannot make: not real things of its own" (LotR 6: I). Thus, we see in these arum-flowers yet another example of the power of evil (here represented by Sauron and the Nazgûl) to corrupt nature. Creation is good, reflecting the goodness of the Creator—but sadly, it can be marred and is in need of redemption (see Y. Imbert, "Eru Will Enter Ea: The Creational-Eschatological Hope of J. R. R. Tolkien," in *Representations of Nature in Middle-earth*, 2015, M. Simonson, ed.).

FIGURE 7.75 *White flowers of the Morgul Vale.*

Etymology: No common name is provided for the white flowers of the Morgul Vale since they had not previously been encountered by Frodo and Sam or any other hobbits. The scientific name *Arum* is the Latin form of the classical Greek name for these plants: *aron.*

Distribution and Ecology: *Arum* is a genus of 28 species distributed from Europe and northern Africa eastward across the Middle East to the Himalayan Mountains. These herbs grow in temperate and warm temperate woodlands, in hedges, rocky open areas, along rivers, and in open disturbed areas. The unusual inflorescence has a showy bract, the spathe, which partially surrounds the spadix, a thick axis with an exposed apical clublike structure and a lower zone, hidden by the surrounding spathe, which bears tentaclelike sterile flowers, staminate flowers, and carpellate flowers. The inflorescence emits a scent of carrion and attract small flies, which are temporarily trapped in the lower, chamberlike portion of the spathe. The flies become covered by pollen while they are trapped, and, after leaving (usually the next day), they often are trapped by another inflorescence. In the process, they bring this pollen to the stigmas of the carpellate flowers of that inflorescence.

Economic Uses: Some species are cultivated as ornamentals because of their unusual inflorescences, in which the tiny flowers are borne on a thickened axis (the spadix) that is surrounded by a white or pale yellow, sometimes purple-tinged to red-purple or purple-black leaflike bract (the spathe). The tubers of a few species are used as a starch source, and some are used medicinally (bone setting).

Description (of plant of Morgul Vale, based on *Arum* spp.): Perennial herbs, each arising from an erect, more or less globose tuber. Leaves alternate and spiral, three or four per plant and basal, the blade simple, ovate, with pinnate venation, and the higher-order veins forming a network, the apex acute to acuminate, the base sagittate, the margin entire, and long petiolate; stipules absent. Inflorescence fully exposed above the leaves comprising a leaflike, occasionally irregularly formed and sometimes twisted, sometimes asymmetrical, white spathe, the lower portion of which surrounds and hides the flowers, and the spadix, a thick axis with an elongate, white to pale yellow upper portion (producing a foul odor of rotting flesh) and a shorter lower portion (surrounded by the spathe) that bears two zones of tentaclelike sterile flowers (one above the staminate flowers and another above the carpellate flowers); a zone of inconspicuous staminate flowers, each with three or four stamens; and finally, a lowermost zone of carpellate flowers, each with a globose ovary. Fruit a dense cluster of translucent-white berries (Figure 7.75).

WHITE TREE OF GONDOR

(A species unique to J. R. R. Tolkien's legendarium.)

> *And he climbed to it, and saw that out of the very edge of the snow there sprang a sapling tree no more than three foot high. Already it had put forth young leaves long and shapely, dark above and silver beneath, and upon its slender crown it bore one small cluster of flowers whose white petals shown like the sunlit snow. (LotR 6: V)*

In this quote we see Gandalf and Aragorn in a high hollow upon Mt. Mindolluin, above Minas Tirith, where they discovered a sapling of the White Tree of Gondor

(Figure 7.76), the source of the heraldic symbol of Gondor—a White Tree crowned by seven stars. This discovery is for Aragorn the longed-for sign that the line of Elendil will be re-established, guiding the combined kingdom of Gondor and Arnor. Gandalf had been the Enemy of Sauron, but the task of opposing evil in the world now must be taken up by Aragorn. As Gandalf told him, "The burden must lie now upon you and your kindred." Yet, to remove any possible source of pride, Gandalf also reminded Aragorn that the "line of Nimloth is older far than your line" (LotR 6: V). Indeed, in the depths of time the Vala Queen Yavanna Kementári, knowing that the elves of all things in Valinor loved most the White Tree Telperion (the elder of the Two Trees of Valinor), "made for them a tree like to a lesser image of Telperion, save that it did not give light of its own being" (SILM 5; see also *Akallabêth*). This tree was named Galathilion, and it was planted in Tirion, within the courts beneath the Mindon, the Tower of Ingwë. We are told that its seedlings were many in Eldamar, and one of these, Celeborn, was planted on Tol Eressëa, the Lonely-Isle. A sapling of that tree, Nimloth, was given to Aldarion, the sixth king of Númenor, and was planted in the King's Court in Armenelos. It was long cherished as a remembrance not only of the ancient connection between the Númenóreans and the Eldar but also of the light of Valinor in the West. When Ar-Pharazôn, the last king of Númenor, under the influence of Sauron and the fear of death, turned to the worship of the Melkor (= Morgoth) and the Dark, he for a while resisted Sauron's urgings to kill the White Tree because he considered the fortunes of his house to be bound up with that of Nimloth the Fair. Thus, ironically, "he who now hated the Eldar and the Valar vainly clung to the shadow of the old allegiance of Númenor" (*Akallabêth*). It was then that Amandil, a ship-captain and leader of the elf-friends, realizing that Nimloth was in danger, spoke to his son Elendil and his grandsons Isildur and Anárion; Isildur, alone and in disguise, went into the guarded courts of the king, now forbidden to the elf-friends, and took a fruit from the Tree, escaping with grave injury. Soon after, Ar-Pharazôn ordered that the White Tree be cut down, and its wood was burned by Sauron as a sacrifice on the altar to Morgoth. The stolen seed sprouted that spring, and, upon the destruction of Númenor, the young tree was brought to Middle-earth in Isildur's ship and planted at Minas Ithil. Not long afterward, this tree was destroyed when Sauron captured Minas Ithil, but a seedling had already been planted at Minas Anor (= Minas Tirith). For more than a thousand years, this tree grew in the Court of the Fountain, but it died when King Telemnar died (in the time of the Great Plague). A seedling was planted by his nephew and successor, King Tarondor, and that tree survived until the death of Belecthor II, the twenty-first Steward, then it also died. A seedling could not be found at that time, so the dead tree was left standing in honor, until Aragorn returned to claim the kingship and with Gandalf found the already blossoming sapling described in the preceding quote. Aragorn's honoring of the White Tree and his acceptance of what it represented allowed him, unlike Ar-Pharazôn, to live in right relationship to others and to accept, at the end, his own death. His last words to his wife, as recorded in the "Tale of Aragorn and Arwen" (see LotR Appendix A (v)), were "In sorrow we must go, but not in despair. Behold! We are not bound for ever to the circles of the world, and beyond them is more than memory. Farewell!"

As discussed by Christopher Tolkien in the *Peoples of Middle-earth* (1996, vol. 12, see pages 147–149) and also by W. G. Hammond and C. Scull (2005, *The Lord of the*

FIGURE 7.76 *White Tree of Gondor.*

Rings: A Reader's Companion, p. 637), the words of Gandalf at the end of the chapter "The Steward and the King" concerning the line of Nimloth disagree with the passages from the *Akallabêth* and *Silmarillion* (cited earlier) in that Celeborn is not included in the lineage of the White Tree of Gondor and also that Galathilon is stated to be "a fruit of Telperion of many names, Eldest of Trees" (LotR 6: V). In the other references, Celeborn is included in the lineage of the White Tree and Galathilion is merely an "image" or "memorial" of Telperion and not actually from a fruit of that eldest tree. The conception in *The Silmarillion* (i.e., that Galathilion was an "image" of Telperion) is by us considered to be Tolkien's preferred view.

Etymology: *Galathilion* is Sindarin for "Radiant-Holy-Moon," likely a reference to the fairly large flowers of this tree, which had a moonlike shape due to their partially fused white petals that formed a globular, bell-shaped corolla. These flowers recalled the first rising of the moon, when the elves looked up in delight, and "Fingolfin let blow his silver trumpets and began his march into Middle-earth" (SILM 11). *Celeborn* is another Sindarin name, meaning "Tall-Silver-[Tree]," and this name highlights the fact that the White Trees had a straight, tall trunk, with branches bearing leaves that were pale silver on their lower surface. *Nimloth* is also Sindarin (and is related to the Quenya *ninquelótë*); it can be translated as "White-Blossom," a reference to the tree's white flowers, somewhat similar to those of cherry (*Prunus avium*; see HoM-E vol. 5, part 2: 6, p. 209), which were especially striking since it bloomed during the winter.

Distribution and Ecology: White Trees (descendants of Galathilion, an "image" of Telperion) grew in Eldamar and Tol Eressëa, but only a very few individuals ever grew outside Elven-home. Nimloth once grew in the King's Court in Armenelos in Númenor, and a seedling of this tree was carried by Elendil to Middle-earth and planted in Gondor at Minas Ithil; in turn, a seedling of that tree was planted in the Court of the Fountain in Minas Anor (= Minas Tirith) where the lineage continued into the Fourth Age. These White Trees were all plants of garden habitats—none grew in the wild (see illustrations by Ted Nasmith, opposite p. 315 in the illustrated *Silmarillion*, and by Alan Lee, opposite page 1008, in the illustrated *Lord of the Rings*). The fragrant, evening-blooming flowers likely were pollinated by moths. Although the trees flowered regularly, fruits were only rarely produced.

Economic Uses: Individuals in the lineage of Galathilion—the various White Trees—had no practical uses. They were, however, beautiful—but were more than mere ornamentals. Such trees represented the deep connection of the Númenóreans with the Eldar and also served as a reminder of the need to maintain right relationships with others and the divine, as represented by the Valar. This is tragically seen in the tale of *Aldarion and Erendis: The Mariner's Wife* (in the *Unfinished Tales*). Aldarion, who was to become the sixth King of Númenor, was presented by the Eldar with the sapling Nimloth as a gift at his wedding to Erendis. Its bark was snow-white and its stem straight, strong, and pliant "as it were of steel." Aldarion, who was much interested in forestry because of his passion for sea-faring and ship-building, said "I thank you . . . The wood of such a tree must be precious indeed." The elves replied that they did not know the qualities of the White Tree's wood, stating that "None has ever been hewn. It bears cool leaves in summer, and flowers in winter. It is for this that we prize it" (UT 2: II). Nimloth was indeed beautiful, and its beauty, if valued, had the power to enrich Aldarion's life. However, only much later, after his relationship with his father Tar-Meneldur and his wife Erendis were damaged beyond repair, and he in bitterness had commanded that all the trees of his gardens at Armenelos be felled except for Nimloth, did he for the first time see that this tree, now standing alone in his desolated garden "was in itself beautiful" (UT 2: II). The Númenórean kings later held that the fortunes of their house were bound with that of Nimloth the Fair, and its descendants, successively, became the chief symbols of the royalty of Gondor.

Description: Tardily deciduous tree with straight trunk, snow-white bark, and upheld, strong yet pliant branches. Leaves alternate, simple, ovate to elliptic, or narrowly so, pinnately veined, with secondary veins smoothly arching toward the margins, dark green above and silver-white beneath, the apex acute to slightly acuminate, the base acute, the margin entire; petiolate, dropping in the early winter, soon after the tree has come into bloom; stipules absent. Inflorescences axillary, few-flowered clusters (the showy flowers produced from leaf axils in early winter but continuing to bloom even after the leaves have fallen). Flowers radially symmetrical, bisexual, pendulous, and fragrant in the evening, the sepals five, triangular and fused, forming a crownlike calyx, the petals five, basally fused, white, forming a bell-like corolla with well-developed, elliptic, and overlapping lobes; the stamens 15, hidden within the corolla, the ovary superior, with five fused carpels. Fruit medium-sized, globose, nutlike, and containing only a single, long-lived seed (Figure 7.76).

WILLOWS (*SALIX* SPP.)

(In the Willow family [Salicaceae].)

> *In the midst of it there wound lazily a dark river of brown water [the Withywindle], bordered with ancient willows, arched over with willows, blocked with fallen willows, and flecked with thousands of faded willow-leaves. The air was thick with them, fluttering yellow from the branches; for there was a warm and gentle breeze blowing softly in the valley, and the reeds were rustling, and the willow-boughs were creaking. (LotR 1: VI)*

Although willows were widespread in the Middle-earth of the Third Age, they probably figure most prominently in this description of the valley of the Withywindle River in the Old Forest, where Merry and Pippin were trapped by an old willow tree (Figure 7.77). They were rescued only with the help of Tom Bombadil. Frodo was almost drowned by one of its roots (and rescued by Sam, who heard him fall into the water and saw him close to the stream's edge, "and a great tree-root seemed to be over him and holding him down," LotR 1: VI). The forest along the river is described in beautiful detail and is dominated by willows—they arched over the stream, their dead trunks and branches partially blocked the stream's flow, and thousands of willow leaves fluttered in the air and floated on the stream. It is here that the hobbits had become overpowered by an urge to sleep—perhaps explained by the warm breeze, rustling leaves, the creaking willow-boughs, and their tiring trek through the Old Forest. But it was more than that—the old willow tree was actually singing about sleep—at least that was how Sam interpreted the sounds. The willow tree, called Old Man Willow by Tom, is described as huge, old, and hoary, with a knotted, twisted, and fissured

FIGURE 7.77 *Willows.*

trunk and sprawling branches "going up like reaching arms with many long-fingered hands" (LotR 1: VI). The tree also is shown in a colored pencil drawing (titled *Old Man Willow*) made by Tolkien (and published in *J. R. R. Tolkien: Artist and Illustrator*; see fig. 147), and this image also highlights the tree's pendant branches; yellow, narrowly elliptic leaves; and writhing roots. It serves, along with the text, to establish a clear picture of this ancient willow tree and is botanically accurate (representing either the white willow, *Salix alba*, or the black willow, *S. fragilis*). Although the heart of this Entish willow was rotten (and turned to evil), Tolkien makes it clear that this is not to be taken as a characteristic of all willows, which are common in the wetland habitats that Tolkien especially loved. As Treebeard told Merry and Pippin, whether or not a tree has a bad heart has nothing to do with its wood, and he explained this by relating that he had known "some good old willows down the Entwash . . . they were quite hollow, indeed they were falling all to pieces, but as quiet and sweet-spoken as a young leaf" (LotR 3: IV). Finally, the centrality of willows in the first meeting of the hobbits and Tom Bombadil is reflected in the name Withywindle itself. In this name, "withy" is a reference to willows (see "Etymology") and "windle" is an old word for spindle (used for winding yarn), derived from Old English *windel* (basket), from *windan* (to wind or twist). So Withywindle could be translated as "twisted or wound willow branches" (as in a basket), although the name could also suggest the twisting and curving of the willow-lined river.

In the First Age, willows figure most prominently in events occurring in Nan-tathren (a Sindarin word meaning "willow-vale"), a swampy region where the Narog flowed into the Sirion River in southern Beleriand. As the elven mariner Voronwë told Tuor, as he described the plants of this region, the "fairest of all are the willows of Nan-tathren, pale green, or silver in the wind, and the rustle of their innumerable leaves is a spell of music: day and night would flicker by uncounted, while still I stood knee-deep in grass and listened" (UT 1: I). The description of these willow-lands is amazingly similar to that of the Withywindle valley, a slowly flowing river lined by beautiful willows (their characteristics matching, most likely, those of *S. alba*). Both forests have the ability to cast a spell, although the enchantment of Nan-tathren is beautiful and good while that of Old Man Willow (and the trees under his domination) has been twisted to evil. It is significant that in his pencil drawing of *Old Man Willow*, the willow forest lining the Withywindle is pictured as beautiful and tranquil—matching his description of the willow-lands of Nan-tathren—even though the description of the Old Forest in *The Lord of the Rings* is dark and frightening. Finally, Tolkien's very positive view of willows also is seen in that they grow in Valinor and add their beauty to the refreshing and peaceful gardens of Lórien (the home of the Vala Irmo, master of dreams and visions). We are told in *The Silmarillion* that Finwë went often to these gardens and there sat "beneath the silver willows beside the body of his wife" Míriel (SILM 6), but in weariness her spirit had already departed, and he alone in all the Blessed Realm lived in sorrow.

Etymology: The name "willow" comes from Old English *welig* (willow), from Proto-Germanic **wel-*, and probably Proto-Indo-European **wel-* (to turn, roll), but a more typical Germanic word for this tree is the English "withy" (willow, especially the osier willow, *S. viminalis*; or a long, flexible willow branch used in thatching, gardening, weaving, etc.), which is derived from Old English *withig* and Proto-Germanic **with-* (willow), from the Proto-Indo-European root **wei-*, meaning to bend or twist.

Both names likely refer to the very flexible character of willow shoots. The scientific generic name, *Salix*, is the classical name of these trees and may have been borrowed from another language. It is related to the English noun "sallow," which refers to several shrubby willows, and this willow name (as well as withy) is used by Tolkien in the poem "Bombadil Goes Boating," in which Tom hauls his boat "through reed and sallow-break, under leaning alder" and then goes down the river singing "Silly-sallow, flow withy-willow-stream over deep and shallow!" (TATB 2: lines 22–24). Willows are so significant to Tolkien's view of the Withywindle valley that he uses all three of these traditional English names for this tree (i.e., willow, withy, and sallow) in his descriptions of this beautiful river. The Sindarin name for willows is *tathren* or *tathar*; the Quenya name may be *tasarë* or *tasarin*.

Distribution and Ecology: Willows are a very species-rich group, with some 450 species. They are widely distributed in arctic to warm temperate regions and especially diverse in the northern hemisphere—from North America to Europe and across Asia. More than 100 species occur in North America (north of Mexico), ca. 70 occur in Europe, some 275 in China, and about 40 in Japan. Willows occur in a wide variety of usually moist to wet habitats, growing in bogs, fens, heaths, prairies, dunes, tundra, alpine habitats, various upland forests, and disturbed habitats, but they are especially characteristic of wetlands, occurring in swamps, marshes, along streams or rivers, and around lakes. Species are often very difficult to identify. Hybridization is common among willows, and the species thus form hybrids (which are more or less intermediate in their characteristics) where they come into contact. Polyploidy also is common, and many species of willows are polyploids, having arisen through hybridization between two parental species, followed by chromosome doubling. These polyploid entities, since they are morphologically intermediate, obscure the distinguishing characteristics of their parental species. Additionally, a single individual cannot provide the full range of structures needed for identification. This is because willows are dioecious (so staminate and carpellate flowers are on different plants); they often flower well before they leaf out. Proper identification, therefore, requires that the population be visited several times because ideal specimens for identification include both staminate and carpellate flowers, mature leaves, mature fruits, and even the winter twigs. Their flowers are inconspicuous (and densely clustered in catkins) but are not wind-pollinated as would be expected; each flower has a nectar gland (or glands), and the flowers attract insects (especially bees) as pollinators. The small capsules open to release tiny seeds, each with a tuft of hairs, leading to dispersal by wind. If they land in the water, they will also float and can be dispersed by currents. Many species also spread vegetatively by means of the rooting of broken branches.

Economic Uses: Willow branches (withies) are extremely pliable and are therefore used in baskets, to cane chairs, and in woven fences and lattices. Willows have a long history of medicinal use because their bark contains the glucoside salicin (converted to salicylic acid in our bodies), which is used in the manufacture of acetylsalicylic acid (aspirin), an economically important analgesic used in the treatment of pain, fever, inflammation, and in the "thinning" of the blood. They are used as honey plants because their early-blooming flowers are an important nectar source for bees. Willows are ecologically significant, providing browse for mammals; they are early successional and important in the recovery of vegetation after disturbance and are thus used both in erosion control and revegetation projects. They are rapidly growing, and some species are used for biomass production (wood for pulp, use as fuel wood or charcoal). Finally, several species of willows provide horticulturally important trees or shrubs, such as weeping willows (*S. babylonica*, or hybrids of this species with *S. alba* or the crack willow, *S. fragilis*), corkscrew willow (*S. matsudana* 'Tortuosa'), pussy willows (*S. caprea, S. discolor*, and relatives), white willow (*S. alba*), shining willow (*S. lucida*), and purple-osier willow (*S. purpurea*).

Description: Deciduous trees to low shrubs, often forming clones by root shoots, rhizomes, or stem fragmentation; the bark various, smooth to strongly vertically furrowed. Winter buds with only a single protective scale. Leaves alternate, spirally arranged, simple, ovate to obovate, sometimes narrowly so, pinnately veined, with secondary veins curving toward the margin, forming clear to obscure loops; the apex acute or acuminate to rounded; the base acute to rounded (or cordate); the margin entire, crenate, to variously serrate; and each tooth with a globose, glandular tip, with hairs various; the leaves petiolate with the early- and late-formed often quite different; stipules usually present, minute to leaflike. Flowers inconspicuous, radially symmetrical, unisexual, and with staminate and carpellate flowers on different plants; lacking a perianth, each with one or two nectar glands and in the axil of a bract, densely clustered in terminal or axillary, dangling or erect catkins; staminate flowers with 1–10 stamens, their filaments either distinct or fused; carpellate flowers with two fused carpels, with a single short style and two stigmas, the ovary nude (but actually superior, based on comparison with related genera having a perianth), sometimes stalked. Fruit a capsule opening by two valves; seeds small, with a tuft of hairs (Figure 7.77).

WOOD-PARSLEY (*ANTHRISCUS SYLVESTRIS*)

(The Carrot family [Apiaceae or Umbelliferae].)

> *No tree grew there, only rough grass and many tall plants: stalky and faded hemlocks and wood-parsley, fire-weed seeding into fluffy ashes, and rampant nettles and thistles. (LotR 1: VI)*

It is of interest that in this quote, which is the only reference to wood-parsley in J. R. R. Tolkien's legendarium, we find this plant growing along with hemlocks in the Old Forest. Presumably wood-parsley refers to *Anthriscus sylvestris*, although this species is usually called cow parsley or Queen Anne's lace. Perhaps Tolkien here coined a name more to his liking and better fitting the circumstances of his story merely by translating the specific epithet, *sylvestris*, which means "pertaining to woods," into English and thus arriving at the name "wood-parsley" (fitting its occurrence in the Old Forest) instead of cow-parsley (which would have been inappropriate, since the hobbits had left the pastures of the Shire behind). Hemlocks (*Conium maculatum*) are very similar plants, and, like wood-parsley, they belong to the umbel-bearing family Apiaceae. Tolkien, as narrator, notes their similarities, calling both "stalky and faded." Wood-parsley (*A. sylvestris*) differs from hemlock (*C. maculatum*) mainly in lacking purple spots on the stem, having hairy leaves, and in its longer and unribbed fruits. In addition, wood-parsley is only mildly resinous when crushed, whereas hemlock has a strong and unpleasant odor. Yet even though these two species are often confused, it is significant that they are clearly distinguished—and given individual mention—in the quoted passage. This fact bears on the ongoing discussion relating to the identity of the "hemlocks" in the woodland clearing in Doriath in which Beren first saw Lúthien dancing (see treatment of hemlock or hemlock-umbels) and the relationship of this mythological element of the First Age to Tolkien's actual visit to Roos in Yorkshire with his wife Edith in 1917. As has been discussed, it is likely that the tall umbellifer at Roos was actually *A. sylvestris*, yet, within the context of his legendarium, *A. sylvestris* (wood-parsley) clearly is a different plant from *C. maculatum* (hemlocks or hemlock-umbels), and only the latter is associated with the woodland clearings of Doriath. It seems likely that Tolkien did not consistently distinguish these two related species when he encountered them; however, within his legendarium, they are clearly distinguished through the application of different names—a distinction made all the more apparent by their side-by-side occurrence in the Bonfire Glade. Finally, the garden parsley (*Petroselinum crispum*) is yet another related and similar umbellifer, and it is mentioned as one of the fragrant herbs of Ithilien.

Etymology: The word "wood" is derived from Old English *wudu*, from Proto-Germanic **widu-*, from Proto-Indo-European **widhu-* (tree, wood), and the word "parsley" is the result of the merger of Old English *petersilie* and Old French *peresil*, both derived from Latin *petroselinum*, and Greek *petroselinon* (rock-parsley), from *petros* (rock, stone) + *selinon* (celery), the ancient name of the plant now called *Petroselinum crispum* (parsley). The scientific Latin name *Anthricus* is derived from the Greek *anthriskos*, an ancient name for a member of this genus.

Distribution and Ecology: *Anthricus sylvestris* (one of nine species of the genus) occurs in Europe, adjacent western Asia, and northern Africa; it is a common plant of woodlands, meadows, hedgerows, and roadsides in England. Its white flowers are pollinated by various insects, and the dry fruits are wind-dispersed. It has been introduced into North America, Iceland, and a few other areas, where it has the potential to become invasive.

Economic Uses: None, although the related species *Anthricus cereifolium* is used as a culinary herb. There is an old English superstition that bringing *A. sylvestris* (cow-parsley) into a house will lead to the death of the mother (giving rise to one of the English common names, "mother-die").

Description (of *Anthriscus sylvestris*): Biennial to short-lived perennial herb to 1.7 m tall, with secretory canals and thus a resinous odor; stems hollow, lacking purple spots. Leaves alternate, spirally arranged, twice- to thrice-pinnately compound with the numerous leaflets ovate, deeply dissected, pinnately veined, with simple hairs; the apex acute; the base attenuate; the margin lobate-serrate; and the leaf base petiolate and sheathing. Inflorescences terminal and axillary, stalked compound umbels (i.e., umbels of umbels). Flowers radially to slightly bilaterally symmetrical, bisexual, tiny; the sepals five, extremely tiny; the petals five, obovate, the tip inflexed, white; and the stamens five, with globose anthers, all borne at apex of the ovary, which is composed of two fused carpels; the styles two, recurved and basally thickened and nectar-producing; the ovary inferior. Fruit a small, dry, two-parted schizocarp, long-ovoid, each segment smooth, unridged, attached to a central stalk; the seed body with oil canals.

YEWS (*TAXUS* SPP.)

(In the Yew family [Taxaceae].)

> *Then Gwindor roused Túrin to aid him in the burial of Beleg, and he rose as one that walked in sleep; and together they laid Beleg in a shallow grave, and placed beside him Belthronding his great bow, that was made of black yew-wood. (SILM 21 and CoH 9)*

From prehistoric times until the late 1500s, yew was the wood of choice in the British Isles (and elsewhere in northern Europe) in the construction of bows. In fact, its use for longbows led to the depletion of mature individuals of the European yew (*T. baccata*) from British forests, requiring importation of yew-wood from other European forests. Most references to yew in J. R. R. Tolkien's writings relate to this important historical use of the wood. Beleg's bow (as related in *The Silmarillion*) was made of yew-wood, as was the great bow of Bard, with which he shot the black arrow killing Smaug (Hobbit XIV) (Figure 7.78). Orcs, however, especially the larger ones, could also use such weapons. Aragorn, for example, encountered "four goblin-soldiers of greater stature" that had bows of yew among those slain from the group that had ambushed Merry and Pippin near Parth Galen (LotR 3: I). There are fewer references to yews as living trees, but one of the most striking is the description of the place at the edge of Taur-nu-Fuin where Beleg was buried, where

> There black unfriendly
> was a dark thicket, a dell of thorn-trees
> with yews mingled that the years had fretted.
> The leafless limbs they lifted hopeless
> were blotched and blackened, barkless, naked
> . . . charred chill fingers changeless pointing
> to the cold twilight.
> (Lays I: lines 1676–1683)

Túrin saw this thicket in a dream as he called on Beleg, asking where his body is buried; in response, he heard a veiled voice: "Seek no longer. My bow is rotten in the barrow ruinous; my grove is burned by grim lightening" but "my life has winged to the long waiting in the halls of the Moon o'er the hills of the sea. Courage be thy comfort, comrade lonely!" (Lays I: lines 1690–1698). It is then that Túrin awoke by the lovely, marsh-lined shores of Ivrin and was healed of his grief and madness.

Etymology: The English name "yew" is derived from the Old English *iw* or *eow*, which may have been derived from the Proto-Germanic **iwa-*, and the Proto-Indo-European **ei-wo-*, and perhaps originally referred to the motley, reddish coloration of the bark. The Latin name, *Taxus*, is the classical name for this plant, which may be a loan word of Scythian origin; it is related to the Greek *toxon* (bow), from which also was derived the Latin *toxicum* (poison). Both *taxus* and *toxon* likely were derived from the same word, and their logical connection probably is the preferred use of yew-wood in bows. The Quenya name for yew may be *tamuril*.

FIGURE 7.78 *Yews.*

Distribution and Ecology: Yews (a genus of about 10 species, although specific distinctions are often difficult and some botanists recognize more than double this number) are widely distributed in the Northern Hemisphere, occurring from boreal to tropical montane habitats, and extending into the Southern Hemisphere in Sumatra and Sulawesi. They usually occur in moist to dry, deciduous or evergreen woods, swamps or bogs, in a wide range of soils, and are very shade-tolerant. The European yew (*T. baccata*) occurs in British forests and is widespread across Europe. Other major species include the American yew (*T. canadensis*), Chinese yew (*T. wallichiana*), Japanese yew (*T. cuspidata*), and Pacific yew (*T. brevifolia*). Species are distinguished by characters such as stature, leaf coloration and anatomical characters, and sexual condition. Their ovules are pollinated by wind and the seeds dispersed by birds, which are attracted to the colorful arils.

Economic Uses: Yews are widely grown as ornamentals because of their shade tolerance, evergreen habit, and ability to be severely pruned. They also are exceptionally long-lived; the European yew holds the record for the oldest tree in Europe. They also provide one of our finest coniferous woods; it is hard, durable, and elastic, and has been extensively used in furniture, for agricultural implements, and for weapons (especially bows). Yews contain highly poisonous diterpene alkaloids called taxanes in the leaves, stems, and seeds (but not the arils); these chemicals inhibit cell division and, therefore, are used in the treatment of some cancers. Because of the presence of these alkaloids, yews are extremely poisonous.

Description: Evergreen trees or shrubs, the bark scaly, often reddish, the wood without resin canals. Leaves alternate, spirally arranged but often twisted so as to appear two-ranked, simple, linear and needlelike, flattened, lacking resin canals, with a single midvein, the apex acute to mucronate, but not sharp to the touch, the base cuneate to ± acute, the margin entire, without hairs, petiolate and with decurrent bases, which form longitudinal ridges on the stems; stipules absent. Plants usually with pollen cones and seeds produced on separate individuals. Pollen cones globose, tiny, with each pollen-bearing structure umbrellalike, bearing 2–9 sporangia. Seed cones absent, and the seeds solitary, brown, surrounded by a fleshy and mucilaginous cup-shaped scarlet to orange-scarlet aril, associated with several bracts, and borne in leaf axils (along branches) (Figure 7.78).

PLANTS OF ITHILIEN

(Various plant families.)

> *Primeroles and anemones were awake in the filbert-brakes; and asphodel and many lily-flowers nodded their half-opened heads in the grass: deep green grass beside the pools, where falling streams halted in cool hollows on their journey down to Anduin. (LotR 4: IV)*

Even the most casual reader of *The Lord of the Rings* picks up on the fact that the forests and meadows of Ithilien are especially diverse and floristically quite different from the Shire. J. R. R. Tolkien mentions about 50 plants as occurring in Ithilien, and most of these are recorded in the single chapter "Of Herbs and Stewed Rabbit"!

Frodo and Sam had scarcely entered this region when Tolkien (as narrator) says that they saw "shrubs that they did not know" (LotR 4: IV), and, just a page later, after listing numerous fragrant shrubs and herbs, he states that they saw "many herbs of forms and scents beyond the garden-lore of Sam" (LotR 4: IV). The hobbits were surrounded by a forest of resinous trees, including fir (*Abies*), cedar (*Cedrus*), and cypress (*Cupressus*) "and other kinds unknown in the Shire" (LotR 4: IV). They were now far south of their home, and since they were in a region with a warmer climate, they naturally encountered many unfamiliar plants. When these plants of Ithilien are mapped, their distributions are focused on the eastern Mediterranean region: especially the southern Balkans, Greece, and Turkey. This connection of Ithilien with the eastern Mediterranean also fits given Tolkien's map of Middle-earth and the distance traveled by Sam and Frodo. The flora of Ithilien in the Third Age, therefore, should be envisioned as similar to that of Greece and Turkey today (although some of its species now occur only in more westward Mediterranean regions; e.g., the holm oak, *Quercus ilex*). Seventeen of these Ithilien plants are listed here (along with their economic uses and brief descriptions); these are mainly minor species—most are only mentioned once, and they are not as significantly connected to events in Tolkien's legendarium as those species given more detailed presentations. Several other characteristic species of this region, such as bay (*Laurus nobilis*), cedar (*Cedrus libani*), filberts (*Corylus maxima*), and olive (*Olea europaea*), are covered in more detail (see their individual treatments), while, naturally, many other plants growing in Ithilien represent widespread species, and these are also recorded from several other regions of Middle-earth. For more information, see individual treatments of ashes (*Fraxinus*), beeches (*Fagus*), bracken (*Pteridium aquilinum*), briars (*Rubus*, possibly also *Rosa*), fir (*Abies*), gorse (*Ulex europaeus*), heather or ling (*Calluna vulgaris*), irises (*Iris pseudacorus*), larches (*Larix*), lilies (*Lilium*), oaks (*Quercus*), pines (*Pinus*), thorns (*Crataegus*), and whortleberry (*Vaccinium*).

In addition to this emphasis on a Mediterranean flora, Tolkien stresses that these plants are remarkably aromatic—the air of Ithilien is "fresh and fragrant," the terebinth is described as "pungent," and "everywhere there was a wealth of sweet-smelling herbs and shrubs" (LotR 4: IV). Among these species of Ithilien, the ones best known for their fragrance are various trees: bay (*Laurus nobilis*), cedars (*Cedrus libani*), cypress (*Cupressus sempervirens*), junipers (*Juniperus* spp.), myrtles (*Myrtus communis*), and terebinths (*Pistacia terebinthus*), and the culinary herbs marjoram (*Origanum majorana*), parsley (*Pteroselinum crispum*), sages (*Salvia officinalis*, among others), and thyme (*Thymus vulgaris*). All of these have specialized canals or cavities in their tissues that contain aromatic compounds, such as terpenes, terpenoids, resins, and ethereal oils. The fragrances of Ithilien are diverse and refreshing—lifting the hearts of Frodo and Sam (Figure 7.79). In contrast, the air along the approach to Sauron's Black Gate had been dry and harsh, "filled with a bitter reek" (LotR 4: II), and there they had nearly lost hope. Before the gates of Mordor nothing lived, but, in stark contrast, Ithilien was nearly bursting with life—as illustrated by its diverse array of plants, described in loving detail. (For example, we especially appreciate the description of thyme, which with its "woody creeping stems mantled in deep tapestries the hidden stones" (LotR 4: IV) upon which it grew.) Unlike the Dagorlad, Ithilien had been under the environmentally destructive dominion of Sauron for only a few years. It was still a place of beauty—both of sight and scent.

Plant List

Asphodel (*Asphodelus* spp.): The Aloe family (Asphodelaceae). Asphodels (i.e., the 12 species of the genus *Asphodelus*) are broadly distributed in the Mediterranean region and central to western Asia; five species occur in southern Europe and a single species is naturalized in the southwestern United States and Mexico (and also naturalized in Australia and New Zealand). Asphodels also are mentioned in *The Lay of Leithian* (see Lays III: line 3235) and likely grew in Doriath during the First Age. They are occasionally grown today as ornamentals. Asphodels are herbs with basal, linear leaves with parallel venation, showy lilylike flowers with six tepals, six stamens, and three fused carpels (with superior ovary), and capsular fruits with black seeds.

Box or boxwood (*Buxus sempervirens*, and relatives): The Boxwood family (Buxaceae). The ca. 70 species of *Buxus* are broadly distributed, occurring in Europe, Asia, Africa, Madagascar, northern South America, Mexico and Central America, and the Caribbean region. *Buxus sempervirens* (the European box) grows widely in Europe and southwestern Asia and also occurs in northwestern Africa. They are used in hedges or topiaries, and the wood is prized by carvers. The boxwoods are evergreen shrubs or small trees with opposite simple leaves (pinnately veined), axillary and inconspicuous, but fragrant flowers that have an inconspicuous perianth of tepals and three-lobed capsules.

Broom (*Cytisus scoparius*): The Legume family (Fabaceae or Leguminosae). The common broom (*C. scoparius*) is widespread in western and central Europe (and has become invasive in North America, Australia, New Zealand, and India); it is one of ca. 50 species of the genus *Cytisus*. They are sometimes used as ornamental shrub and contain toxic alkaloids. Brooms are shrubs with green twigs, small alternate and compound leaves, each with only three leaflets, with axillary, bright yellow and pealike flowers and legume fruits that explosively open to disperse the seeds.

Celandine (probably *Chelidonium majus*): The Poppy family (Papaveraceae). Although this common name is also applied to *Ranunculus ficaria* (of the Ranunculaceae; see entry for buttercups), *C. majus* is native to Europe and western Asia and has naturalized in North America. It is poisonous, containing various isoquinoline alkaloids, so it is not surprising that it often has been used medicinally. Celandine is a perennial herb with a bright yellow-orange sap and alternate, deeply pinnately lobed leaves; its bisexual and radially symmetrical flowers have two sepals (quickly falling), four slightly wrinkled, yellow petals, numerous stamens, and two fused carpels; the fruit is an elongated capsule with the two halves falling away from a central rimlike and seed-bearing structure; seeds each with a fleshy aril.

Clematis (*Clematis* spp.): The Buttercup family (Ranunculaceae). The species of *Clematis* number ca. 300 and are broadly distributed in the temperate regions of the Northern Hemisphere (although a few are subarctic, while others grow in tropical mountains). Ten species occur in Europe, while 32 are reported for North America (north of Mexico) and 147 are known from China. In many regions, they are common wildflowers, and, because of their showy flowers, they are also popular as ornamentals. *Clematis* species are vines, climbing by their tendril-like petioles, or, less commonly, perennial herbs, with opposite, simple, or compound leaves;

FIGURE 7.79 *Plants of Ithilien.*

sometimes the leaflets are also lobed; their flowers are bisexual or unisexual, radially symmetrical, with four colorful (white, blue, purple, red, yellow, green) tepals, these forming a spreading to bell-shaped perianth, numerous stamens, and usually also numerous distinct carpels; the fruits are a cluster of achenes, each often with a persistent plumose style.

Cornel (*Cornus mas*): The Dogwood family (Cornaceae). The cornel (or cornelian cherry or dogwood) is one of ca. 60 species of the genus *Cornus*, which is widely distributed in the Northern Hemisphere. The cornel grows in southern Europe, from France to Ukraine, and also occurs in adjacent regions of southwestern Asia. The tree is used as an ornamental, and its fruits are edible. Cornels are shrubs or trees with simple, opposite leaves with pinnate venation and an entire margin; their flowers are borne in terminal clusters (appearing before the leaves), and each bisexual and radially symmetrical flower has four tiny sepals, four yellow petals, four stamens, and two fused carpels (the ovary inferior); the fruit is a bright red drupe with a single, ridged pit.

Cypress (*Cupressus sempervirens*, and relatives): The Cypress family (Cupressaceae). The eight species of cypress are distributed from the Mediterranean region to southern China, and the Mediterranean cypress (*C. sempervirens*) grows from the eastern Mediterranean to southwestern Asia. Nine American species also are frequently also included within *Cupressus*, although they perhaps should be segregated as species of *Callitropsis*. These trees are grown as ornamentals and also valued for their scented wood. Cypress trees are evergreens with cylindrical or four-angled branches bearing opposite and decussate scale leaves; their pollen strobili and ovulate cones are borne on the same plant: pollen strobili solitary, at the tips of short branchlets; seed cones spherical or oblong, woody, with a few, centrally stalked, hexagonal scales, sometimes not opening until after fire, with two to numerous angular and narrowly winged seeds per scale (and the wings derived from the seed coat).

Hyacinth (*Hyacinthus orientalis*): The Hyacinth family (Hyacinthaceae). The three species of hyacinth grow in western and central Asia, with *H. orientalis* occurring from Turkey south to Lebanon, Syria, and Israel and west to Iran (and also naturalized in Europe). Hyacinths are popular garden ornamentals because of their showy flowers. Their Sindarin name may be *lingui*. They are herbs, arising from bulbs, with alternate, basal, linear, parallel-veined leaves; their bisexual and radially symmetrical flowers are in terminal racemes and have six fused, usually blue, purple, pink, or white tepals, six stamens, and a gynoecium of three fused carpels that develop into a globose capsule, opening to release black seeds.

Junipers (*Juniperus* spp.): The Cypress family (Cupressaceae). The 54 species of junipers (i.e., *Juniperus*) are widely distributed in temperate regions of the Northern Hemisphere but extend southward into the tropical regions in the Caribbean, Mexico, and Central America; eastern Africa; and southern China. Junipers are closely related (and very similar) to cypress trees, differing from them chiefly in usually having the pollen strobili and seed cones on different plants and, most importantly, in having the mature seed cones much reduced, having only one to three pairs (or groups of three) of scales, which are fused, and the entire cone is ± fleshy, berrylike, and remains closed at maturity; the seeds are dispersed with the fleshy cone and are unwinged.

Marjoram (*Origanum majorana*): The Mint family (Lamiaceae or Labiatae). The nearly 40 species of *Origanum* are Eurasian in distribution; marjoram is native to Cypress and southern Turkey and very early was taken westward into Europe, cultivated by the ancient Greeks and Romans, and used as a flavoring, especially in meats. A related species, oregano (*O. vulgare*), of southern Europe, is an important spice (and especially used in pizza). These herbs have small, opposite, and decussate leaves, with pinnate veins and entire to obscurely toothed margins; their flowers are in dense, axillary, reduced cymes that are clustered distally along the shoots, with each inflorescence more or less three-branched and each flower bisexual and bilaterally symmetrical, with five fused sepals, five fused petals, with a bilabiate, white to pink corolla, with two petals forming an upper lip and three a lower lip; four stamens, with two shorter than the others; and two fused carpels with the ovary deeply four-lobed; fruits schizocarps, of four nutlets, surrounded by the persistent calyx.

Myrtle (*Myrtus communis*): The Myrtle family (Myrtaceae). Myrtles are native to the Mediterranean region and are grown as ornamentals, the fragrant leaves and fruits are used in scent-making and also in flavoring alcohol and meats; the branches were used in garlands in ancient Greece and Rome (and associated with the goddess Aphrodite), and, in Jewish ritual, it is one of the "four species" used on the first day of Tabernacles. Myrtles are trees or shrubs with the leaves opposite and decussate, ovate to elliptic, aromatic (due to aromatic oils in spherical cavities, forming pellucid dots), with pinnate venation, and an entire margin; the flowers solitary and axillary, bisexual, radially symmetrical, each with five sepals, five white petals, and numerous spreading stamens; the ovary is inferior, developing into a blue-black berry and crowned by the persistent calyx.

Primeroles, primroses (*Primula* spp.): The Primrose family (Primulaceae). The primroses (i.e., members of the large genus *Primula*, a group of some 450 species) are widely distributed, but are most characteristic of northern temperate regions; 34 species occur in Europe, 37 in North America (incl. *Dodecatheon*), and ca. 300 in China. Primroses are often grown as ornamentals. They are perennial herbs with leaves alternate and spirally arranged, usually in a basal rosette; simple, variously shaped, the veins pinnate, with margin entire to variously toothed; the flowers usually borne in umbels (or reduced to a solitary flower) atop an elongated scape and each flower bisexual, radially symmetrical, with 4–5 fused sepals and 4–5 fused petals, forming a bell-shaped to tubular variously colored corolla with the well-developed lobes reflexed or not, 4–5 stamens that have their filaments attached to the corolla tube, and 4–5 fused carpels; the ovary superior; fruits are capsules with seeds attached on a free-central axis.

Tamarisks (*Tamarix* spp.): The Tamarisk family (Tamaricaceae). *Tamarix*, a group of 54 species, is widely distributed in Eurasia and Africa, and these arid-adapted plants are characteristic of saline soils and river margins; several species have naturalized in North America. They are shrubs or small trees with alternate and spirally arranged scalelike leaves with salt-excreting glands and often densely clustered small flowers, each with 4–5 sepals, 4–5 white to rose petals, four to numerous stamens, 3–4 fused carpels, with a superior ovary; fruits are capsules, opening to release hairy seeds.

Terebinth (*Pistacia terebinthus*): The Sumac or Poison Ivy family (Anacardiaceae). The terebinth is a member of the genus *Pistacia* (a group of nine species) and grows in the Mediterranean region from Greece and Turkey westward to Morocco and Portugal; the closely related species, *P. palaestina*, grows in Syria, Lebanon, and Israel. The tree is a source of turpentine, and the fruits are occasionally used in flavoring alcohol. Terebinths are deciduous trees with alternate and spirally arranged pinnately compound leaves, with resin canals in their leaves, stems, and bark; their small flowers are unisexual (with staminate flowers and carpellate flowers on different trees) and radially symmetrical, with inconspicuous sepals, no petals, five stamens (in staminate flowers), and three fused carpels with a superior ovary (in carpellate flowers) that develop into reddish drupes (each with a single pit).

Thyme (*Thymus vulgaris*, and relatives): The Mint family (Lamiaceae or Labiatae). Thyme is native to southern Europe but is now widely cultivated as a culinary herb (especially used in flavoring meat dishes) and is also used medicinally (for respiratory problems). The genus *Thymus* is large, with some 220 species, and is widely distributed in temperate Eurasia and northern Africa. Common thyme is an aromatic, perennial herb to low shrub with opposite and decussate leaves that are quite small, ovate, with pinnate venation and an entire margin; their flowers are in reduced cymes, clustered distally on the shoots; each flower is either bisexual or carpellate (with the bisexual and carpellate flowers on different plants), with five fused sepals and five fused petals forming a white to rose or purple, two-lipped corolla, with the upper lip with two lobes and the larger lower lip with three lobes, with four stamens (two long and two shorter), and two fused carpels; the ovary superior and deeply four-lobed, maturing into a schizocarp of four nutlets (surrounded by the persistent calyx).

Parsley (*Petroselinium crispum*): The Carrot family (Apiaceae or Umbelliferae). Parsley is native to the central Mediterranean (southern Italy, Algeria, and Tunisia) but is widely cultivated as a culinary herb and has occasionally naturalized elsewhere in Europe and in North America. Parsley is a biennial herb with alternate and spirally arranged, three-times pinnately compound leaves that are aromatic (with oil canals); the leaflets are flat or curly and with serrate to lobed margins; the flowers are borne in terminal, compound umbels, and each tiny flower has five minute sepals, five green to yellow curled petals, five stamens, and two fused carpels, with an inferior ovary; the fruits are two-parted, grooved schizocarps.

FOOD PLANTS OF MIDDLE-EARTH

(Various plant families.)

And just bring out the cold chicken and pickles! (Hobbit I)

In addition to the 11 edible plants treated briefly here, several other food and seasoning plants are mentioned in J. R. R. Tolkien's writings—these are provided more detailed, individual treatments (because of their importance in the legendarium) and include apples, barley, blackberries, blueberries, cherries, coffee, grapes, plums, potatoes, raspberries, sage, sorrel, tea, turnips, and wheat. In addition, olives are mentioned as occurring in Ithilien, but only as wild trees, not as cultivated plants, providing edible oil or fruits. Finally, the herbs thyme and marjoram are also in Tolkien's legendarium, and these are briefly discussed in the section on Plants of Ithilien. Nearly all of these food plants are native to Europe and/or the Mediterranean region, or to adjacent areas of southwestern to central Asia. Exceptions include cardamom, pepper, oranges, tea, and cucumbers (i.e., pickles), which are native to southern or eastern Asia, and most of these—but not tea or oranges—were very early brought to Europe. Coffee is native to the mountains of Ethiopia and, like tea, did not become popular in England until seventeenth century. New World species (i.e., American food plants) are conspicuously absent from the foods recorded in Tolkien's legendarium: the sole exceptions being potatoes and tobacco (pipeweed), and Tolkien clearly struggled with how he should deal with these two, which brings up the reason that pickles are mentioned in the *Hobbit.* The inclusion of pickles (i.e., cucumbers) in the first chapter of *The Hobbit* (see introductory quote) is interesting because it is a modification, originating in the third edition (in 1966), of the original (1937) wording at this point, which was "cold chicken and tomatoes!" (as discussed in Tom Shippey's, 1992, *The Road to Middle-earth,* and Douglas Anderson's, 2002, *The Annotated Hobbit*). Why were tomatoes (*Solanum lycopersicum*) removed and replaced by pickles? This change was surely motivated by the fact that tomatoes are native to the Americas and thus would be a discordant element, both botanically and linguistically, in the Shire—which was conceived by Tolkien as basically English in nature (Figure 7.80). The same reasoning lies behind his use of pipeweed for tobacco (*Nicotiana tabacum*) and taters for potatoes (*Solanum tuberosum*) in *The Lord of the Rings*—changes that downplay the linguistic dissonance that would have arisen through use of the names tobacco and potato—both of which are derived from Taino, an extinct Amerindian language. These three aside, the foods of Tolkien's Middle-earth (see Figure 7.80, which shows the common-room of the Prancing Pony) are remarkable in that they largely reflect a diet that would have been recognized by those living in the Europe of the Middle Ages, in contrast to our modern diet, which is based on foods that have been brought together from around the globe. Finally, it is surprising that oats (*Avena sativa*), an old and important temperate grain (native to Europe and long used in gruels) are not mentioned as occurring in Middle-earth (although Tolkien does mention oats in the Gnomish lexicon).

Plant list: Cabbage (*Brassica oleracea,* Brassicaceae or Cruciferae): *Brassica oleracea* is native to western and southern Europe and adjacent Asia (and it is one of ca. 35 species in the genus). However, it has been domesticated several times, and the species is now cultivated (as a garden vegetable) in all temperate regions and has naturalized in Asia, North America, and Australia. The cultivated varieties of *B. oleracea* are diverse in form, including cabbage, broccoli, cauliflower, Brussels sprouts, kohlrabi, Chinese broccoli, kale, and collard greens. Of these, only cabbages are mentioned in J. R. R. Tolkien's writings, and this entity, *B. oleracea* var. *capitata,* was very early (perhaps ca. 8,000 years ago) derived from coastal populations of var. *oleracea* in Western Europe. Cabbages are biennial herbs containing mustard oils, in which the leathery leaves form a dense head; their flowers are produced in the second year and borne in racemes, with each flower having four sepals, four white or yellow petals, held in the shape of a cross, six stamens (with four long and two short), and two fused carpels (with a superior ovary); the fruits are elongated siliques in which the two halves fall away from a persistent structure that bears the seeds.

FIGURE 7.80 *The common-room at the Prancing Pony (Food plants of Middle-earth).*

Cardamon (*Elettaria cardamomum*, Zingiberaceae): Cardamon is one of seven species of *Elettaria*, and it is native to southern India; the species is an important spice, was imported into Europe as early as the time of the Romans, and is now widely cultivated in tropical regions. Cardamon is an aromatic, perennial herb with alternate and two-ranked simple leaves, each with a sheath (the lower portion, which surrounds the stem), ligule (i.e., flap of tissue at junction of sheath and blade), and elliptical blade, with pinnate venation and an entire margin; the flowers are borne on erect inflorescences that arise from rhizomes and thus are separated from the vegetative stems; each flower is bilaterally symmetrical, bisexual, with three fused sepals, three fused petals with well-developed lobes, two large and petaloid sterile stamens, which are fused to each other and form a white and red liplike structure, a single functional stamen that appears to be grabbing the style and stigma, and three fused carpels, with an inferior ovary; the fruit is a capsule.

Carrots (*Daucus carota*, Apiaceae or Umbelliferae): Carrots, one of ca. 22 species of *Daucus*, grow wild in Europe and southwestern Asia, but the species was domesticated in central Asia; the fleshy roots may be white, yellow, orange, red, or purple (and contain carotene and/or anthocyanins). Carrots (and their wild relative, Queen Anne's lace) are aromatic biennial herbs, from a taproot, with alternate, spirally arranged much dissected leaves, the petioles of which are sheathing; their flowers are in terminal, compound umbels, and each flower has five tiny sepals, five curled, usually white petals (but often purple-red in the central flower or flowers of the umbel), and five stamens, all arising atop the inferior ovary (of two fused carpels, with the style bases conspicuously expanded, producing nectar); the fruits are two-parted schizocarps, and each segment is covered with rows of stiff hairs.

Lettuce (*Lactuca sativa,* Asteraceae or Compositae): Lettuce is one of ca. 75 species of *Lactuca* and native to the eastern Mediterranean region. It is grown for its edible leaves (used in salads) and was domesticated ca. 4,500 years ago in Egypt, perhaps from ancestral populations closely related to *L. serriola*; it is now cultivated nearly worldwide and widely naturalized. Lettuce is an annual or biennial herb with milky sap, alternate and spirally arranged, simple, ovate to orbicular leaves, with pinnate venation (the midvein usually not prickly) and an entire to variously toothed or lobed margin; the flowers are in clustered heads, with each head having all ligulate flowers (i.e., the flowers bilaterally symmetrical, bisexual, with the sepals modified into a pappus of numerous bristles), the petals fused, yellow, their lower portion forming a tube but distally flexed to one side of the flower, forming an elongated, tonguelike structure with five minute teeth at the apex, five stamens, fused into a tube, and two fused carpels; the ovary inferior; the fruit is a flattened achene, with a narrowed portion separating the fruit body from the persistent pappus.

Hops (*Humulus lupulus,* Cannabaceae): Hops are one of six species of *Humulus; H. lupulus* is widespread in Eurasia and has naturalized in North America. The closely related species *H. neomexicanus, H. pubescens,* and *H. lupuloides* grow in North America. Hops are economically most important as a flavoring in brewing beer (due to their resins and ethereal oils), but the plant also is used medicinally (especially to induce sleep, and for breast, uterus, or bladder problems). Hops are twining vines, with two-branched nonglandular hairs and gland-headed hairs, with alternate, spirally arranged, simple, usually lobed, and ovate leaves, with palmate venation; the apex is acuminate, the base cordate, and margin serrate. Flowers are unisexual (with staminate and carpellate flowers usually on different plants) and radially symmetrical and in axillary or terminal cymes; the staminate flowers are inconspicuous, with five tepals and five stamens, the carpellate flowers have a reduced perianth and two fused carpels; fruits are achenes, with associated perianth tissue and bracts.

Onions (*Allium cepa,* Alliaceae): Onions are a large group (probably more than 700 species, of wide distribution), and, of these species, the onion of commerce (*A. cepa*) is the most valuable, although garlic (*A. sativum*) is also an economically important crop. Onions probably were domesticated about 7,000 years ago in central Asia (perhaps from *A. oschanini*), and their various cultivars are only known from cultivation. In addition to being edible, onions are used medicinally, containing antibacterial and antifungal compounds. Their name in Sindarin may be *nínholch*. Onions are biennial herbs, arising from a bulb that contains aliphatic disulfides; the leaves are alternate, two-ranked, basal, simple, semicircular in cross-section, with parallel venation and entire margins; their flowers are borne in an umbel atop a long scape, and each flower is radially symmetrical, bisexual, with six white to pink tepals, six stamens, and three fused carpels, with a superior ovary; the fruit is a capsule, opening to release black seeds.

Oranges (*Citrus* × *aurantium,* Sweet Orange Group, Rutaceae): Oranges are the most widely cultivated member of the genus *Citrus,* a group of some 25 species native to southern and southeastern Asia, northern Australia, and New Caledonia. The sweet orange originated in southern China more than 2,300 years ago and is of hybrid origin (as indicated by the multiplication sign before its specific epithet). It is a backcrossed hybrid between the pomelo (*C. maxima*) and mandarin (*C. reticulata*), with about 25 percent of its genes coming from the former and 75 percent of its genes coming from the latter species. Other major genetic groups within *C.* × *aurantium* include the sour oranges (including the bergamot oranges), grapefruits (which resulted from backcrossing with *C. maxima*), and the tangelos (resulting from a cross of grapefruit and *C. reticulata*). As can be seen, the history of citrus fruits is highly reticulate, and the asexual formation of seeds, which are genetically identical to the parent tree, is common. The Sindarin name for orange-tree may be *culforn,* while the Quenya name possibly is *culmarin.* Oranges are aromatic trees with alternate and spirally arranged leaves that have the pellucid-dotted, pinnately veined, and crenate-margined blade separated from the winged petiole by a distinct joint (and thus represent a reduction from an ancestral compound leaf, as in *C. trifoliata*); the juvenile trees can be thorny. The axillary and fragrant flowers are radially symmetrical, bisexual, and have five small sepals, five showy white petals, numerous stamens, and several fused carpels (with the ovary superior); fruits are large orange berries with a leathery rind and segments (the flesh of which is derived from juice-filled hairs).

Peas (*Psium sativum,* Fabaceae or Leguminosae): Peas are native to the Mediterranean region and western Asia and were domesticated ca. 9,000 years ago; they are now widely grown in cool temperate regions for their nutritious and flavorful seeds and young pods. The Quenya name for pea is *orivaine.* Peas are annual vines with alternate, spirally arranged, pinnately compound leaves, the apices of which are modified into twining tendrils, and each leaf is associated with a pair of well-developed leafy stipules. The axillary, white, pink, or purple flowers are pealike (i.e., bilaterally symmetrical, with a dorsal, banner petal, two lateral wing petals, and the two ventral petals fused, forming a keel-like structure), with ten stamens, nine of which are fused by their filaments, but with the dorsal stamen distinct, surrounding the single, elongated carpel, which develops into a legume.

Pepper (*Piper nigrum,* Piperaceae): The important spice pepper is made from the fruits (drupes) of *P. nigrum,* a species native to humid tropical forests of southern India; black pepper is made from immature green drupes, which are briefly boiled and then dried, causing them to turn black; white pepper is made from the mature, red drupes, with the red flesh removed and the white pits then dried. The hotness of this spice comes from piperine, with associated terpenes contributing to the odor of the spice. The fruits have been used for at least 4,000 years and were very early taken to Europe in trade. *Piper* is a large pantropical genus consisting of ca. 1,050 species. Pepper is a liana with alternate, simple leaves, with venation intermediate between pinnate and palmate, an acuminate apex, and entire margin; the bisexual or unisexual flowers are borne in spikes positioned opposite the leaves, and each flower is extremely tiny, lacking a perianth; the 3–5 fused carpels develop into globose, red drupes (each with a single pit).

Pickles or cucumbers (*Cucumis sativa,* Cucurbitaceae): Cucumber is one of nearly 70 species of *Cucumis,* and it is native to the Himalayan region of northern India; the species was domesticated more than 3,000 years ago and is now a widely cultivated and important vegetable (with numerous cultivars—some for slicing, others for pickling). The fruits are frequently used raw in salads and are also pickled in vinegar, often with dill (*Anethum graveolens,* Apiaceae) as a flavoring. Their Sindarin name may be *colost,* and the Quenya name *colosta.* Cucumbers are vines with tendrils arising at the nodes (and slightly displaced from the leaf axil) and alternate and spirally arranged, simple leaves, with palmate venation and toothed and often lobed margins. The axillary flowers are unisexual (with staminate and carpellate flowers on the same plant), radially symmetrical, with five small sepals, five fused yellow or orange-yellow petals, five twisted and fused stamens, and three fused carpels, with an inferior ovary; the fruits are leathery, elongated berries, often with a few prickles.

Strawberries (*Fragaria vesca*, and close relatives, Rosaceae): The modern strawberry is *Fragaria* × *ananassa*, the result of an accidental hybridization event (around 1750, in Europe) between the Chilean strawberry (*F. chiloensis*, which also has native populations in western North America) and the North American strawberry (*F. virginiana*). The Chilean strawberry, North American strawberry, and the European strawberry (*F. vesca*, which also has three geographical races native in North America) are also sometimes cultivated. Strawberries, a group of 24 widely distributed species, are occasionally considered within the genus *Potentilla*. Strawberries are perennial, rhizomatous herbs with a very short stem bearing alternate and spirally arranged compound, stipulate leaves, each with three leaflets, which have pinnate veins and serrate margins. The axillary flowers are bisexual and radially symmetrical, with associated bracts, five sepals, five white petals, and 15 stamens, all borne on a saucer-shaped floral cup (= hypanthium), and with numerous distinct carpels; the fruits are small achenes, borne on a bright red, expanded, floral receptacle.

HOBBIT NAMES

(Various plant families.)

> *"Well, Sam," said Frodo, "what's wrong with the old customs? Choose a flower name like Rose. Half the maidchildren in the Shire are called by such names, and what could be better?" (LotR 6: IX)*

As this quote makes clear, flower names were frequently used in the Shire for girls and women, reflecting, we think, J. R. R. Tolkien's love of floral beauty and gardening—and in this we see his similarity to Bilbo, Frodo, and Sam (see Figure 7.81, showing Frodo's garden). Sam and Rose's love for their first-born daughter is also clear—as is evident from Sam's comment: "But if it's to be a flower-name . . . it must be a beautiful flower, because, you see, I think she is very beautiful, and is going to be beautifuller still." They end up choosing the name Elanor, the sun-star, with its golden flowers, which Sam remembered from the grass of Lothlórien. Sam and Rose are typical of hobbit parents, as we see in the wonderful array of flower-names in Appendix C of *The Lord of the Rings*—31 in all! All are beautiful plants, and they have meanings connected with concepts such as love, purity, fidelity, virtue, humility, bashfulness, imagination, and remembrance. Only a few are negative in connotation, for example, Lobelia, in the "language of flowers," is associated with malevolence—and this certainly connects with aspects of the personality of Lobelia Sackville-Baggins! In addition to the 16 plant names briefly described in this section, an additional 15 names (representing 11 genera) are the source of hobbit names. These 15 names are listed here along with the relevant individual (or individuals) carrying the name: Asphodel (*Asphodelus*, Asphodelaceae; Asphodel Burrows, née Brandybuck),

FIGURE 7.81 *Frodo's garden (Hobbit names).*

Celandine (*Chelidonium*, Papaveraceae; Celandine Brandybuck), Daisy (*Bellis perennis*, Asteraceae; Daisy Baggins, Daisy Gamgee), Eglantine (*Rosa rubiginosa*, Rosaceae; Eglantine Took, née Banks), Elanor (*Anagallis*, Primulaceae; Elanor Gamgee), Filbert (*Corylus maxima*, Betulaceae; Filibert Bolger), Goldilocks (based on *Ranunculus*, a yellow wildflower, Ranunculaceae; Goldilocks Took, née Gamgee), Lily (*Lilium*, Liliaceae; Lily Goodbody, née Baggins, Lily Cotton, née Brown), Myrtle (*Myrtus*, Myrtaceae; Myrtle Burrows), Primrose and Primula (*Primula*, Primulaceae; Primula Baggins, née Brandybuck, i.e., mother of Frodo, Primrose Gamgee), Rose and Rosa (*Rosa*, Rosaceae; Rosa Took, née Baggins, Rose Gamgee, née Cotton), Rowan (*Sorbus aucuparia*, Rosaceae; Rowan, daughter of Holman the Greenhanded), and Salvia (*Salvia*, Lamiaceae; Salvia Bolger, née Brandybuck). These 11 genera are also considered in more detailed in individual treatments or in the treatment of the plants of Ithilien because they are mentioned in J. R. R. Tolkien's writings as occurring in the landscape—not merely represented in the legendarium as names of particular hobbits. All 31 of these plants are native to Europe and, in this regard, are similar to the food plants that are part of Tolkien's legendarium—except for *Camellia*, which originated in southern China and did not become popular as an ornamental in England until the nineteenth century. Finally, for more information, especially relating to the history and folklore associated with these hobbit names, we recommend D. Hazell's book *The Plants of Middle-earth: Botany and Sub-creation* (2006).

Amaranth (*Amaranthus* spp., Amaranthaceae; Amaranth Brandybuck): The amaranths are a group of ca. 70 tropical and temperate species of wide distribution. Many are weedy plants, and *A. caudatus* is a cultivated grainlike species and popular ornamental; the southern Asian *A. tricolor* is also used as an ornamental and edible herb. Amaranths are usually annual herbs, with or without hairs, with alternate and spirally arranged leaves that are simple, variously shaped, with pinnate venation, with the apex acute to emarginate, and the margin entire, sometimes undulate. The flowers are densely clustered and tiny, usually unisexual (and the staminate and carpellate flowers on the same or different plants), radially symmetrical, with a perianth of 3–5 distinct to slightly fused tepals (or absent), 3–5 stamens, and usually 2–3 fused carpels; the small, one-seeded fruits are surrounded by the persistent tepals.

Angelica (*Angelica sylvestris*, Apiaceae; Angelica Baggins): Angelica, one of ca. 60 species of the genus *Angelica*, is widely distributed across Europe and adjacent Asia (and has been introduced into North America), occurring in open habitats and woodlands. Angelica is a biennial, aromatic herb with alternate and spirally arranged twice-pinnately compound leaves, with broad, pinnately veined leaflets with serrate margins, and the petioles broadly expanded and sheathing; the inflorescences are terminal compound umbels; the flowers are small, with five minute sepals, five white, curled petals, five stamens, and two fused carpels with expanded, nectar-producing style bases and an inferior ovary; the fruits are schizocarps, the two segments of which are flattened and winged.

Belladonna (*Atropa belladonna*, Solanaceae; Belladonna Baggins, née Took, i.e., mother of Bilbo): Belladonna, native to Europe, western Asia, and North Africa (and occasionally naturalized in North America), is extremely poisonous, containing various tropane alkaloids. It was once used cosmetically, the specific epithet derived from Italian and meaning "beautiful lady" in an allusion to its dangerous use to increase pupil size. Belladonna is a perennial herb or subshrub with alternate, simple, pinnately veined leaves and axillary inflorescences. The flowers are bisexual and radially symmetrical, with five fused, triangular sepals, five purple and strongly fused petals, forming a bell-shaped corolla; five stamens, which have their filaments fused to the lower part of the corolla tube; and two fused carpels, with a superior, globose ovary and elongate style; the fruits are black berries with many small seeds.

Camellia (*Camellia* spp., Theaceae; Camellia Baggins, née Sackville): *Camellia* is a genus of ca. 120 species of southern and eastern Asia. Species of *Camilla* are shrubs or small trees frequently grown for their beautiful white, red, or pink flowers, and the most commonly grown ornamental species are *C. japonica* and *C. sasanqua*. These species are quite similar in morphological features to *C. sinensis* (the source of tea), but their flowers are larger (see also treatment of tea).

Cotton (*Gossypium* spp., Malvaceae, Holman Cotton, Tolman Cotton, Wilcome Cotton, Rose Cotton, etc., see *The Longfather-tree of Master Samwise*; we also note that the name Gamgee was based on Gamgee tissue, the cotton wool used in Tolkien's youth, see Letters, No. 257): The ca. 50 species of cotton grow in warm temperate to tropical regions of the Americas, Africa, Asia (Middle East), and Australia. Several species of *Gossypium* are economically important as sources of fiber (elongate hairs on the seeds) and cooking oil. The Quenya name for cotton may be *line*. Cotton plants are shrubs with slime canals and alternate and spirally arranged, stipulate, simple, ovate, usually lobed and palmately veined leaves, with apex acute to acuminate, base cordate, margin entire; flowers axillary, radially symmetrical, bisexual, large, associated with bracts, with five fused sepals, five distinct, cream, yellow, or rose-colored petals, with or without a dark spot at base, numerous stamens, fused into a hollow column, and 3–5 fused carpels, with a superior ovary and elongate style (running through center of staminal tube); the fruits are capsules, opening to release usually hairy seeds.

Gilly, gillyflower (*Dianthus caryophyllus*, Caryophyllaceae; Gilly Baggins, née Brownlock): *Dianthus* is a large genus (including ca. 320 species) that is mainly Eurasian in distribution. Several species are used as ornamentals, and *D. caryophyllus* also is used as a source of oil for soap and scent. *Dianthus* species are annual to perennial herbs with opposite and decussate, simple, basally sheathing, linear to ovate or oblong, one-veined leaves with an acute apex. The flowers are in cymose clusters or solitary; they are radially symmetrical, bisexual, and have five fused sepals, forming a tube, five distinct petals (but more in many cultivars), each pink or red, sometimes white or purple, with an erect narrowed base and spreading, distally broadened, toothed to fringed apical portion, 10 stamens, and two fused carpels, with a superior ovary and two styles; fruits are ovoid to cylindric capsules, opening by four apical teeth to release numerous small seeds.

Lobelia (*Lobelia* spp., Campanulaceae; Lobelia Sackville-Baggins, née Bracegirdle): The ca. 400 species of *Lobelia* are widely distributed in both temperate and tropical regions; several species are used as ornamentals. Lobelias are herbs to small trees with milky latex, with leaves alternate and spirally arranged, simple, variously shaped, pinnately veined leaves, with margin entire to variously toothed; their inflorescences are terminal or axillary, and the flowers are bilaterally symmetrical, bisexual, with five fused sepals, five fused petals, with a variously developed dorsal slit in the corolla and usually with two upper lobes and three lower lobes; five stamens, fused together by their anthers and distal portion of the filaments, sheading pollen toward the center of the anther-tube; and two fused carpels, with an inferior ovary and long style, with apical pollen-collecting hair-tuft; the fruits are capsules.

Malva (*Malva* spp., Malvaceae; Malva Brandybuck, née Headstrong): The ca. 30 species of *Malva* are native to Eurasia and northern Africa, but some species are weedy and are widely naturalized. They are occasionally grown as ornamentals (especially *M. alcea*) and several species are eaten (e.g., *M. parviflora, M. neglecta,* in salads, or as greens). The species of *Malva* are annual to perennial herbs or shrubs, with or without simple or stellate hairs, and with mucilage canals, with alternate and spirally arranged stipulate, simple, sometimes lobed, palmately veined leaves with margin crenate to dentate. The usually axillary flowers are radially symmetrical, bisexual, associated with three bracts, with five fused sepals, five distinct, white, pink, rose, or purple petals, numerous stamens fused into a tube by their filaments, and six to numerous fused carpels, with a superior ovary. The fruits are schizocarps, circular and flattened, separating into triangular units.

Marigold (*Calendula officinalis,* Asteraceae; Marigold Cotton, née Gamgee): Commonly called the pot marigold (in contrast to *Tagetes erecta,* the French marigold, a native of Mexico and Central America): The 15 species of *Calendula* mainly occur in the Mediterranean region (eastward to Iran). *Calendula officinalis* is only known from cultivation and may be of hybrid origin; it is used as an ornamental (because of its showy yellow to orange flower-heads) and is also edible (flowers used as garnish in salads, leaves used in soups) and used medicinally (fevers, possible antibacterial and antiviral activity). Pot marigold is a short-lived perennial with alternate and spirally arranged simple, oblong, elliptic to obovate, pinnately veined leaves with margin entire to sparsely dentate; the flowers are borne in terminal heads, with 30 to more than 100 ray flowers surrounding 60 to more than 150 disk flowers, with ray flowers bilaterally symmetrical, carpellate, yellow to orange, with tonguelike extension of the corolla with three apical lobes, and disk flowers radially symmetrical, minutely trumpet-shaped, staminate, yellow, reddish, or purple; the fruits are curved achenes.

Melilot or sweet clover (*Melilotus* spp., Fabaceae; Melilot Brandybuck): The 20 species of sweet clovers are distributed in temperate to subtropical regions of Eurasia and northern Africa. Their sweet odor is due to the presence of coumarins in their tissues, which (in contrast to their odor) taste bitter and probably protect against herbivory. Melilot is used to enrich soil nitrogen, as livestock feed, and as a bee plant. Warfarin, a commonly used anticoagulant, is derived from *M. officinalis.* Species of *Melilotus* are annual or biennial herbs with erect stems and alternate, spirally arranged, stipulate leaves, each with three ovate to obovate leaflets and serrate margins. The inflorescences are axillary racemes; the flowers are bilaterally symmetrical, bisexual and pealike, with five fused sepals, five white or yellow petals, with the dorsal petal larger than the others forming a standard, the two lateral petals winglike, and the two innermost petals fused, forming a keel, with ten stamens, nine of which are fused, and a single carpel, developing into a short, 1–4 seeded pod that does not open.

Mentha (*Mentha* spp., Lamiaceae; Mentha Brandybuck): The 19 species of *Mentha* are widely distributed, aromatic herbs; they are commonly used as flavorings; such as spearmint (*M. spicata*) and peppermint (*M.* × *piperita*). Species of *Mentha* are perennial, rhizomatous herbs with opposite and decussate simple, pinnately veined leaves with serrate margins. The flowers are in dense cymose clusters forming pseudo-whorls at each node; flowers are bisexual and only slightly bilaterally symmetrical, with five fused sepals forming a tubular to bell-shaped calyx, five fused, white to pink petals forming an apparently four-lobed corolla, the seeming upper lobe wider than the others and actually two fused petals, four stamens, all the same length, and two fused carpels, with an elongate style and four-lobed, superior ovary; the fruits are schizocarps of four-nutlets.

Mimosa (*Mimosa* spp., Fabaceae; Mimosa Baggins, née Bunce): *Mimosa* is a large genus (ca. 500 spp.) that is primarily distributed in tropical to warm temperate regions of the Americas. They are occasionally grown as ornamentals, and *M. pudica* is grown as a novelty because it is sensitive, folding its leaflets when touched. Mimosas are perennial herbs or shrubs, sometimes with prickles, with alternate, spirally arranged, stipulate, twice-pinnately compound leaves, with small leaflets having entire margins. Flowers are in dense heads, and each flower is radially symmetrical and bisexual, with a minute calyx, 4–5 fused to nearly free, white to pink or lavender petals, 4–10 stamens, which are colorful, showy, and more conspicuous than the tiny petals, and a single carpel; fruits are elongated, compressed, often prickly, usually constricted between the seeds, the valves separating from the sutures as a unit, or the fruit breaks into one-seeded segments.

Pansy (*Viola tricolor*, Violaceae; Pansy Bolger, née Baggins): The ca. 400 species of violets are distributed nearly worldwide, especially in temperate regions. Pansies are native to Eurasia and are popular ornamentals due to their showy multicolored flowers; they are widely cultivated and naturalized. They are annual herbs with prostrate to erect stems with alternate and spirally arranged simple, ovate to elliptic or oblong, pinnately veined leaves, with margin crenate-serrate, and each associated with a pair of conspicuous, dissected stipules. Flowers axillary, bilaterally symmetrical, bisexual, with five sepals and five petals: these usually violet, the two lateral ones white, and the lowermost yellow or white, or all violet, white, yellow, or red, and the lowermost and lateral petals purple-veined, the two laterals bearded near base, and lowermost with a nectar spur; five stamens form a ring around the ovary, the two lowermost with nectar-producing projections (that fit into the petal-spur), and three fused carpels, with a superior ovary; the fruits are capsules.

Peony (*Paeonia* spp., Paeoniaceae; Peony Burrows, née Baggins): The 25 species of peonies grow in temperate Eurasia and Western North America; they are popular ornamentals because of their showy, many petaled flowers. Many species are of hybrid origin. Peonies are perennial herbs to shrubs with alternate, spirally arranged, compound or dissected leaves; the flowers are often solitary, radially symmetrical, bisexual, with five sepals, 5–13 white to pink, red, purple, or occasionally yellow petals (or more numerous in some cultivars), numerous stamens, and three to numerous distinct carpels, each developing into a follicle, opening to release the arillate seeds.

Pimpernel (*Anagallis* spp., Primulaceae; Pimpernel Took): The pimpernels (a group of ca. 20 species) are widely distributed across Eurasia, Africa, and the Americas. They are ornamental herbs and occasionally used medicinally. Species of *Anagallis* are annual to perennial herbs, with opposite, alternate, or whorled simple, pinnately veined leaves, with the apex acute to obtuse and the margin entire. The axillary flowers are radially symmetrical, bisexual, with five slightly fused sepals; five fused, white, pink, red, orange, or blue petals, with the corolla more or less bell-shaped, the base tubular and expanding distally into ovate-triangular lobes; five stamens, these fused to the corolla tube; and with five carpels, these fused, with a superior ovary; the fruits are capsules, each dehiscing around its rim. As is discussed earlier, it is nearly certain that elanor is also a species of *Anagallis,* so can also be considered as a kind of pimpernel.

Poppy (*Papaver* spp., Papaveraceae; Poppy Bolger, née Chubb-Baggins): The ca. 80 species of poppies grow in North America, Eurasia, Africa, and Australia. They are rich in alkaloids, especially opiates, and several species are grown as ornamental herbs; *Papaver somniferum* is cultivated as the source of opium (and derivatives such as morphine and codeine) and also for its edible seeds. The Quenya name for these flowers may be *fúmella* or *húmella.* Poppies are annual to perennial herbs with white, orange, or red latex, with alternate and spirally arranged leaves in a basal rosette or distributed along the stem, each leaf simple, pinnately veined, often lobed or dissected, with margin entire to variously toothed. Flowers are on long scapes or with a long peduncle, radially symmetrical, bisexual, with two sepals that quickly drop off, four wrinkled, pink, red, lilac, orange, or yellow petals, numerous stamens, and three to numerous fused carpels; the ovary is superior; fruits are capsules, opening by pores or short valves, releasing numerous tiny seeds.

BREE NAMES

(Various plant families.)

> *The men of Bree seemed all to have rather botanical (and to the Shire-folk rather odd) names, like Rushlight, Goatleaf, Heathertoes, Appledore, Thistlewool and Ferny (not to mention Butterbur). Some of the hobbits had similar names. The Mugworts, for instance, seemed numerous. (LotR 1; IX)*

Several of these "rather botanical" names of men and hobbits of Bree are recorded in *The Lord of the Rings.* Some of the plants on which these names are based are treated in detail in individual treatments, such as Appledore, based on an old name for apple tree (*Malus domestica,* Rosaceae); Ferny, based on ferns (species of leptosporangiate ferns, especially the Polypodiales); Heathertoes, based on heather (*Calluna vulgaris,* Ericaceae, and perhaps, as suggested by J. R. R. Tolkien [see *Nomenclature of the Lord of the Rings,* p. 759, in W. G. Hammond and C. Scull, 2005, *The Lord of the Rings: A Reader's Companion*] a joke of the Big Folk, suggesting that hobbits, walking barefoot, collected heather, twigs, and leaves between their toes); Rushlight, based on a candle with a rush stem as the wick (*Juncus,* Juncaceae); and Thistlewool, based on the fluffy, hair-tufted, and wind-dispersed fruits of thistles (*Cirsium,* Asteraceae). The remaining names, Butterbur, Goatleaf, and Mugwort, are discussed here (and are briefly described). These names, along with the "natural" surnames characteristic of many hobbits (e.g., Banks, Brockhouse, Longholes, etc.) indicate that both the humans and hobbits of Bree were closely connected to the natural environment that surrounded them.

Among these names, the most familiar to readers of *The Lord of the Rings* is certainly Butterbur because it is the surname of the innkeeper of the Prancing Pony in Bree—Barliman Butterbur. His name is doubly botanical, as his first name is based on barley (*Hordeum vulgare*), the most important grain used in brewing beer, and it is thus appropriate for the owner of an establishment where beer is the main beverage served. The food provided at the Prancing Pony is delicious, perhaps bringing to mind thoughts of butter—on taters, turnips, carrots, or bread (see also Figure 7.80). The surname Butterbur is based on the medicinal plant of the same name, *Petasites hybridus*, which has large leaves that were once used to wrap butter.

The surname Goatleaf is based on the goatleaf honeysuckle (*Lonicera caprifolium*), an ornamental liana and native of Europe, which has leaves that imaginatively resemble the hoof prints of a goat. The name Mugwort, which was common among Bree hobbits, is based on members of the large genus *Artemisia*, aromatic herbs that long have been used to repel insects. The name is derived from Old English *mugwyrt*, meaning "midge-wort," as it was used to repel flies and midges. The plant surely would have come in handy when Aragorn and the four hobbits passed through the Midgewater Marshes just east of Bree!

Butterbur (*Petasites hybridus*, and relatives, Asteraceae): Butterburs (i.e., the 19 species of *Petasites*) are distributed across temperate and arctic regions of Eurasia and North America. A few are used as ornamentals, such as *P. hybridus* (common butterbur) and *P. japonicus* (giant butterbur), and the former is also medicinally useful (especially in treatment of migraine headaches). Butterburs are perennial, rhizomatous herbs with leaves in a basal rosette and also along the erect stem, these alternate and spirally arranged, simple, ovate to orbicular, with venation palmate or intermediate between palmate and pinnate, and margin entire to variously toothed. The flower heads are in clusters, with the predominantly staminate and predominantly carpellate heads on different plants and with both ray (bilaterally symmetrical, positioned marginally in head) and disk flowers (radially symmetrical, positioned centrally in head) or merely with disk flowers. The tiny flowers have a calyx of numerous pappus bristles and a white to pink or purple corolla. The two fused carpels (with inferior ovary) develop into achenes.

Goatleaf (*Lonicera caprifolium*, Caprifoliaceae): The ca. 180 species of *Lonicera*, the honeysuckles, are widely distributed in the Northern Hemisphere; many are used as ornamental lianas or shrubs. *Lonicera caprifolium* is Eurasian in distribution and is grown ornamentally, but not as frequently as *L.* × *italica*: its hybrid with *L. etrusca*. Goatleaf honeysuckle is a liana with opposite and decussate, simple, pinnately veined leaves, with margins entire. The flowers are terminal and axillary, associated with a pair of fused leaves, bisexual, and bilaterally symmetrical; they have five tiny sepals, five showy, fused, white to rose petals that form two lips, the upper with four lobes and the lower consisting of a single petal lobe; the five stamens are fused to the corolla; the two fused carpels have an inferior ovary and an elongate style with a capitate stigma; fruits are red berries.

Mugwort (*Artemisia* spp., Asteraceae): *Artemisia* is a genus of some 400 species, and these are mainly distributed in the Northern Hemisphere, especially in regions where rainfall is limited; they are also called sagebrush, wormwood, and felon-herbs. These aromatic and bitter herbs and shrubs, containing terpenoids and sesquiterpene lactones, are of medicinal value (being used to repel insects, as a vermifuge or disinfectant, in antimalarial preparations, and to ease childbirth). Some species are also used as flavorings (especially of alcohol) or as scent plants. Mugworts are annual to perennial herbs or shrubs with erect branches and alternate and spirally arranged, simple and variously shaped, but usually pinnately and/or palmately lobed leaves, with margins entire to toothed and with or without hairs. The flowers are arranged in numerous small heads, usually entirely composed of disk flowers; each flower is bisexual, radially symmetrical, with sepals lacking, five fused, petals, forming a cylindrical, yellow corolla, five stamens, with their anthers fused into a tube, and two fused carpels, with an inferior ovary; the fruits are achenes, often gland-dotted.

UNIDENTIFIED AND EXCLUDED MIDDLE-EARTH PLANTS

It must be kept in mind that many passages of *The Lord of the Rings* are written from the perspective of the hobbits, and they do not have a perfect knowledge of the plants of Middle-earth. They are knowledgeable, of course, especially Sam, who is a gardener, but they know best those plants of their own gardens, fields, and forests. Thus, we should not be surprised to read that, when in the Old Forest, Frodo, Sam, Merry, and Pippin saw to the left of the path a drier forest dominated by pines and firs, which replaced "the oaks and ashes and other strange and nameless trees of the denser wood" (LotR 1: VI). Of the four hobbits, only Merry had been in the Old Forest before, and he had been only a little way in—so none of them had ever seen some of the trees that occurred in the "denser wood." We do not know, therefore, what these species were, but perhaps *Carpinus betulus* (a common species of Europe and western Asia) and *Acer campestre* (a widespread species of Europe and southwestern Asia and Africa) were among them—as these are the only two tree species native to England that are not known from J. R. R. Tolkien's legendarium. Later in their journey, the hobbits probably also encountered *Picea abies* (Norway spruce) as this species is common in European mountains, so should have been seen, for example, in the White Mountains of Gondor. Perhaps they confused this coniferous tree with the more common and superficially similar fir trees. Likewise, we are informed in the chapter "Of Herbs and Stewed Rabbit" that Sam and Frodo encountered in Ithilien many trees that were "unknown in the Shire," as well as "many herbs of forms and scents beyond the garden-lore of Sam" (LotR 4: IV). Of course, this makes sense because the flora of Ithilien is quite different from that of the Shire. The hobbits' limited knowledge adds interest to the story—we are discovering Middle-earth along with them as we read! Still, it means that we cannot easily discover the names of these unknown plants. However, a few guesses can be made—at least satisfying our curiosity. Distinctive trees and shrubs of the Balkans and/or Turkey (regions that are probably near the location of Ithilien), which could have been seen by Frodo and Sam include species such as *Acer sempervirens* and *A. monospessulanum* (Cretan maple, Montpellier maple), *Arbutus andrachne* (Greek strawberry-tree), *Carpinus orientalis* (oriental hornbeam), *Celtis tournefortii* (Oriental hackberry), *Cercis siliquastrum* (Judas-tree), *Erica arborea* (tree heather), *Ostrya carpinifolia* (hop hornbeam), *Picea omorika* (Serbian spruce), *Platanus orientalis* (oriental plane), *Rhododendron luteum* (yellow azalea), *Styrax officinalis* (storax), *Syringa vulgaris* (common lilac), and *Zelkova abelicea* (Cretan zelkova). It is interesting that most of these are biogeographical relics, with close relatives in either eastern Asia or North America. For a quite imaginative presentation of additional plants that may also occur in Middle-earth, the reader should consult *Beyond Middle Earth* by G. Williams (2014).

We sometimes are even given brief sketches of plants but do not have sufficient information to actually identify them. For example, when the company is about ready to leave Lothlórien, an elf put ropes in each of the boats. Sam asked him what they were made of, since he was interested in rope-making. The elf told him that they were made from fibers of *hithlain* (using its Sindarin name), but we are not told anything about that plant, so its botanical identity remains a mystery. When Frodo and Sam came to the cross-roads in Ithilien they saw a beautiful white-flowered vine, which had wound around the head of the fallen king, causing Frodo to exclaim, "The king has got a crown again!" Could this be a species of *Ipomoea* or *Convolvulus*, in the Convolvulaceae? Possibly—but we just are given too little information to be sure. After the defeat of Sauron, the celebration of the forces of Gondor and Rohan was held in an opening in the woods of Ithilien, and we are told that the field was "bordered by stately dark-leaved trees laden with scarlet blossom" (LotR 6: IV). Their identity is unknown, although they suggest the beauty and diversity of the place. In yet another instance, Voronwë told Tuor about the marshes along the Sirion (in the forest of Nan-Tathren), where "the grass is filled with flowers, like gems, like bells, like flames of red and gold, like a waste of many-coloured stars in a firmament of green" (UT 1: I). So what were these bell-like flowers? It is tempting to suggest some species of *Campanula* (in the family Campanulaceae). What were these flowers "like flames of red and gold"—could they perhaps represent *Gloriosa*, in the Colchicaceae? Were the many-colored stars actually species of *Hypoxis, Rhodohypoxis*, and *Spiloxene*, in the Hypoxidaceae? In the end, we rejected all of these suggestions; sadly, they are just too speculative. Even the reference to Bilbo's embroidered silk waistcoat (see LotR 1:I) leads us to speculation about plants. Silk is made by silkworms (*Bombyx mori*), and these larvae only feed on the leaves of mulberry (species of *Morus*, especially the white mulberry, *Morus alba*) and paper mulberry (*Broussonetia papyrifera*), both members of the plant family Moraceae (which also includes figs, breadfruit, and jackfruit). Thus, although neither *Morus* nor *Broussonetia* is mentioned in Tolkien's writings, some kind of mulberry must have occurred in the Middle-earth of the Third Age! We know too little, however, about its identity to include it based on so indirect a reference.

Finally, we have excluded some species through our circumscription of Middle-earth. We include those plants mentioned in J. R. R. Tolkien's major writings, but we (for example) do not include the mythical tales related in the *Book of Lost Tales* (see HoM-E, vols. 1 and 2, edited by Christopher Tolkien) or in *The Notion Club Papers* (see HoM-E, vol. 9, part 2). For the purposes of this book, the plants mentioned in such works are not considered within our rather limited circumscription of Middle-earth. Thus, for example, ten plants are included in the various versions of "Kortirion Among the Trees" (final title: "The Trees of Kortirion"; see H-o-M-E, vol. 1), and of these harebells (*Campanula rotundifolia*, of the Campanulaceae) and maples (*Acer* spp., of the Sapindaceae) are not included in this book because neither occurs in *The Hobbit, Lord of the Rings*, or *Silmarillion*. Thus, the inclusion of this poem within our coverage would have corrected the mysterious absence of maples from Tolkien's Middle-earth. Also, poplars (although included in our book) are barely part of Middle-Earth—Tolkien only used the name of these trees so that the readers of *The Lord of the Rings* would understand the sound of mallorn leaves. In contrast, in this poem, poplars are actually part of the imaginative legendarium:

> Your trees in summer you remember still:
> The willow by the spring, the beech on hill;
> The rainy poplars, and the frowning yews. ("The Trees of Kortirion," lines 23–25)

Thus, we are led from the imaginative world of Tolkien's Middle-earth, to our own world—the Middle-earth seen in the landscapes all around us. There we will find the full measure of floristic diversity. Tolkien's writings help us to regain a clear view of the natural world—including its green plants.

EIGHT

A Note from the Illustrator

I THINK IT IS IMPORTANT TO DISCUSS THE ILLUSTRATIONS IN THIS BOOK, not only because they are integral to what we seek to accomplish, but also because J. R. R. Tolkien, whose created world is the subject of this book, was in his own right a great illustrator. In the art community, unlike the scientific one, there is little that does not come down to conjecture or opinion. Artistic critics and philosophers argue over all kinds of minutiae, searching for the faintest of inklings in order to persuade others to believe their own opinion regarding a piece; there are no facts or figures, no data or analyses—simply feelings, emotions, and belief. But, in much the same way as scientists, artists toil away at their craft, researching theories and hunches late into the night, hoping to find truth. In this way, I have always found a connection with my peers in fields of research far less whimsical than art. It is also a blessing that in this book we get to discuss not simply the merits of excellent aesthetics but the functionality that art can have, especially its ability to improve our ability to enjoy—and function in—both imaginary and real worlds.

In the beginning of all botanical illustration, there was printmaking. Not really, of course—there were drawings, paintings, and even a few Egyptian hieroglyphics and temple bas-relief. But for all intents and purposes, our story, and the dissemination of scientific information, hinged on the ability of the printing press to replicate images and text, which then could be distributed to the masses or at least to the wealthy and educated (who, at this point in history, were often physicians interested in the use of plants in treating diseases). This great explosion of information flowing out into the world needed to be illustrated, and relief printmaking was perfectly suited for the job—as well as the inspiration for the artistic styling of this book. It should be noted here that, to avoid any confusion, relief wood carving is a completely different technique from its fancier sister, wood engraving. Not only does it not share any of the same tools, but also the matrix in which the image is carved is different as well. The most common engravings (with which we are all familiar) are our currencies, most of which have the engraved faces of dead heroes on them. In an engraving, you will notice the fine line quality and exquisite control of form and value. This process is created in wood engraving by running sharp blades over the surface of a piece of wood that has been cut horizontally across the lumber, thus allowing the artist to make marks into the grain of the wood instead of along it, which is how relief printmaking works. It received its name because the matrix itself has much the same look as a relief sculpture because the artist carves large sections of material away, allowing for higher regions. Ink is then rolled with a firm rubber or leather roller over the wood to create an image that can be run through a press along with text. This is what gave relief its strength in early printing, unlike intaglio (i.e., etching in metal) or other techniques that would provide greater detail. Later, relief techniques worked in tandem with movable type and the letter press printing equipment (that would follow). Relief carving requires much less skill in carving than its counterpart, wood engraving, and, because the images could be carved from planks of wood instead of into the grain of the wood, it was significantly cheaper and faster to produce.

Enough with the medium itself—relief carving became the dominant force in printed illustration due to its accessibility, affordability, and ease of use—not for its precision, amazing detail, or scientific accuracy. These flaws, however, have been corrected by the modern practices used in this book (while I fantasize about the resulting damage to my wrists that I could have sustained carving and editing the 160 odd illustrations for this book). Instead, however, I have resorted to—dare I say it—digital techniques. However, in these digital images, I have strived to maintain the aesthetic sensibilities achieved through relief carving and printmaking. These digital techniques have allowed me to create very accurate botanical illustrations that can be used to identify the Middle-earth plants in your own parks and gardens. Despite the nod to the history of scientific illustration and the brief description of the method, my personal love of relief printmaking results from the fact that the act of illustrating with relief techniques reduces the image to its foundations. Through the use of digital technology, we are able to edit and produce botanical illustrations that don't just illuminate the form of the plant but also highlight the characteristics that are most important for its identification. From personal experience, I have found my botanically untrained eye incapable of distinguishing a plant I was studying for an illustration from another similar plant (or even from its surroundings) in photographs. With the help of my co-author and artistic editor, however, I have been able to reduce each plant down to just what is minimally needed to identify it—employing the salient and most diagnostic characteristics of each. This is why you may notice that the book is devoid of photographs, although they have become the staple of many botanical field guides. Such photographs, while accessible, just don't provide the level of simplification and ease of use that traditional illustration offers. They also create a disjointed aesthetic in the consumption of the material, pulling us back out of the land of fantasy and instead placing us firmly within a "reality" that at times can be quite off-putting. Instead, in this book, we seek to evoke the ancient European herbals with their woodcut illustrations.

But our inspiration for the art of this book did not start with the history or principles of botanical illustration; it truly started long before I had any inclination that I would become an artist or illustrator. It started with the literary and artistic works of J. R. R. Tolkien and, secondarily, with the works of many other talented artists, especially those of Alan Lee. Tolkien's line-based illustrations, often with intense undulating lines (see *The Mountain-path*, a view of the Misty Mountains, fig. 109 in W. G. Hammond and C. Scull, 1995, and first published in *The Hobbit*) harken back to an age of relief illustration in which the beautiful free-flowing value shifts offered by contemporary lithographic processes (i.e., techniques using the principle of water's repulsion of oil in order to create an image) were just a twinkle in Alois Senefelder's eye. Additionally, Tolkien's own artwork so breaks the mold of contemporary artistic inclinations that it refreshes and rejuvenates my own mark-making processes. Tolkien was not restricted by what has now become a canonized, certified, licensed property. Alan Lee, while not offering Tolkien's vibrant use of line, provided my first glimpse (as a youth) into what the naturalistic world of Tolkien was like. I remember melting into the thick undergrowth of Fangorn Forest as I wandered off into the woods behind our house, imagining far-off worlds. Alan Lee's work, in this way, is nearly unique because it breaks with the traditions that hold a firm grasp on contemporary fantasy illustration. Much fantasy illustration is rooted in classical European romantic and historical painting, in which we imagine bold characters charging forth on frothing steeds, crashing into their foes, and captured by the artist in magnificent oils, with every saber sparkling and each beam of light creating the bold silhouettes of our honored heroes. As magnificent as these images are, the stages on which they take place are just that: stages—flat, empty planes largely devoid of any of meaningful detail, which is instead given to the valiant figures who typically undulate across the entire surface of the image. Trees and bushes are summed up in the flick of a paintbrush and not much thought is given to anything besides what is considered important: characters, drama, and the historical accuracy of the figures. In contrast, the landscape, not the characters, is the artistic focus in many of the paintings of Alan Lee; for example, see the painting of Merry and Pippin entering Fangorn (opposite p. 480) or of the steps up to the high seat of Amon Hen (opposite p. 416 in the illustrated *Lord of the Rings*). Thus, in the works of Alan Lee, I remember (for the first time) seeing this mold broken in Tolkien-themed fantasy illustrations. And these works also provided a launching point for how I approach my own illustrations. As stated, the characters and the drama take center stage in most contemporary fantasy illustrations because, for all intents and purposes, the viewer is most interested in these things. Because this is a botanical book, however, we have taken a plant-centered view of the world, intentionally de-emphasizing the human, hobbit, elven, or dwarven characters, and this has led us to some very unique and interesting images. Our approach has the advantage that the illustrations purposely seek not to impose on the imagination of the reader, reflecting the concerns of Tolkien. As he stated, the reader should retain full power to give to each character "a peculiar personal embodiment in his imagination" (see *On Fairy-Stories*, p. 70).

It was difficult to design and create images that allow the viewer to interpret the unfolding scene while at the same time prioritizing the flora over the fauna—to give precedence to the plants over the drama of the unfolding scene stands staunchly against the fundamentals of my classical arts education. But, as the work developed, it became an intriguing and enlightening exercise and made me realize the need for (and importance of) this short chapter. We learn quite quickly after our birth that the world can be simply too much to take in all at once: as babies, we liked to be swaddled so that we couldn't move and to be set down in a quiet room with not very much to look at so as not to be visually overstimulated. Only slowly do we learn to compartmentalize and reduce things down into little bites of information that we can process quickly and appropriately. If an instructor asked you to draw a person or a face, you might quickly scribble down a circle some eyes, a nose, and mouth. Everything you need to make a face, but this is not a face—it is a symbol of a face. And so our brain works, simplifying, reducing, and compartmentalizing our life because it's hard to focus on all things and truly see them. So we reduce our world to symbols, allowing us to move quickly and not become overwhelmed—but sacrificing our ability to accurately see our surroundings. This process of relearning to look (and to actually "see") is, of course, the process of learning to draw. And it is how we learn to see plants. As you read this book and engage with the images, you are learning to look and to really see. You are training your mind to look for what is important, to decompartmentalize the world around you and see the plants of Middle-earth in your own environment. This process gives the world of Middle-earth the weight that Tolkien intended in its creation. As he noted "we should look at green again, and be startled anew (but not blinded) by blue and yellow and red" (see *On Fairy-Stories*), and it is part of the recovery that is the goal of fantasy. Tolkien could have quite easily filled his world completely with fantastical plants, and, if you were to ask most fans, I think the assumption would be that he created more unique plants than he actually did. His imagination was great enough to do so, but instead the world of Middle-earth is filled in large part with the same plants with which we interact on a daily basis, even if we do so unknowingly. The use of the ordinary—the plants that we see every day—in such an extraordinary setting allows us to really see plants in a fresh light. Thus, we hope that you will use this book to increase your appreciation of both the natural world around us and of Tolkien's works. No longer will characters act out dramas for you on a blank stage; instead, it will be filled and satisfying.

Glossary

This glossary is mainly composed of specialized terms used in the plant descriptions.

Achene. Small, indehiscent, dry fruit with a thin and close-fitting wall surrounding, but free from, the single seed

Acorn. Fruit of species of oaks (*Quercus*); a nut associated with a variously scaly cup-like structure.

Acuminate apex. Apex with somewhat concave sides that taper to a sharp point.

Acute apex. Apex ending in a point of less than 90 degrees, with the sides of the tapered apex more or less straight to slightly convex.

Acute base. Base ending in a point of less than 90 degrees, with the sides of the tapered base slightly convex.

Aerial root. Root naturally occurring above ground.

Alternate leaves. Leaves borne one leaf per note along the stem; such leaves may be spiral, two-ranked, etc.

Androecium. Collective term for all the stamens of a flower.

Annual. Plant growing from seed to maturity within a single growing season.

Anther. Pollen-bearing part of the stamen, located at the top of the filament.

Aril. A hard to juicy often brightly colored outgrowth of the seed.

Asymmetrical flower. Flower lacking a plane of symmetry; i.e., neither radial nor bilateral.

Asymmetrical leaf. Leaf lacking a plane of symmetry; often wider on one side than the other, and with widest point differently positioned on each side of the blade.

Attenuate apex. Apex tapering gradually to a narrow tip.

Axillary bud. Bud located in the leaf axil.

Banner petal. Distinctive uppermost petal of flowers, usually the largest, especially those of Fabaceae subfam. Faboideae; also called the *standard* or *flag* petal.

Berry. Indehiscent, fleshy fruit with (one or) a few to many seeds; the flesh may be more or less homogeneous or the outer part of the fruit may be firm, hard, or leathery.

Biennial. Plant that lives for two years, growing vegetatively during the first and flowering and fruiting in the second.
Bilateral symmetry. Divisible into equal halves by only one plane of symmetry.
Bisexual flower. Flower with both androecium (stamens) and gynoecium (carpels).
Blade. Flat, photosynthetic portion of a leaf or leaflets of a compound leaf.
Bract (associated with a flower). A reduced leaf, in the axil of which arises a flower or an inflorescence branch; such reduced leaves may also be positioned on the floral pedicel (and these often called *bractlets*).
Bract (of a cone). One of the reduced leaves associated with the cone axis; in its axil arises the ovulate (or cone) scale. Such bracts may be free from the ovulate scale or fused to it.
Bud. Small embryonic shoot, often protected by modified leaves (bud scales), stipules, or hairs.
Bulb. Short, erect, underground stem surrounded by thick, fleshy leaves or leaf bases.
Calyx. Collective term for all the sepals of a flower.
Capsule. Usually dry fruit from a two- to many-carpellate gynoecium that opens in various ways to release the seed or seeds. (Occasionally also used for the sporangium of a moss.)
Carpel. Ovule-bearing unit(s) that make up the gynoecium of a flower.
Carpellate flower. Flower with a gynoecium (carpel or carpels) but no functional androecium (stamens).
Catkin. Inflorescence consisting of a dense, elongated mass of inconspicuous, usually wind-pollinated flowers.
Compound leaf. Leaf with two or more blades (leaflets).
Connective. Portion of the stamen connecting the two pollen sacs of an anther.
Cordate base. Base heart-shaped, with rounded lobes.
Corm. Short, erect, underground, more or less fleshy stem covered with thin, dry leaves or leaf bases.
Corolla. Collective term for all the petals of a flower.
Cuneate base. Base that is more or less triangular and tapered to a point.
Cyme. Determinate, compound inflorescence composed of repeating units of a pedicel bearing a terminal flower and, below it, one or two bractlets; each bractlet is associated with an axillary flower, further bractlets, and so on.
Deciduous leaves. Leaves that fall at the end of the growing season.
Decurrent base. Blade base tapering gradually to the petiole, also used for leaf base that extends down the stem.
Decussate leaves. Opposite leaves in which those of adjacent nodes are rotated 90 degrees from each other.
Dehiscence. Method or process of opening of a structure, such as a fruit or anther.
Dendritic. Branching like a miniature tree.
Dentate margin. With coarse teeth that are perpendicular to the margin.
Determinate inflorescence. Inflorescence in which the axis is developmentally converted into a flower, resulting in the cessation of growth of that axis.
Distinct. Term describing the condition of like parts (e.g., all the petals of a flower) being unfused.

Drupe. Indehiscent, fleshy fruit in which the outer part is soft or fleshy and the center contains one or more hard pits or stones consisting of a bony endocarp surrounding a seed or seeds.

Elliptic. Widest near the middle.

Entire margin. With a smooth margin, lacking any teeth.

Epigynous. With perianth and stamens apparently borne on the ovary and the ovary, therefore, inferior.

Epiphyte. Plant growing upon another plant, which is used as a support.

Evergreen plant. Plant that is leafy throughout the year.

Fertilization. Fusion of the sperm nucleus and the egg nucleus.

Fiber. Any long, narrow, thick-walled, and lignified cell.

Fibrous roots. Root system with roots of more or less equal thickness and often well-branched; the tap root absent or not obvious.

Filament. Stalk of a stamen.

Flora. Enumeration of the plants occurring in a particular geographical area, usually with keys, descriptions, illustrations, and distribution maps; also the plants occurring within a designated region.

Floral cup (or floral tube). Flat, cup-like, or tubular structure on which the sepals, petals, and stamens are borne; usually formed from the fused base of the perianth parts and stamens or from a modified receptacle; also the *hypanthium*.

Flower. Reproductive structure of the flowering plants or angiosperms consisting of a modified shoot, the floral axis or receptacle bearing modified leaves (i.e., the perianth parts, stamens, and/or carpels).

Follicle. Usually dry fruit derived from a single carpel that opens along a single longitudinal slit.

Free. Term describing the condition of unlike parts (e.g., the petals and stamens of a flower) not being fused together.

Fruit. Mature ovary, sometimes with associated accessory parts (such as a fused floral cup, as in an apple).

Fused. Flower parts developmentally joined to form a single entity, such as petals fused to form a cup-shaped corolla, or carpels fused to form a gynoecium with a single ovary and several styles or style branches.

Glabrous. Lacking hairs.

Glaucous. With a waxy covering and thus often blue or white in appearance.

Grain. Small, indehiscent, dry fruit with a thin wall surrounding and more or less fused to a single seed; characteristic fruit of the grass family (Poaceae or Gramineae).

Gynoecium. Collective term for all the carpels of a flower.

Habit. General appearance of a plant, e.g., whether a tree, shrub, vine, etc.

Hair. Epidermal outgrowth of diverse form, structure, and function.

Head. Compact determinate or indeterminate inflorescence with a very short, often disk-like axis and usually flowers without pedicels (stalks).

Herb. Plant without a persistent above-ground woody stem and either dying (in annuals) or dying back to ground level (in herbaceous perennials) at the end of the growing season.

Hypanthium. See *Floral cup*.

Hypogenous. With perianth parts and stamens arising from below the ovary, and the ovary thus superior.

Indeterminate inflorescence. Inflorescence in which the main axis produces only lateral flowers, branches, or groups of flowers so that the lowermost or outermost flowers usually open first, with the main axis (and lateral axes, if present) often elongating as the flowers develop, without the production of terminal flowers.

Inferior ovary. Ovary that is positioned beneath the point of attachment of the other floral parts, which appear to arise from its apex.

Inflorescence. The shoot system that serves for the formation of flowers; or, more simply, a cluster of flowers.

Infructescence. Mature inflorescence, with flowers replaced by fruits.

Leaf. Usually flat, determinate, photosynthetic part of a plant, borne on a branch or stem.

Leaf axil. Space in the angle between a leaf and the stem that bears it.

Leaflet. One of the blades of a compound leaf.

Legume. Dry, more or less elongated fruit derived from a single carpel that opens, often explosively, along two longitudinal slits; the most common fruit type of the legume family (Fabaceae or Leguminosae).

Lenticel. Wartlike protuberance on the stem surface that is involved in gas exchange.

Liana. Woody climbing plant.

Ligule. Projection from the top of the leaf sheath, as in gingers (Zingiberaceae) or grasses (Poaceae).

Linear. Long and very narrow (i.e., very narrowly oblong).

Lobed. Having large projections along margin.

Long shoot. Stem with long internodes; this term is applied only in plants in which internode length is clearly bimodal, and both long and short shoots are present.

Midvein. Central vein of a leaf, often called the midrib.

Mucronate apex. Apex terminated by a distinct, short, and abrupt point.

Nectary. Nectar-producing gland.

Nut. Fairly large, indehiscent dry fruit with a thick bony wall surrounding a single seed.

Oblong. With the sides nearly or quite parallel for most of their length.

Obovate. Widest near the apex.

Obtuse apex. Apex blunt, having an angle greater than 90 degrees, with straight to slightly convex sides.

Obtuse base. Base blunt, having an angle greater than 90 degrees, with slightly convex sides.

Opposite leaves. Pair of leaves borne along a stem, the members of which are positioned on opposing sides of the stem.

Ovary. Ovule-bearing part of a carpel (or several fused carpels).

Ovate. Widest near the base.

Ovulate scale. The structure of a seed cone that bears the ovules (and eventually the seeds), derived from a short shoot and arising in the axil of a bract.

Ovule. Structure in seed plants comprising the female gamete-bearing plant, usually one or two integuments, and a stalk; after fertilization, it develops into the seed.

Palmate venation. Three or more primary veins (or well-developed secondary veins) arising at the base of the blade or near the base of the blade.

Palmately compound leaf. Leaf with more than three leaflets attached to a common point, like the fingers of a hand.

Panicle. Indeterminate inflorescence with two or more orders of branching, each axis bearing flowers or higher order axes.

Parallel venation. Several to many parallel veins running the length of the blade.

Pedicel. Stalk of a single flower in an inflorescence.

Peltate. Flat structures attached by their surfaces rather than the base or margin (e.g., like an umbrella).

Perennial. Plant that lives for three or more years and usually flowers and fruits repeatedly.

Perianth. Collective term for calyx and corolla, or all tepals (when calyx and corolla are not distinguished), of a flower.

Perigynous. With perianth parts and stamens borne on a floral cup or tube (= hypanthium) that surrounds, but is not fused to, the superior ovary.

Petal. Member of the inner perianth whorl, usually colorful and assisting in attracting pollinators to the flower.

Petiole. Stalk of a leaf.

Pinnate venation. Where secondary veins arise along the length of the single primary vein, like the teeth of a comb or the lateral units of a feather.

Pinnately compound leaf. Compound leaf with leaflets more than three and attached along two sides of an axis (i.e., feather-like).

Pollen. Small spore containing the male gametophyte; in the flowering plants, germinating to form a pollen tube, which rapidly transports the sperm to the ovule.

Pollination. Transference of pollen from the anther to the stigma.

Pome. Indehiscent, fleshy fruit in which the outer part is soft and the center contains papery to cartilaginous structures enclosing the seeds; characteristic fruit of apples, pears, and their relatives (Rosaceae tribe Maleae).

Prickle. Sharp-pointed hair (involving only epidermal tissue) or emergence (involving epidermal and subepidermal tissues, but not vascularized).

Pubescent. With hairs of any type.

Pulvinus. Swollen portion of the petiole involved in movement, usually positioned at the petiole base.

Raceme. Simple, indeterminate inflorescence with a single axis bearing stalked (pedicellate) flowers.

Radial symmetry. Divisible into equal halves by two or more planes of symmetry.

Receptacle. Floral axis that bears the flower parts.

Retuse (or emarginate) apex. Slightly to strongly notched apex.

Rhizome. Horizontal stem, often underground or on the surface of the ground, usually bearing scale-like leaves.

Root. Portion of plant axis lacking nodes and leaves, usually branching irregularly and found below ground; it holds the plant in place and is involved in water and nutrient uptake.

Rounded apex. Round in outline (at apex).

Rounded base. Round in outline (at base).

Sagittate base. Arrowhead-shaped base.

Samara. Winged, indehiscent, dry fruit containing a single seed (or rarely two or a few seeds).

Schizocarp. Dry to rarely fleshy fruit breaking into one-seeded (occasionally few-seeded) segments.

Scorpioid cyme. Coiled cyme in which the lateral branches (and flowers) develop alternately on opposite sides of the axis.

Secondary vein. Vein that branches from a primary vein.

Seed cone scale. The matured ovulate scale, bearing one or more seeds; one of the basic units composing a seed cone (in the conifers, such as pines and firs).

Sepal. One member of the outer perianth whorl, when floral whorls are differentiated; usually green and protecting the inner flower parts in bud.

Serrate margin. Margin with saw-teeth (i.e., the teeth point forward).

Short shoot. Stem with short internodes, other shoots of the same plant having distinctly longer internodes; compare *Long shoot.*

Shrub. Woody plant that is usually shorter than a tree and produces several stems or trunks from the base.

Silique. Fruit derived from a two-carpellate gynoecium in which the two halves of the fruit split away from a persistent partition (around the rim of which the seeds are attached).

Simple leaf. Leaf with a single blade.

Spadix. Spike with a thickened, fleshy axis; characteristic of members of the arum family (Araceae).

Spathe. Large bract surrounding or subtending an inflorescence; as in the often showy inflorescence bract of members of the arum family (Araceae).

Spike. Simple, indeterminate inflorescence with a single axis bearing stalk-less flowers.

Spine. Reduced, sharp-pointed leaf or stipule, or sharp-pointed marginal tooth.

Spirally arranged. Leaves (or other organs) borne in a spiral or ascending coil along and around an axis.

Stamen. Pollen-bearing part of a flower, composed of a filament (stalk) and anther (pollen-sacs).

Staminate flower. Flower with an androecium (stamen or stamens) but not a functional gynoecium (carpel or carpels).

Stellate. With branches radiating in a star-like pattern.

Stem. The plant axis bearing leaves with axillary buds at the nodes (with are separated by internodes); such axes are usually above ground.

Stigma. Part of the carpel (or several fused carpels) that receives and facilitates the germination of the pollen.

Stipule. One, usually of a pair of appendages located on either side of (or on) the petiole base.

Style. More or less elongated part of the carpel (or of several fused carpels) between the stigma (or stigmas) and the ovary.

Superior ovary. Ovary that arises above the point of insertion of the other flower parts.

Taproot. Major root, usually enlarged and growing downward.

Tendril. Elongated and twining structure (modified from an inflorescence, leaf, or stem) assisting in climbing.

Tepal. One of a series of perianth parts that are not differentiated into a calyx and corolla.

Terete. Cylindrical.

Thorn. Reduced, sharp-pointed stem.

Tooth. Small, usually pointed and triangular projection of the leaf margin.

Tree. A tall, woody plant with usually a single main trunk; the term also used as an abbreviation for "evolutionary tree").

Trifoliolate leaf. Compound leaf with three leaflets.

Truncate apex. Apex appearing as if cut off at the end, nearly straight across.

Tuber. Swollen fleshy portion of a rhizome involved in water or carbohydrate storage.

Two-ranked leaves. Leaves borne along just two sides of a stem (i.e., they are all in the same plane).

Umbel. Determinate or indeterminate inflorescence in which all flowers have stalks of equal or unequal length that arise from a single region of the apex of the inflorescence axis.

Unisexual flower. Flower lacking either an androecium (stamens) or a gynoecium (carpels).

Vein. Vascular bundle (containing specialized cells for conduction of water and sucrose), usually visible externally, as in a leaf.

Vine. Herbaceous climbing plant.

Whorled leaves. Three or more leaves at a single node.

Bibliography

Allen, W. 2003. Plant blindness. *BioScience* 53(10): 926.

Anderson, D. A. 2002. *The Annotated Hobbit*. Houghton Mifflin, Boston.

Bernthal, C. 2014. *Tolkien's Sacramental Vision: Discerning the Holy in Middle Earth*. Angelico Press, Kettering, Ohio.

Carpenter, J. M. 2014. *Sindarin-English and English-Sindarin Dictionary*, 2nd edition. Published by the author.

Dickerson, M. and J. Evans. 2006. *Ents, Elves, and Eriador: The Environmental Vision of J.R.R. Tolkien*. University Press of Kentucky.

Faegri, K. and L. van der Pijl. 1979. *The Principles of Pollination Ecology*, 3rd edition. Pergamon, Oxford.

Fauskanger, H. K. 2008. *English—Quenya*. www.ambar-eldaron.com/eng-quen.pdf.

Fisher, M. 2016. *The Encyclopedia of Arda*. www.glyphweb.com/arda/.

Flieger, V. 2000. Taking the Part of Trees: Eco-conflict in Middle-earth. In *J.R.R. Tolkien and his Literary Resonances: Views of Middle-earth*, edited by G. Clark and D. Timmons (pp. 147–158). Greenwood Press, Westport, Connecticut.

Flieger, V. 2002. *Splintered Light: Logos and Language in Tolkien's World*, 2nd edition. Kent State University Press, Ohio.

Flora of North America Editorial Committee. 1993–2017. *Flora of North America north of Mexico*. Vols. 1–9, 12, 19–28. Oxford University Press, New York.

Fonstad, K. W. 1991. *The Atlas of Middle-earth*, revised edition. Houghton Mifflin, Boston.

Frodin, D. G. 2001. *Guide to Standard Floras of the World: An Annotated, Geographically Arranged Systematic Bibliography of the Principal Floras, Enumerations, Checklists, and Chorological Atlases of Different Areas*, 2nd edition. Cambridge University Press, Cambridge.

Garth, J. 2003. *Tolkien and the Great War: The Threshold of Middle-earth*. Houghton Mifflin, Boston.

Giraudeau, D. 2011. Sindarin Corpus. *Parma Eldalamberon* 17: 1–69. lambenore.free.fr/downloads/PE17_S.pdf.

Gray, A. 1880. *Natural Science and Religion—Two Lectures Delivered to the Theological School of Yale College*. C. Scribner's Sons, New York.

Hammond, W. G., and C. Scull. 1995. *J.R.R. Tolkien: Artist and Illustrator*. Houghton Mifflin, Boston.

Hammond, W. G., and C. Scull. 2005. *The Lord of the Rings: A Reader's Companion*. Houghton Mifflin, Boston.

Hammond, W. G., and C. Scull. 2012. *The Art of the Hobbit by J.R.R. Tolkien*. Houghton Mifflin, Boston.

Harper, D. 2016. *Online Etymology Dictionary*. Ohio University. http://www.etymonline.com.

Hazell, D. 2006. *The Plants of Middle-earth: Botany and Sub-creation*. Kent State University Press, Ohio.

Jeffers, S. 2014. *Arda Inhabited: Environmental Relationships in The Lord of the Rings*. Kent State University Press, Ohio.

Judd, W. S., C. S. Campbell, E. A. Kellogg, P. F. Stevens, and M. J. Donoghue. 2016. *Plant Systematics: A Phylogenetic Approach*, 4th edition. Sinauer Assoc., Sunderland, Massachusetts.

Keeling, P. J. 2004. Diversity and evolutionary history of plastids and their hosts. *Amer. J. Bot.* 91: 1481–1493.

Liberman, A. 2008. *An Analytic Dictionary of English Etymology*. University of Minnesota Press, Minneapolis.

Mabberley, J. D. 2008. *Mabberley's Plant Book*, 3rd edition. Cambridge University Press.

Murrell, Z. E. 2010. *Vascular Plant Taxonomy*, 6th edition. Kendall Hunt, Dubuque, Iowa.

Palmer, J. D., D. E. Soltis, and M. W. Chase. 2004. The plant tree of life: an overview and some points of view. *Amer. J. Bot.* 91: 1437–1445.

Proctor, M., P. Yeo, and A. Lack. 1996. *The Natural History of Pollination*. Timber Press, Portland, Oregon.

Rateliff, J. D. 2007. *The History of the Hobbit*, 2 vols. Houghton Mifflin, Boston.

Sallo, D. Accessed 2016. *A Qenya Botany*. folk.uib.no/hnohf/botany.htm.

Shippey, T. A. 1992. *The Road to Middle-earth*, 2nd edition. Grafton, London.

Shippey, T. A. 2002. *J. R. R. Tolkien: Author of the Century*. Houghton Mifflin, Boston.

Simonson, M., ed. 2015. *Representations of Nature in Middle-earth*. Walking Tree Publishers, Zurich and Jena.

Simpson, M. G. 2010. *Plant Systematics*, 2nd edition. Academic Press, Burlington, Massachusetts.

Soltis, D. E., P. S. Soltis, P. K. Endress, and M. W. Chase. 2005. *Phylogeny and Evolution of Angiosperms*. Sinauer Associates, Sunderland, Massachusetts.

Soltis, D. E., P. Soltis, P. K. Endress, M. W. Chase, S. Manchester, W. Judd, L. Majure, and E. Mavrodiev. 2017. *Phylogeny and Evolution of the Angiosperms*, revised and updated edition. University of Chicago Press, Chicago and London.

Spichiger, R. -E., V. Savolainen, M. Figeat, and D. Jeanmonod. 2004. *Systematic Botany of Flowering Plants*. Science Publishers, Enfield, NH and Plymouth, UK.

Tolkien, J. R. R. 1937. *The Hobbit: or There and Back Again*, 1st edition. George Allen and Unwin, London [cited here from 4th edition, George Allen and Unwin, London, 1978, and Houghton Mifflin, Boston, 1978].

Tolkien, J. R. R. 1954. *The Fellowship of the Ring: Being the First Part of The Lord of the Rings*. George Allen and Unwin, London [cited here from the 2nd edition, George Allen and Unwin, London, 1966, and Houghton Mifflin, Boston, 1967].

Tolkien, J. R. R. 1954. *The Two Towers: Being the Second Part of The Lord of the Rings*. George Allen and Unwin, London [cited here from the 2nd edition, George Allen and Unwin, London, 1966, and Houghton Mifflin, Boston, 1967].

Tolkien, J. R. R. 1955. *The Return of the King: Being the Third Part of The Lord of the Rings*. George Allen and Unwin, London [cited here from the 2nd edition, George Allen and Unwin, London, 1966, Houghton Mifflin, Boston, 1967].

Tolkien, J. R. R. 1962. *The Adventures of Tom Bombadil and Other Verses from The Red Book*. George Allen and Unwin, London [cited here from Houghton Mifflin, Boston, 2014, edited by C. Scull and W. G. Hammond].

Tolkien, J. R. R. 1977. *The Silmarillion*. George Allen and Unwin, London [cited here from the 2nd edition, Houghton Mifflin, Boston, 2001].

Tolkien, J. R. R. 1980. *Unfinished Tales of Númenor and Middle-earth*, edited by C. Tolkien. George Allen and Unwin, London, and Houghton Mifflin, Boston.

Tolkien, J. R. R. 1981. *The Letters of J. R. R. Tolkien*, edited by H. Carpenter, with the assistance of C. Tolkien. George Allen and Unwin, London, and Houghton Mifflin, Boston.

Tolkien, J. R. R. 1983. *The Book of Lost Tales, Part I*, edited by C. Tolkien [Vol. I of *The History of Middle-earth*]. George Allen and Unwin, London, and Houghton Mifflin, Boston.

Tolkien, J. R. R. 1984. *The Book of Lost Tales, Part II*, edited by C. Tolkien [Vol. II of *The History of Middle-earth*]. George Allen and Unwin, London, and Houghton Mifflin, Boston.

Tolkien, J. R. R. 1985. *The Lays of Beleriand*, edited by C. Tolkien [Vol. III of *The History of Middle-earth*]. George Allen and Unwin, London, and Houghton Mifflin, Boston.

Tolkien, J. R. R. 1987. The Etymologies. In *The Lost Road and Other Writings*, edited by C. Tolkien [Vol. V of *The History of Middle-earth*] (pp. 341–400). George Allen and Unwin, London, and Houghton Mifflin, Boston.

Tolkien, J. R. R. 1987. *The Lost Road and Other Writings*, edited by C. Tolkien [Vol. V of *The History of Middle-earth*]. George Allen and Unwin, London, and Houghton Mifflin, Boston.

Tolkien, J. R. R. 1992. The Epilogue. In *Sauron Defeated*, edited by C. Tolkien [Vol. IX of *The History of Middle-earth*] (pp. 114–135). George Allen and Unwin, London, and Houghton Mifflin, Boston.

Tolkien, J. R. R. 1992. Part Two—The Notion Club Papers. In *Sauron Defeated*, edited by C. Tolkien [Vol. IX of *The History of Middle-earth*] (pp. 145–327). George Allen and Unwin, London, and Houghton Mifflin, Boston.

Tolkien, J. R. R. 1994a. The Wanderings of Húrin. In *The War of the Jewels*, edited by C. Tolkien [Vol. XI of *The History of Middle-earth*] (pp. 251–310). George Allen and Unwin, London, and Houghton Mifflin, Boston.

Tolkien, J. R. R. 1994b. On Fairy-stories. In *Poems and Stories* (pp. 115–188). Houghton Mifflin, Boston (and first published in 1964 in *Tree and Leaf*, published by George Allen and Unwin, London).

Tolkien, J. R. R. 1996. *The Peoples of Middle-earth*, edited by C. Tolkien [Vol. XII of *The History of Middle-earth*]. George Allen and Unwin, London, and Houghton Mifflin, Boston.

Tolkien, J. R. R. 2007. *Narn I Chîn Húrin: The Tale of the Children of Húrin*, edited by C. Tolkien. HarperCollins, London, and Houghton Mifflin, Boston.

Tutin, T. G., V. H. Heywood, N. A. Burges, D. M. Moore, D. H. Valentine, S. M. Walters, and D. A. Webb. 1964–1980. *Flora Europaea*, Vols. 1–5. Cambridge University Press.

Wandersee, J., and E. Schussler. 2001. Towards a theory of plant blindness. *Plant Science Bull.* 47(1): 2–9.

Williams, G. 2014. *Beyond Middle Earth: A Botanica of the Extinct, Rare and Useful Plants of the Lands of Middle Earth, with Commentaries on the Lands and Peoples of Central Middle Earth.* Published by the Author; Lorien Wildlife Refuge and Conservation Area, Lansdowne, New South Wales, Australia.

Van der Pijl, L. 1972. *Principles of Dispersal in Higher Plants.* McGraw-Hill, New York.

Vickery, R. 1995. *A Dictionary of Plant-lore.* Oxford University Press, London.

Watts, D. C. 2007. *Elsevier's Dictionary of Plant Lore.* Academic Press, Burlington, Massachusetts.

Index